Critical Supplement to the

ATLAS

of the

BRITISH FLORA

INCLUDING

DISTRIBUTION MAPS

OF

500 FLOWERING PLANTS AND FERNS

AND EXPLANATORY TEXT

Critical Supplement to the

ATLAS
of the
BRITISH FLORA

Edited by

F. H. PERRING

Assisted by P. D. SELL

Published for the

BOTANICAL SOCIETY OF THE BRITISH ISLES

by

EP PUBLISHING LIMITED

1978

Originally published by Thomas Nelson & Sons Ltd, 1968

Republished 1978 by EP Publishing Ltd
East Ardsley, Wakefield, England

British Library Cataloguing in Publication Data
Atlas of the British flora. – 2nd ed
Critical supplement
1. Botany – Great Britain – Maps
2. Flowers – Great Britain – Maps
I. Perring, Franklyn Hugh II. Walters, Stuart Max
III. Sell, Peter Derek
IV. Botanical Society of the British Isles
912'.1582'130941 QK306
ISBN 0-7158-1341-2

ISBN 0 7158 1341 2

This is a Supplement to the *Atlas of the British Flora*,
edited by F. H. Perring and S. M. Walters
and published for the Botanical Society of the British Isles
by Thomas Nelson & Sons Ltd. in 1962
and by EP Publishing Ltd (second edition) in 1976

TEXT PRINTED IN MONOPHOTO BEMBO

MADE AND PRINTED IN GREAT BRITAIN BY THE SCOLAR PRESS ILKLEY
EAST PARADE, ILKLEY, WEST YORKSHIRE

Contents

To the memory of the British Botanists of the past without whose critical appreciation of our flora this volume would not have been possible.

Charles Cardale Babington (1808–1895)

James Backhouse jun. (1825–1890)

Arthur Bennett (1843–1929)

John Thomas Irvine Boswell-Syme (1822–1888)

Charles Edward Britton (1872–1944)

Eric Frederic Drabble (1877–1933)

George Claridge Druce (1850–1932)

John Fraser (1854–1935)

Alfred Fryer (1826–1912)

Rex Alan Henry Graham (1915–1958)

Eliza Standerwick Gregory (1840–1932)

Patrick Martin Hall (1894–1941)

Frederick Janson Hanbury (1851–1938)

Augustin Ley (1842–1911)

Edward Francis Linton (1848–1928)

William Richardson Linton (1850–1908)

Edward Shearburn Marshall (1858–1919)

Charles Edward Moss (1872–1930)

Herbert William Pugsley (1868–1947)

William Moyle Rogers (1835–1920)

Charles Edgar Salmon (1872–1930)

Noel Yvri Sandwith (1901–1965)

Edmund Frederic Warburg (1908–1966)

William Charles Richard Watson (1885–1954)

Alfred James Wilmott (1888–1950)

Anthony Hurt Wolley-Dod (1861–1948)

Introduction

IT was towards the end of the work on the *Atlas of the British Flora* that reluctantly the decision had to be taken to omit the major critical genera, *Rubus*, *Alchemilla*, *Sorbus*, *Rhinanthus*, *Euphrasia*, and *Hieracium*. The task of preparing maps of the species in these genera was of a different nature from the majority of the species included in the *Atlas*: a very high percentage of the records had to rest upon the authority of one or more experts in each genus. By delaying the publication of the maps of these genera for six years there has been a much greater opportunity to make use of the services of these experts to increase the completeness of the work. Any inadequacies which remain are due to lack of collectors rather than to inattention by the experts.

Whilst waiting for the maps of the critical genera to ripen it was agreed by the Maps Committee of the Botanical Society of the British Isles that the opportunity should be taken of collecting distribution data on a number of critical species, microspecies, subspecies, varieties and hybrids and, where it seemed practical, of inviting the assistance of members of the B.S.B.I. in the work. The taxa to be considered, with hints on their determination, were listed in a paper published at an early stage (Perring, 1962). The editor has visited the main National Herbaria and personally identified much of the material used in compiling these maps. Thus in almost every case the distribution data are supported by many herbarium records. In cases of extreme difficulty or diversity of taxonomic view the data presented are based entirely on the editor's criteria for delimiting the taxa under consideration. This was particularly necessary in situations where two subspecies are joined by intermediates, e.g. *Juniperus communis* subsp. *communis* and subsp. *nana*, or *Centaurea nigra* subsp. *nigra* and subsp. *nemoralis*.

In the *Atlas of British Flora* it was possible to divide the species into three classes, A, B, and C, to indicate the amount of editing which had been given to each: this is not possible with the critical taxa. Instead the maps are accompanied by brief notes in which the method of compilation, the main sources of information, and the taxonomic criteria used are included. Whenever relevant, reference is made to maps already published in the *Atlas*.

The numbering of the species and the basis of the nomenclature are those of Dandy (1958). Such nomenclatural and taxonomic changes as were made during the preparation of this volume have mainly been published in a separate paper in *Watsonia* (Sell, 1967).

The main sources of the records are listed in abbreviated form after the name of the taxon. The key to the herbarium abbreviations is to be found in an appendix (p. 153). The term "Floras" which appears in some lists of sources means that, for these taxa, all the main Local Floras have been consulted.

The figures which appear in brackets in the text are the Watsonian vice-county numbers. The names of the vice-counties to which they refer may be found in the *Atlas*, pp. 409–10; an overlay showing the vice-county boundaries was supplied with the *Atlas*.

A work of this kind can only appear as a result of the willing collaboration of many people: I have been fortunate to find among botanists a group eager to pool their knowledge and ability. I would like to thank all those people, many of whom are mentioned in the text, who gave generously of their time for the identification of material and the supply of original data. I am also appreciative of the facilities made available for me to collect data myself in the herbaria in Cardiff, Dublin, Edinburgh and London. I am indebted too to the members of the Maps Committee who have advised me, sustained me and finally bullied me during the preparation of this volume. In particular I appreciated the help of the Chairman, J. E. Lousley, and the Secretary, Professor A. R. Clapham, who served in those capacities for fifteen years without a break.

I am grateful to J. E. Dandy for his advice on the nomenclature, even when I did not realize any problem existed. There is also a debt I owe which cannot be repaid, to the late Dr E. F. Warburg, for help with the many critical genera of which he was a master: he revised the section on *Sorbus* only a few weeks before his death. It would not have been possible to publish maps of *Rubus* without the enormous effort made by E. S. Edees and B. Miles, and I must record my thanks to Dr P. F. Yeo for the patience he showed in identifying thousands of *Euphrasia* specimens without unreasonable complaint.

The greatest thanks though must be reserved for the inseparable Sell & West—Peter Sell and Dr Cyril West. This supplement would have been a slim volume without their contribution, which entailed considerable personal sacrifice by them both. I am honoured in having played a small part in the publication of their great work.

I want also to pay tribute to Peter Sell for the invaluable assistance he has given me in editing this supplement, and for his advice on nomenclature, lectures on taxonomy, and wide knowledge of the literature which have made our collaboration both stimulating and pleasurable.

Two people who have helped Peter Sell in the Herbarium of the Botany School, Cambridge shall not go unsung. To Mrs Ann Hill my thanks for typing the manuscript so accurately from such impossible handwriting: to Mary McCallum Webster my thanks for work in preparing the *Hieracium* cards and for brightening our winters. I am also most grateful to my own staff at Monks Wood who have prepared many of the maps, handled much of the data, and kept the office running in my absence with frightening ease.

One name in this list must, to all with a knowledge of the Distribution Maps Scheme, be conspicuous by its absence so far, the name of Dr S. M. Walters. It was Max who launched the scheme and guided it through its formative years towards final publication of the *Atlas*. His influence pervades the *Supplement* also. As my first thanks were to him for having invited me to help with the scheme from the beginning, so my last thanks are to him for having been available with encouragement and advice until the very end.

F. H. Perring

Biological Records Centre,
Monks Wood,
Huntingdon.
1 December 1966

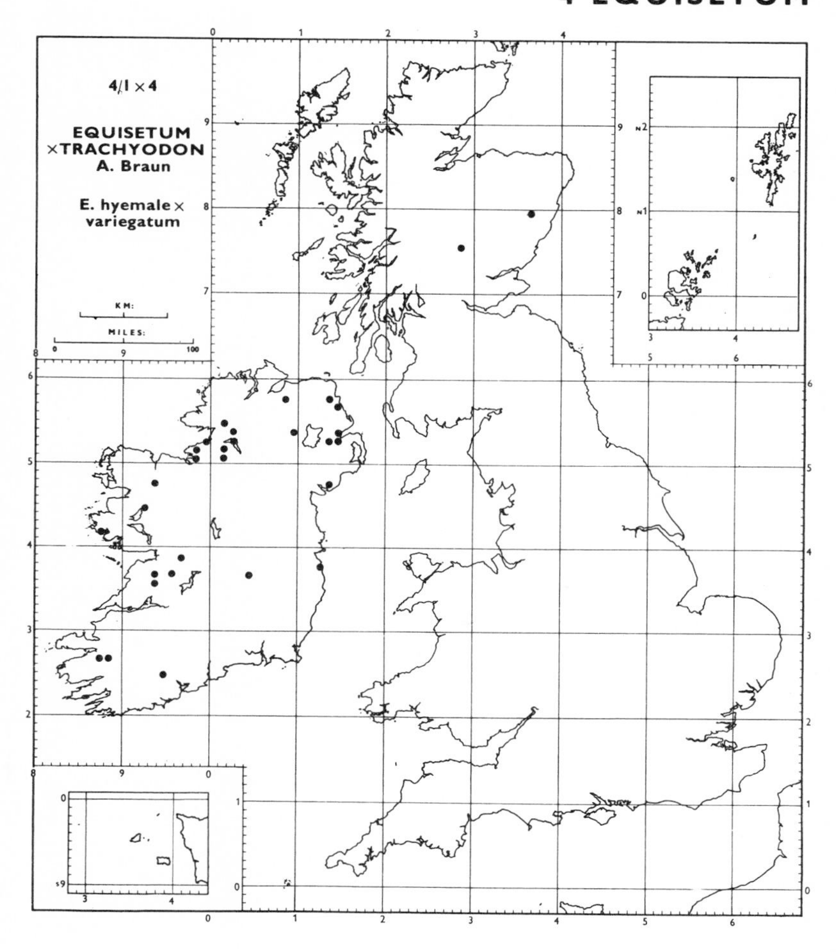

4/1×4 Equisetum ×trachyodon A. Braun
=E. hyemale × variegatum

BFT, DBN: Floras
The parentage of the hybrid is assumed to be that given above; all three taxa have the same chromosome number, i.e. 216. However, *E.* × *trachyodon* often occurs in the absence of either of the putative parents, and in north-west Ireland appears to be the commonest of the three. (*Atlas*, p. 3.)

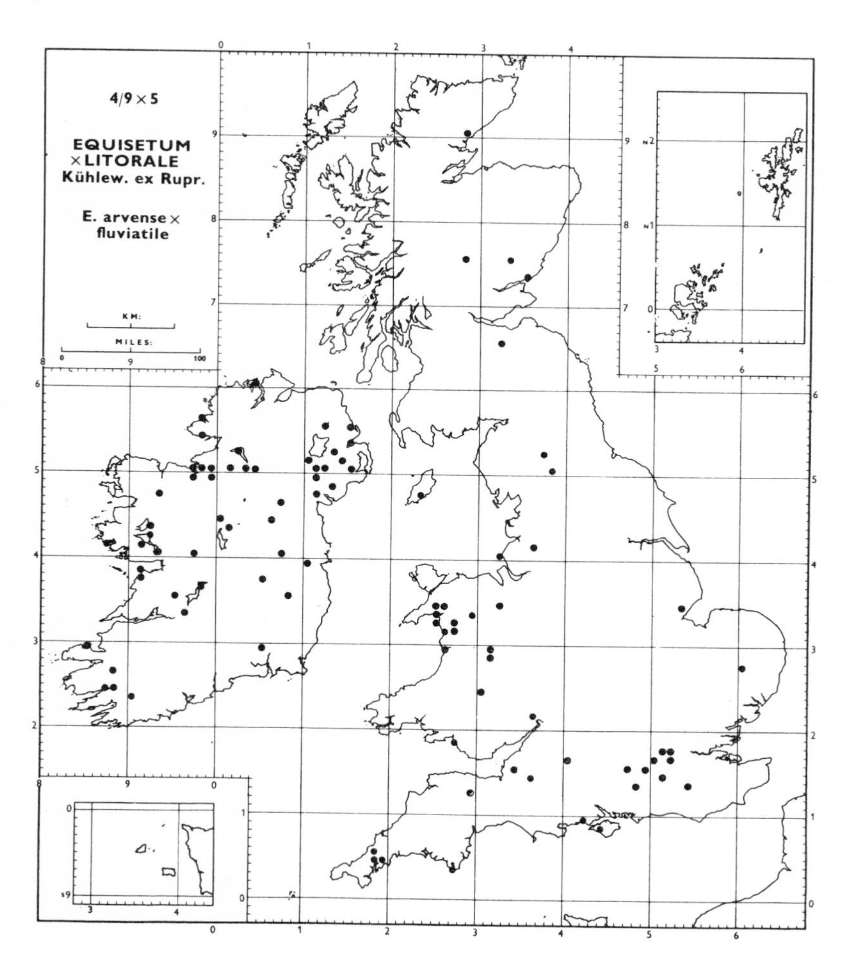

4/9×5 Equisetum ×litorale Kühlew. ex Rupr.
=E. arvense × fluviatile

BM, CGE, E, K, NMW: Floras
The limited occurrence of this hybrid is unexpected as both parents occur throughout the British Isles (*Atlas*, pp. 3–4), and it is probably under-recorded. Nevertheless, its apparently greater abundance in Ireland and its absence from central England may reflect the situation accurately. The Irish records are based mainly on a series of Papers by Praeger (1929, 1932, 1934, 1939, 1946) and Brenan and Simpson (1949): the British records have been supplemented with personal records of P. Taylor.

No information has been discovered to substantiate the statement in *The Flora of Buckinghamshire* (Druce, 1926) that this plant occurs in Berkshire (22).

ASPLENIACEAE

15/1b Asplenium adiantum-nigrum L.
subsp. **onopteris** (L.) Luerss.

BM, CGE, DBN, NMW, OXF, TCD, UCNW
The map is based on herbarium material seen by the editor. Only specimens in which the leaf segments are narrowly lanceolate with *long*, acuminate teeth have been accepted. On this basis this subspecies is confined to south and west Ireland in the British Isles.

Subsp. *onopteris* largely replaces subsp. *adiantum-nigrum* in south and west Europe.

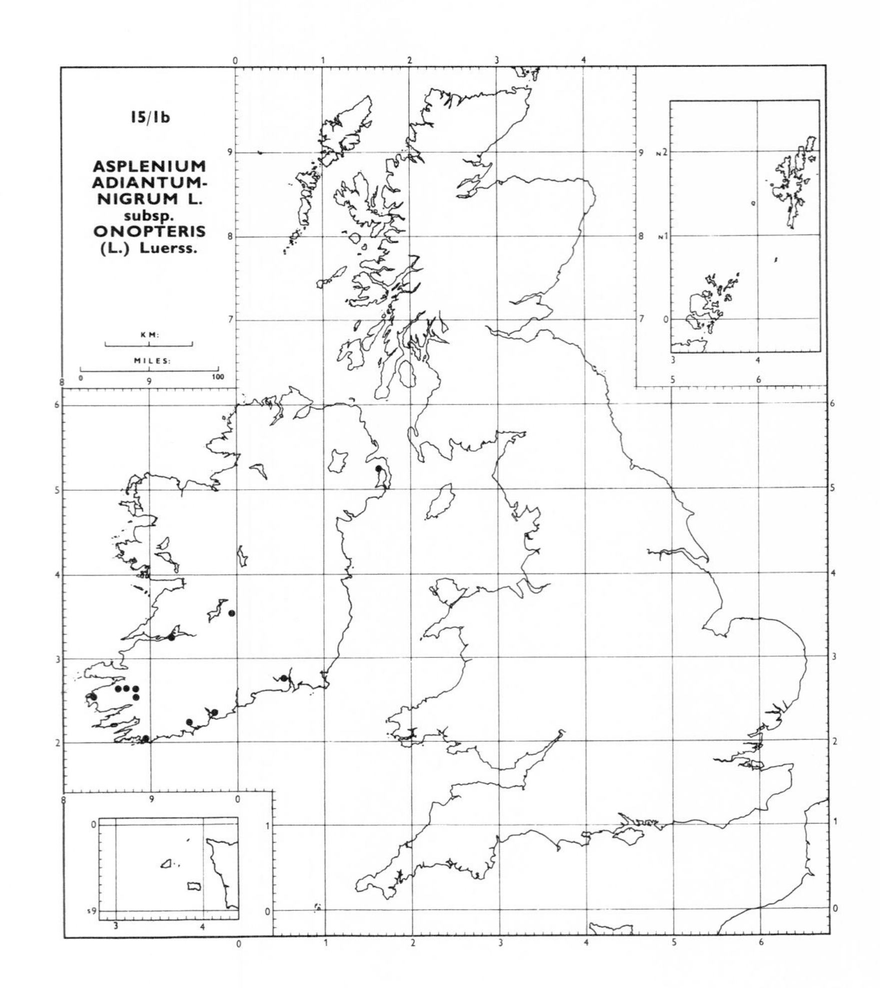

15/8×5 Asplenium ×alternifolium Wulf.
=A. septentrionale × trichomanes

BM, CGE, E, K, NMW, OXF, Lousley, Wallace: Floras
Not known to occur outside the range of overlap of the two parents (*Atlas*, pp. 8–9). The record for Brecon (42), given in Hyde and Wade (1962), requires confirmation.

18/2 Athyrium distentifolium Tausch ex Opiz
var. **flexile** (Newm.) Jermy
A. alpestre var. *flexile* (Newm.) Milde

BM, CGE
This plant of doubtful status is endemic to Scotland. The map is based on herbarium material seen by A. C. Jermy. As far as is known var. *distentifolium* occurs throughout the range of the species as indicated in the *Atlas*, p. 9 (as *A. alpestre*). (Nomenclature, cf. Sell, 1967.)

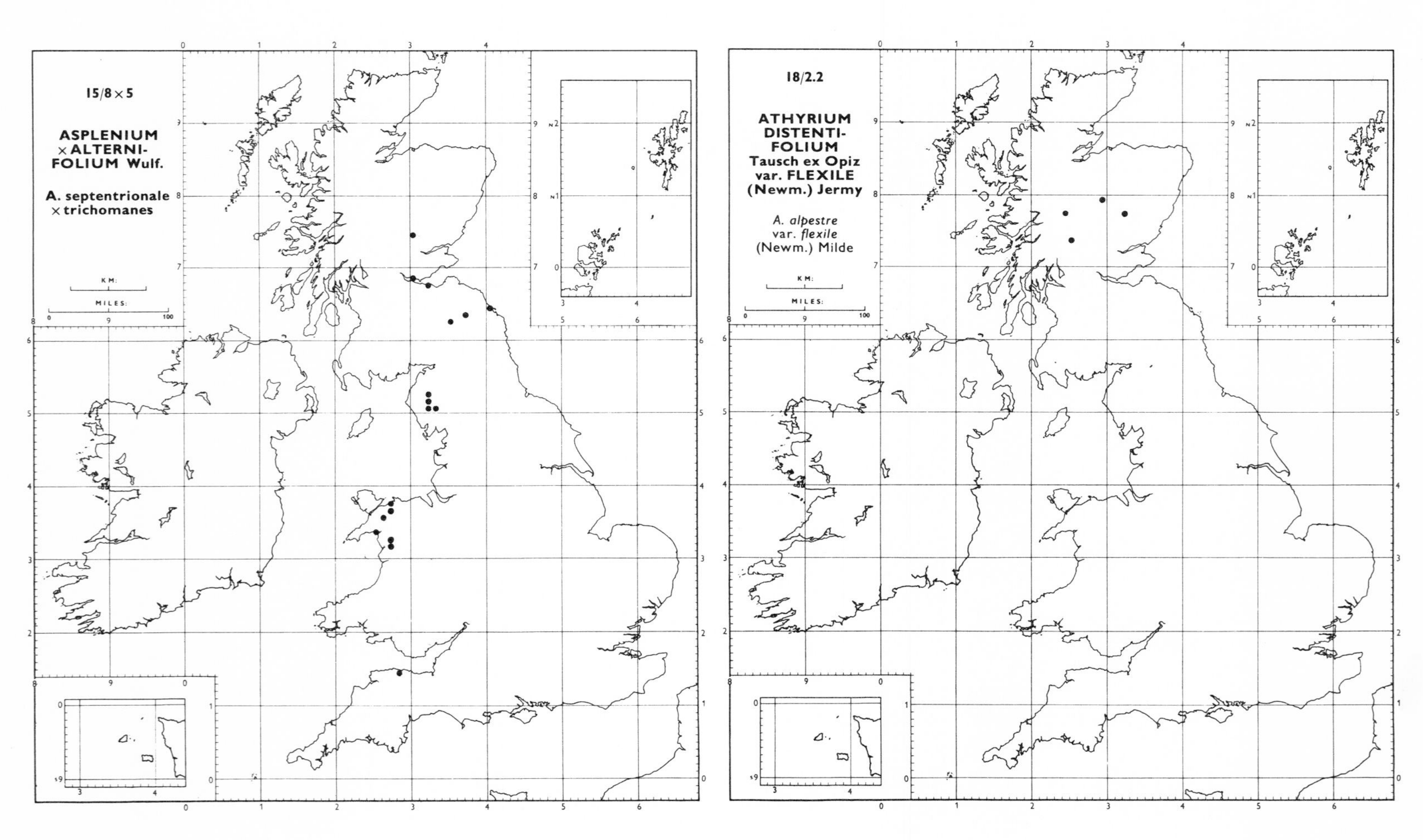

25/1 Polypodium vulgare L.
subsp. **vulgare**
subsp. **serrulatum** Arcangeli
P. australe Fée
subsp. **prionodes** Rothm.
P. interjectum Shivas

BFT, BM, CGE, DBN, E, NMW, NWH, OXF, SHY, TCD

The maps are based almost entirely upon counts of the thick-walled cells of the sporangia carried out on fresh or herbarium material. The majority have been identified by J. A. Crabbe, A. C. Jermy and the editor.

In Clapham, Tutin and Warburg, ed. 2 (1962), specific rank is given to each of these taxa. However, during the examination of several hundred specimens many were found to have sporangia with an intermediate number of thick-walled cells: such specimens were also found to have a number of aborted and empty spores. No records of intermediates are included in the maps.

The three possible hybrids which might arise have been recognized on morphological characters, and inferred from the presence of the parents growing together where the intermediates occur. The occurrence of such intermediates makes field identification of material impossible with certainty. For this reason the editor feels that it is better to regard the three taxa as subspecies rather than species.

The relative distribution of these three subspecies in the British Isles is similar to that shown in Europe. Subsp. *vulgare* occurs throughout most of Europe except the extreme north and south. Subsp. *serrulatum* is the common subspecies of the south and south-west. Subsp. *prionodes* is most frequent in west and west-central Europe. (Nomenclature, cf. Sell, 1967.)

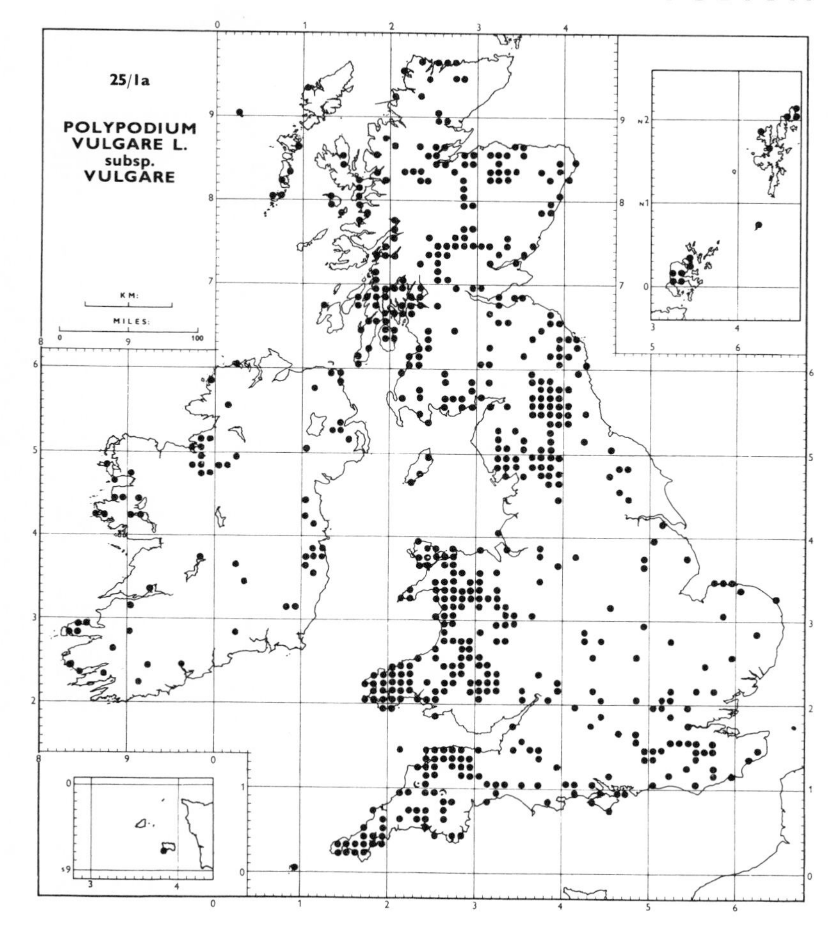

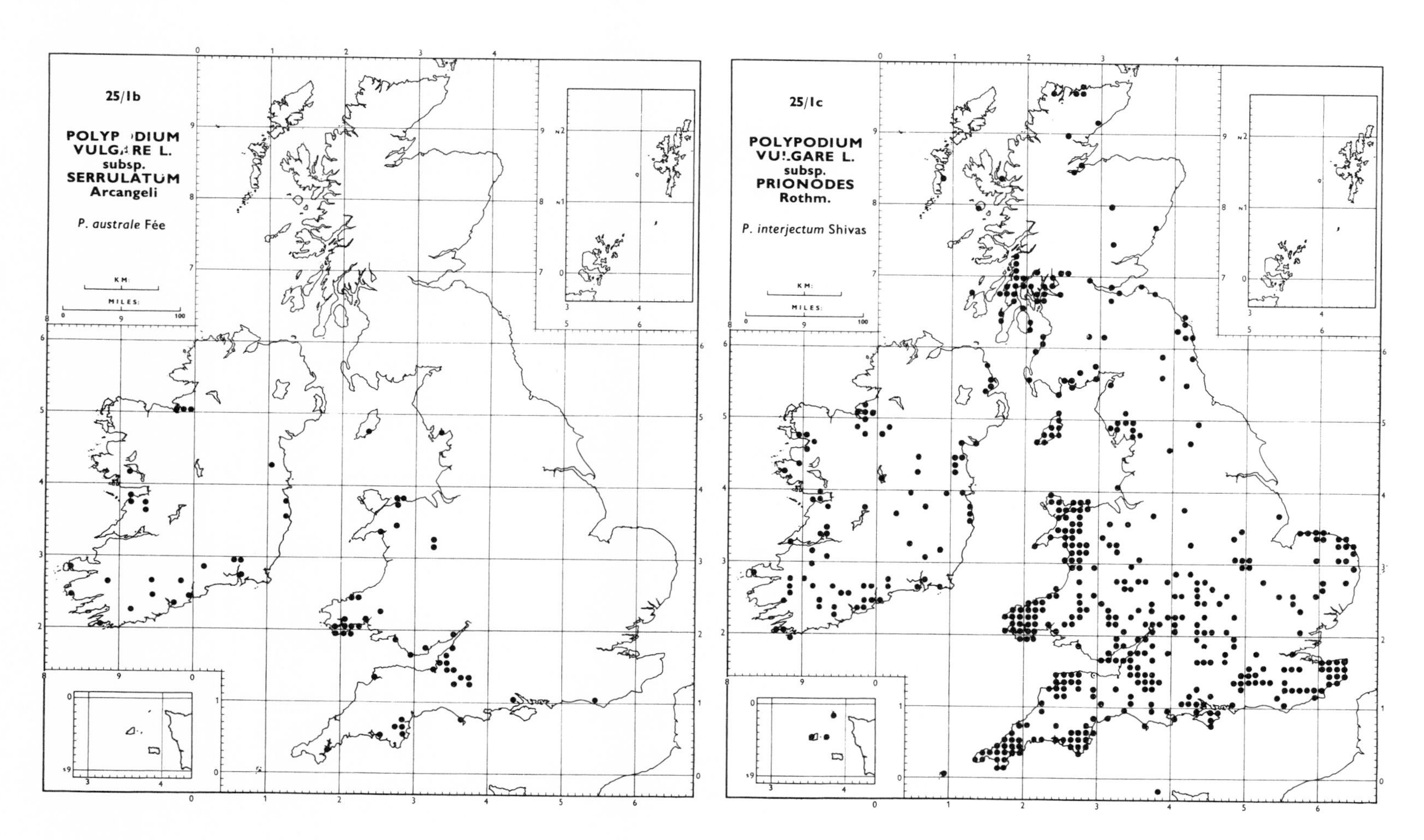

OPHIOGLOSSACEAE

29/1b Ophioglossum vulgatum L.
subsp. **ambiguum** (Coss. & Germ.) E. F. Warb.
O. azoricum C. Presl

BFT, BM, CGE, DBN, E, NMW, TCD

Herbarium material is rarely adequate to determine whether 2–3 leaves occur together as is reported for subsp. *ambiguum*. The map is based on specimens having 14 or fewer sporangia, a sterile blade less than 3·5 cm., and occurring in short turf near the sea. Specimens from Galway and Mayo (H. 16, 17) were recently discovered in the National Museum of Ireland. This subspecies replaces subsp. *vulgatum* in coastal regions of west Europe.

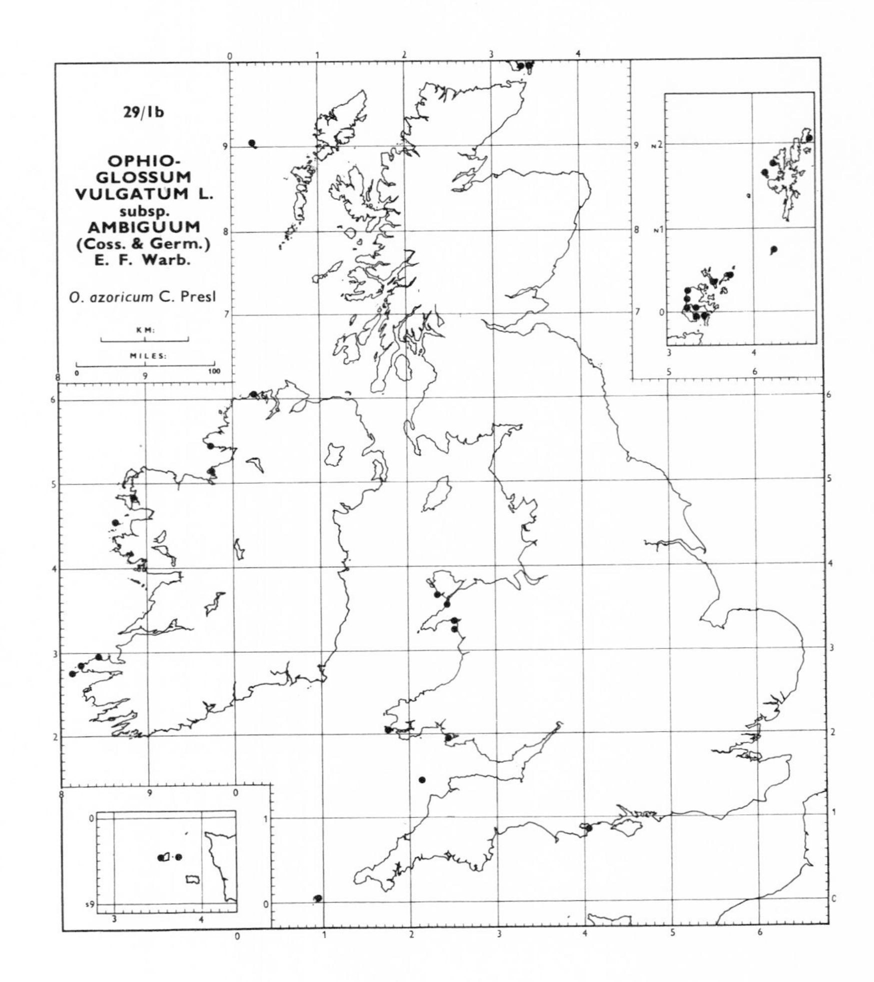

34/1 Juniperus communis L.
subsp. **communis**
subsp. **nana** Syme

BFT, BM, CGE, DBN, E, NMW, SHY, TCD

Although the majority of the material of this species can be referred with ease to one of the two taxa, there is a small percentage of specimens which must be regarded as intermediate in leaf shape and arrangement. A trap for the unwary is that not all prostrate juniper is subsp. *nana*. In the east of Scotland and Orkney, for example, prostrate forms with the leaf shape and arrangement of subsp. *communis* occur: in these maps they are included with upright subsp. *communis*. Records of specimens with intermediate morphology have been omitted; including the isolated record from the Lizard Peninsula, Cornwall (1) (*Atlas*, p. 17) which has a claim to be regarded as a third subspecies.

In the preparation of the map of subsp. *communis* all records received for the species which occur outside the range of subsp. *nana* have been accepted as subsp. *communis*. The Irish distributions of these taxa are based largely on the work of Praeger (Praeger, 1934b). Subsp. *nana* is mainly a plant of higher mountains in Europe, but it descends to sea-level in the north and west.

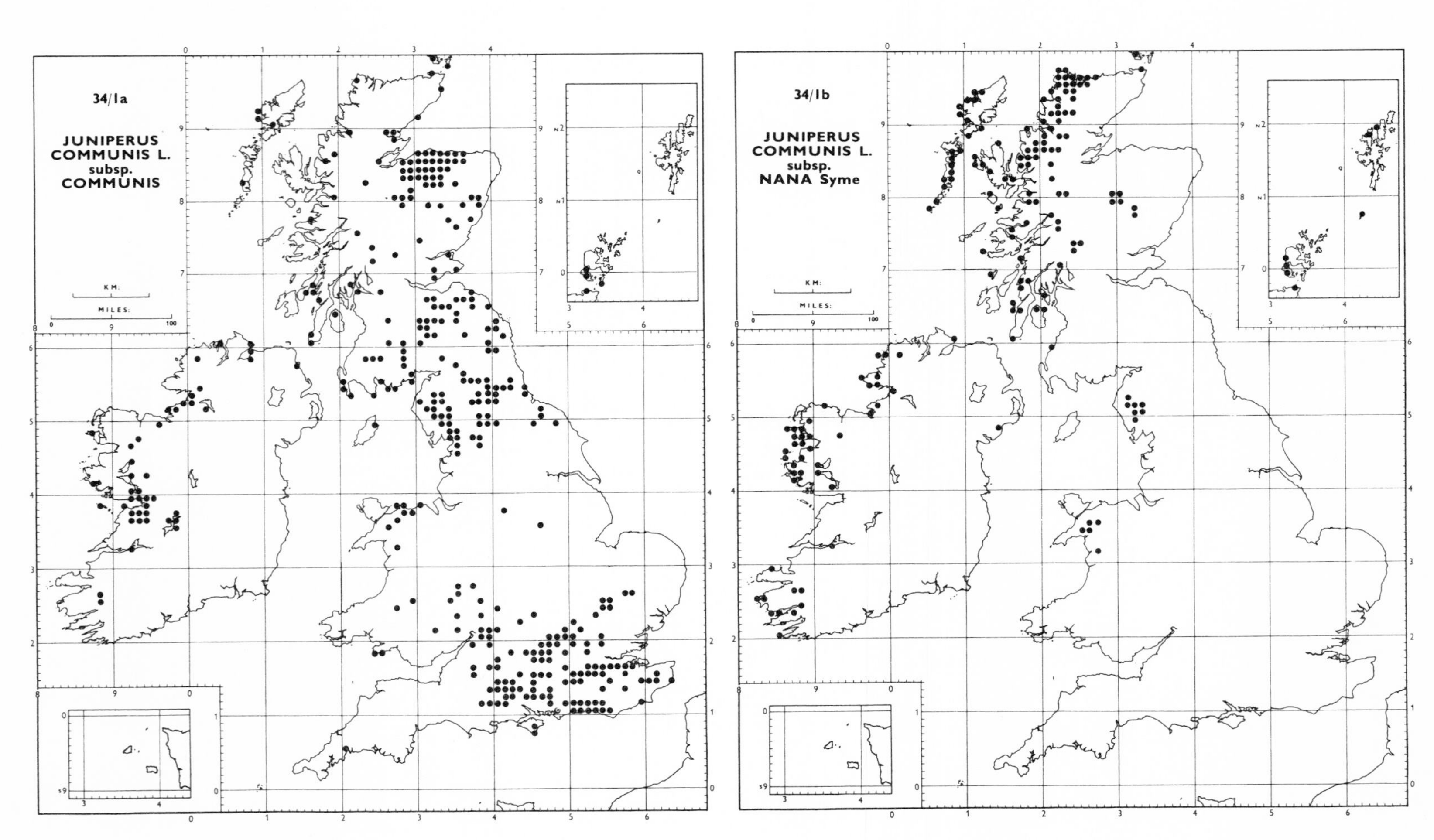

46/24 **Ranunculus ficaria** L.
subsp. **ficaria**
subsp. **bulbifer** Lawalrée

BFT, BM, CGE, DBN, E, NMW, NWH, OXF, SHY, TCD: Floras

Only a small percentage of herbarium material can be used for the determination of these two subspecies. The majority has been collected in too young a state for the development of the distinguishing characters. Herbarium material bearing fully developed ripe achenes and lacking axillary bulbils has been classified as subsp. *ficaria*: material bearing axillary bulbils, and with leaves less than 50 mm. wide, has been classified as subsp. *bulbifer*.

The maps are very incomplete. The species occurs throughout the British Isles, though it is local in the north and west of Britain and in central Ireland (*Atlas*, p. 24). There is considerable observer bias: much more field work has been done in the Lothians (82–84), Northumberland (67, 68), Isle of Man (71), Merioneth (48), Pembrokeshire (45) and the midlands of England than elsewhere: recording in Ireland and north Scotland is extremely poor. However, a difference in the distribution patterns is discernible. Subsp. *ficaria* may be found in all parts of the British Isles, but is less common in eastern Scotland; subsp. *bulbifer* is more easterly in its distribution and is rare on the west coast of Great Britain and in Ireland, and in most of these areas it only occurs as a garden weed. The more continental distribution of subsp. *bulbifer* is not unexpected as this is the only subspecies of *R. ficaria* known in Sweden. Subsp. *ficaria* is known only from western Europe. A map of a third taxon, subsp. **ficariiformis** Rouy & Fouc., has not been included. Without cultivation experiments it is sometimes difficult to distinguish from lush-growing forms of subsp. *ficaria*. It is a plant of the Mediterranean region which is almost certainly native in the Channel Islands (i.e. *R. ficaria* forma *luxurians* Moss).

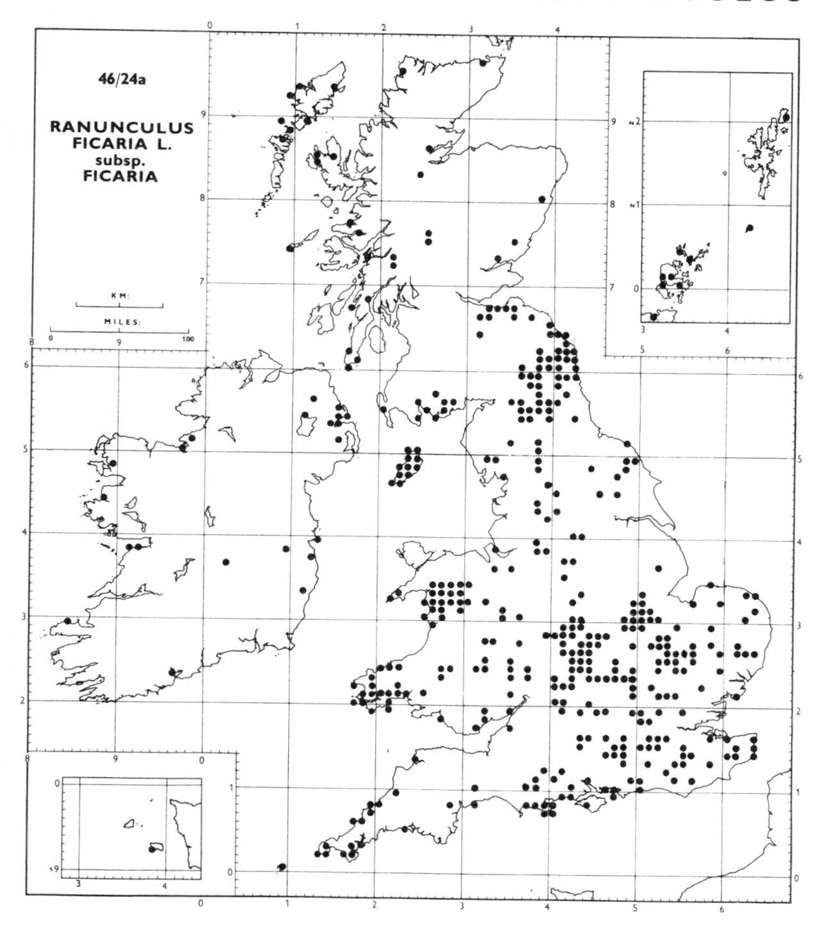

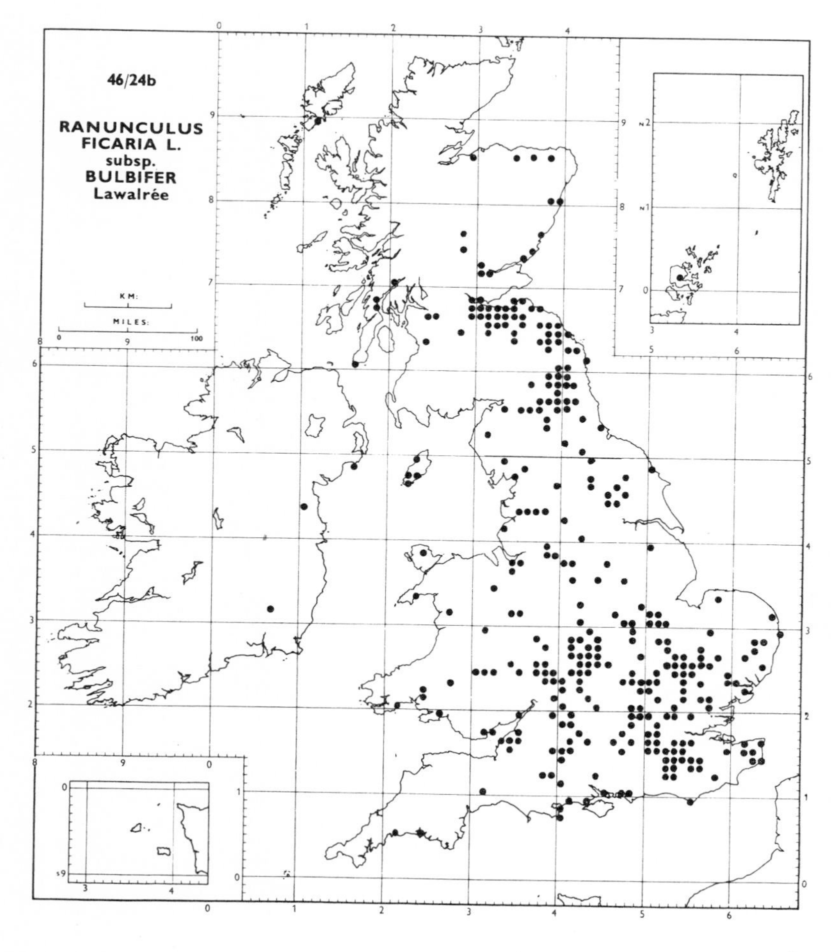

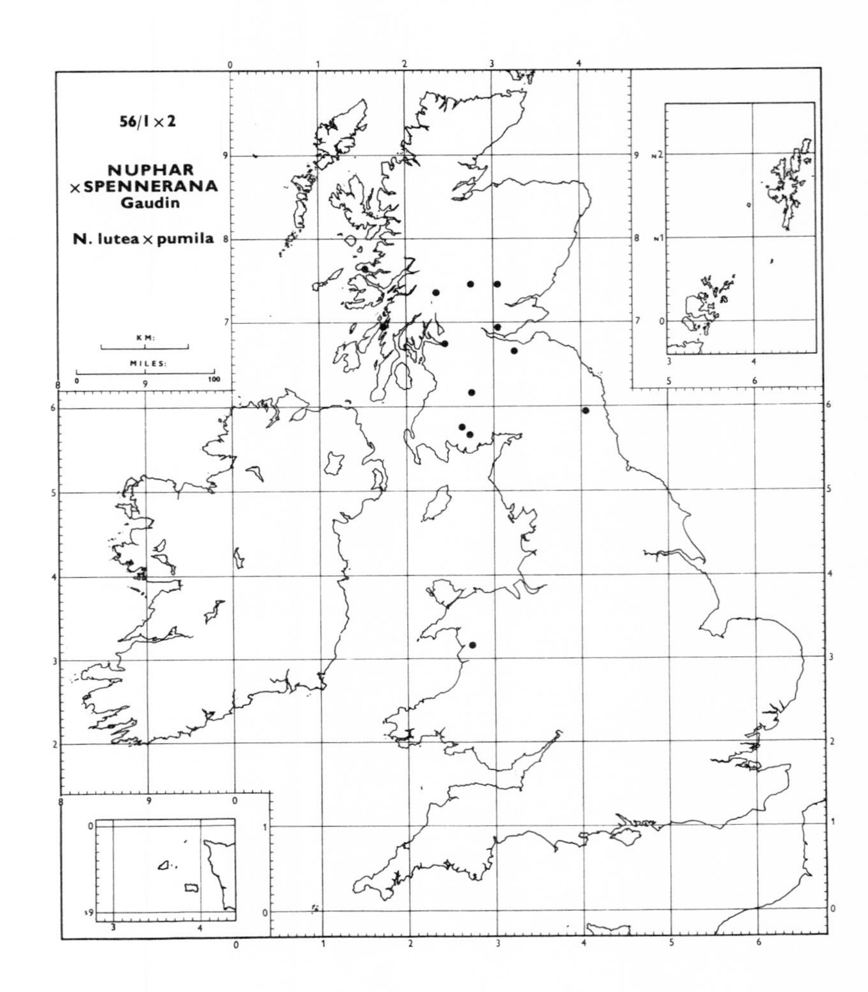

56/1 × 2 Nuphar ×spennerana Gaudin
=N. lutea × pumila

The map is based on the paper of Y. Heslop Harrison (1953). It is interesting that the hybrid occurs in the absence of one of the parents, *N. pumila* (see *Atlas*, p. 27). It is presumed that *N. pumila* did exist in north England at one time during the early post-glacial period, but has become extinct through the climatic amelioration and the gradual elimination of suitable habitats through the growth of peat bogs.

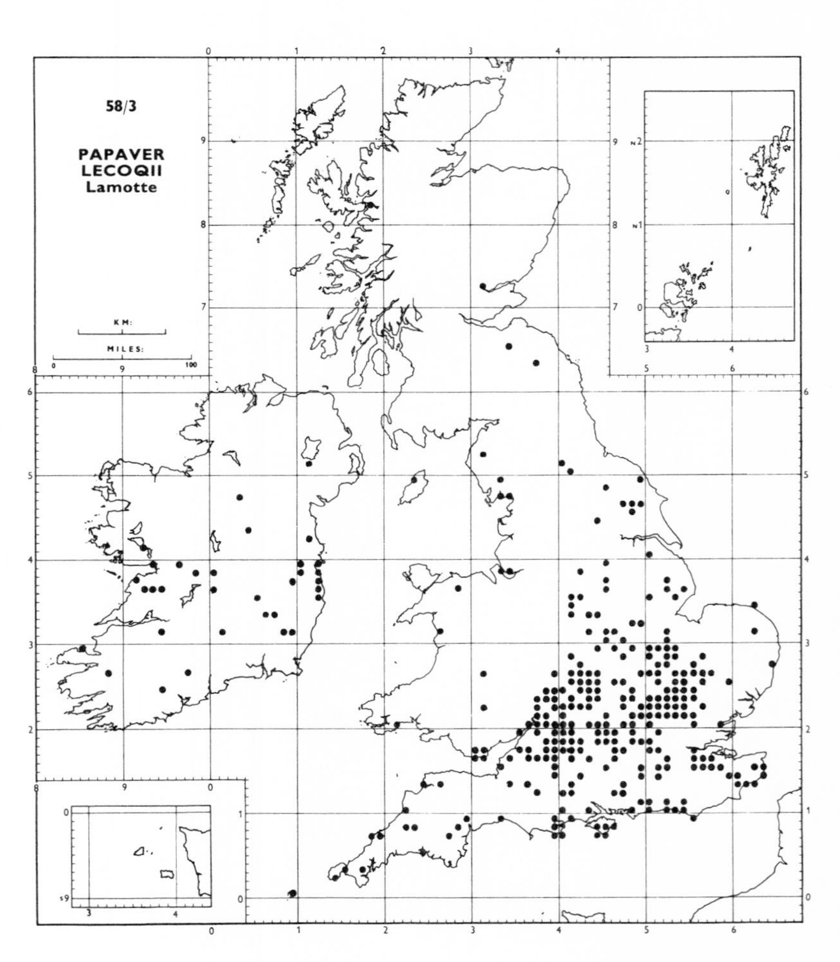

58/3 Papaver lecoqii Lamotte

Floras

In addition to the literature, field records based on the examination of the latex of this species, which turns dark yellow after exposure to the air, have been accepted. Records from the following v.-cs were either not traced or have not been confirmed: 36, 40, 46, 56, 72, 84, 85, 106, H.1, 3, 11, 16, 17, 19, 25, 26, 28, 31, 38–40. As far as is known, *P. dubium* occurs throughout the range of the aggregate (see *Atlas*, p. 28), but it is replaced locally by *P. lecoqii* on the chalk in south and east England.

66/6 Fumaria muralis Sond. ex Koch
subsp. **muralis**
subsp. **neglecta** Pugsl.

CGE, K

This map is based on the lists published in Pugsley (1912), subsequent records in the literature, and herbarium material determined by Pugsley or N. Y. Sandwith. Records of subsp. *muralis* for v.-cs 41, 43, 49, 51 have not been traced, and that for v.-c. 50 was an error. As far as is known the map of the species in the *Atlas* (p. 32) may be taken as subsp. **boraei** (Jord.) Pugsl. Subsp. *neglecta* is endemic to a cultivated field near Penryn, Cornwall (1). Although subsp. *muralis* occurs throughout the range of the species it is extremely rare in Britain, where it may have been introduced.

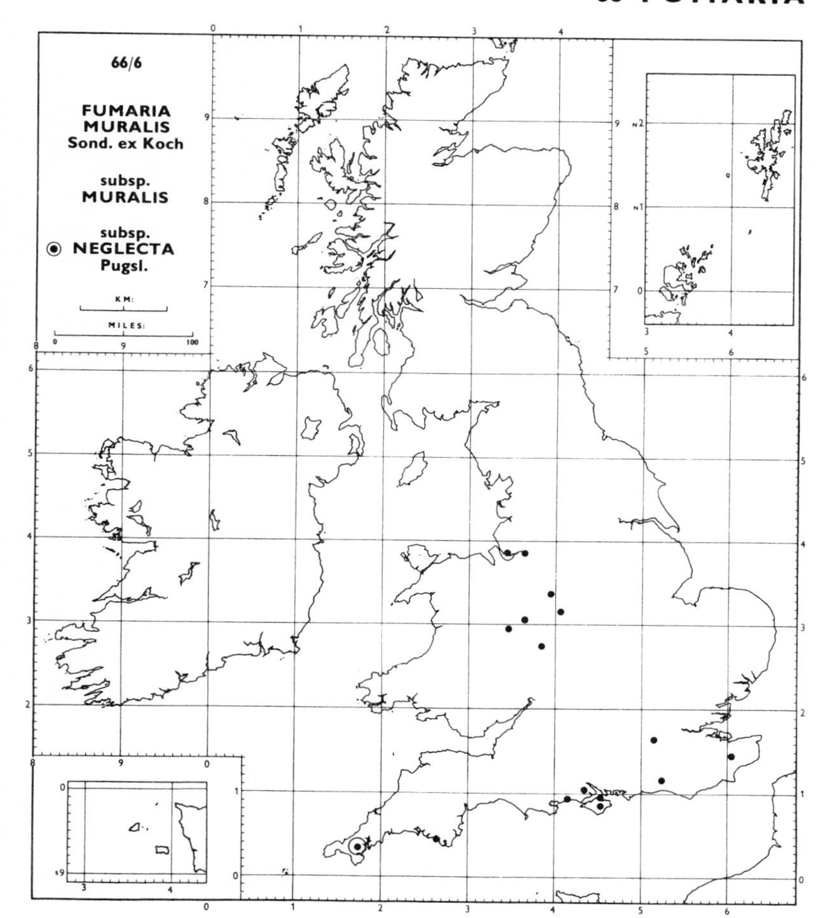

66/6×8 Fumaria ×painteri Pugsl.
=F. muralis × officinalis

CGE, K

An account of the taxon was first published in Pugsley (1912). It is the only fertile hybrid known in the genus and might perhaps prove to be a distinct species. Two records from Shropshire (40) were determined by Pugsley: another record from Kent (16) has been determined by J. E. Lousley.

66/8b Fumaria officinalis L.
subsp. **wirtgenii** (Koch) Arcangeli
F. officinalis var. *minor* Hausskn. et var. *wirtgenii* (Koch) Hausskn.

CGE, K

This map is based on the lists published in Pugsley (1912), and material determined by M. G. Daker, P. D. Sell, or N. Y. Sandwith. Subsp. *wirtgenii* tends to replace subsp. *officinalis* on the chalk of south-east England. It is the more continental subspecies in Europe.

Subsp. *officinalis* has the racemes normally more than 20-flowered and the sepals 2·5–3·5 × 1–1·5 mm. Subsp. *wirtgenii* has the racemes 10- to 20-flowered and the sepals 1·5–2 × *c.* 1 mm. (cf. Sell, 1963a).

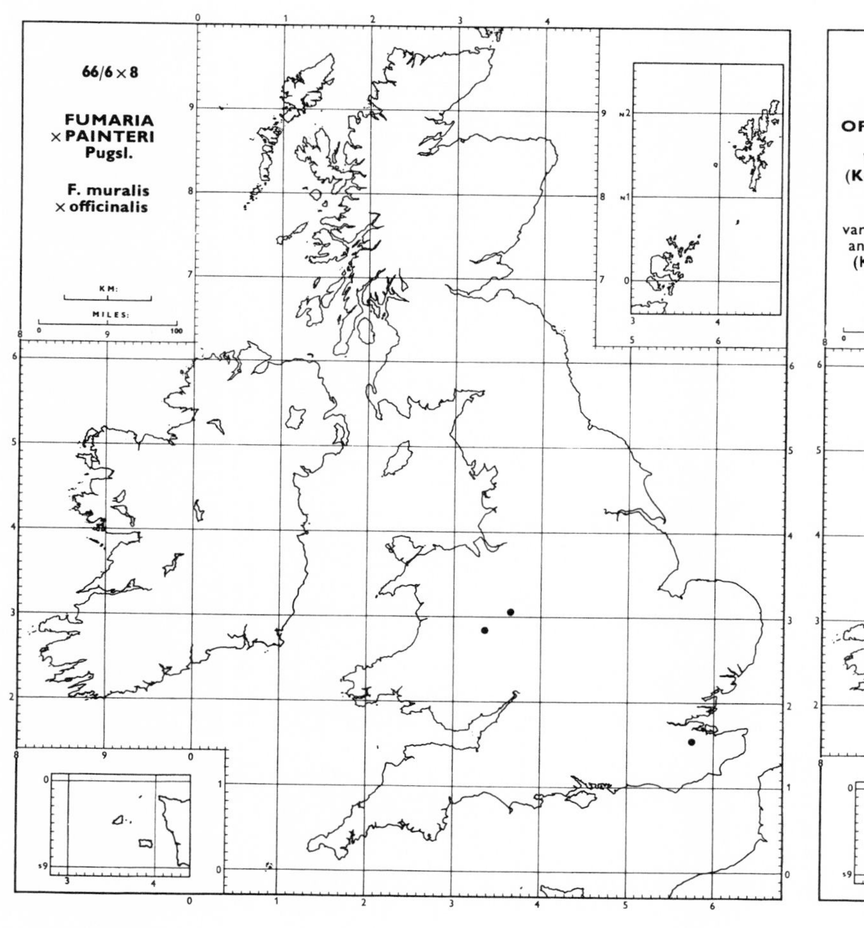

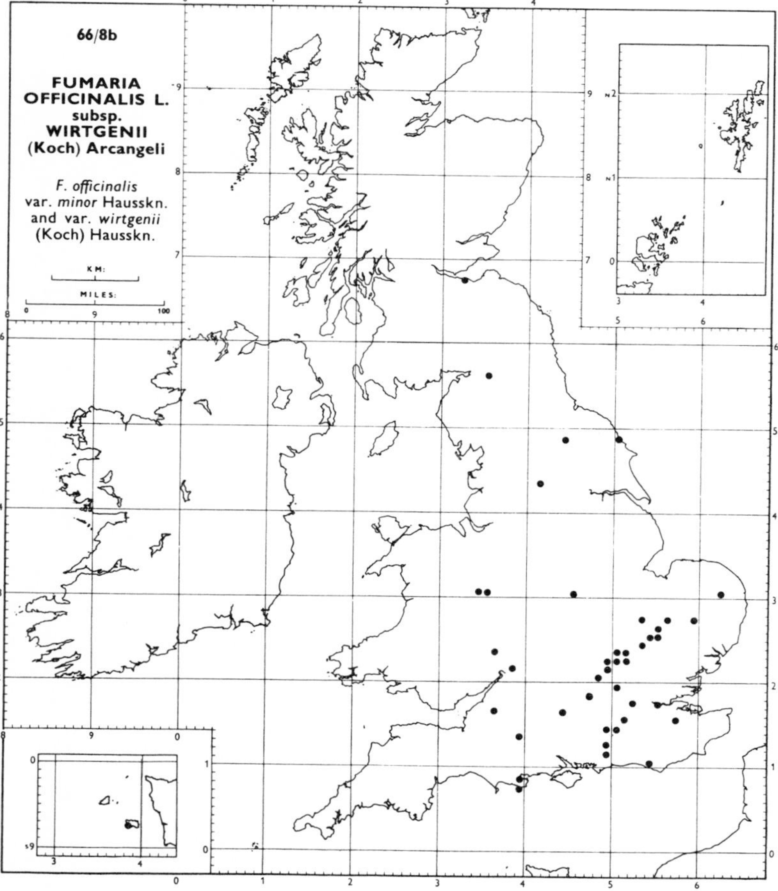

CRUCIFERAE

74/1 **Raphanus raphanistrum** L.

BM, CGE, E, NMW, SHY, TCD

This species exhibits a bewildering variation in petal colours: yellow, lilac or white, often with dark veins, but golden yellow and unveined in some forms. There are many intermediates. There appears to be no geographical distinction between veined and unveined forms, but there is between yellow and white forms. Maps of these are included and are based on field records and herbarium material wherever the colour of the flowers was stated; the colour of the petals cannot safely be determined from the dried plant, unless well prepared and of recent origin. The maps are very incomplete and serve only to indicate the tendency for white forms to be rare in Scotland, whereas yellow forms occur almost anywhere within the range of the species (*Atlas*, p. 36). Pure yellow forms, without prominent veins, are more frequent in the north.

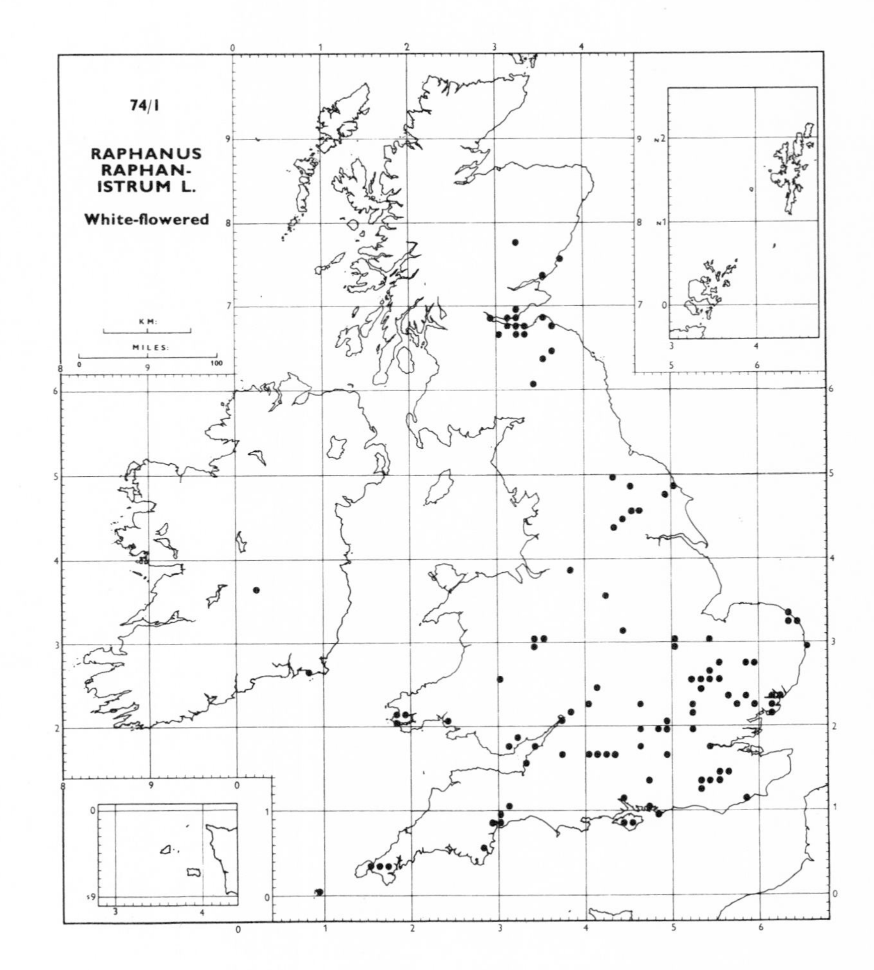

95/2–3 **Erophila verna** (L.) Chevall. subsp. **spathulata** (Láng) Walters

E. spathulata Láng; *E. praecox* auct. brit.

BFT, BM, CGE, E, NMW, TCD

This map is based almost entirely on herbarium material seen by the editor. A preliminary inspection of material suggested that whereas it was not difficult to distinguish *Erophila verna* (L.) Chevall. subsp. *verna*, it was impossible to divide what remained into satisfactory taxa. However, measurements of pod dimensions showed that specimens in which the length is less than 5 mm. and the length/breadth ratio is less than 2 form a group with a distinct geographical distribution and ecological requirements. In its most characteristic form the pods are *c.* 4·0 × 2·5 mm. In the north and west the taxon is almost confined to sand-dunes. This group appears to include most of the specimens referred to *E. spathulata* and *E. praecox* in Britain in the past, with the exception of *E. spathulata* var. *inflata* (Bab.) O. E. Schulz which in pod size and distribution seems more closely related to subsp. *verna*. The taxon here mapped is in accord with the treatment by Walters (Tutin *et al.*, 1964). Subsp. *spathulata* has a similar general distribution to that of subsp. *verna*, but apparently prefers more calcareous substrata. Subsp. *praecox* (Stev.) Walters is mainly a plant of the Mediterranean region and has not been recorded from the British Isles.

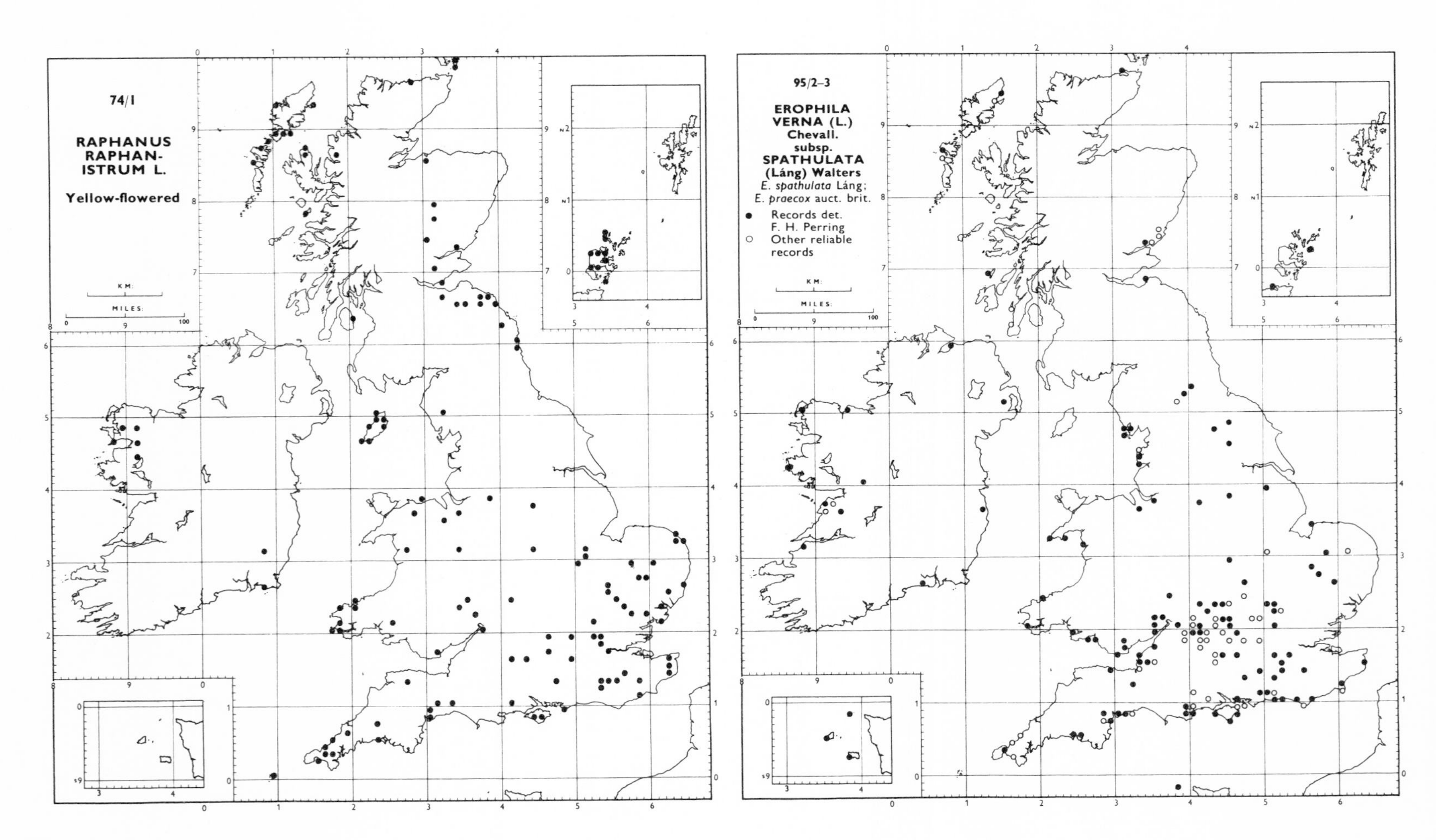

102/1 Rorippa nasturtium-aquaticum (L.) Hayek
Nasturtium officinale R. Br.

102/2 Rorippa microphylla (Boenn.) Hyland.
Nasturtium microphyllum (Boenn.) Reichb.

ABD, ABS, BIRM, BM, CGE, DBN, DHM, E, GL, HAMU, HLU, K, MANCH, OXF, NMW, STA (Pre-1951)
These maps are based on the work of Howard and Lyon (1950, 1951) with the addition of many field records. An aggregate map was published in the *Atlas* (p. 48) because at the time of going to press it was considered that the maps of the two segregates were inadequate. The difficulty has not been completely overcome even now, as in some counties records of neither segregate have been received. The deficiency is particularly noticeable in the east midlands. *R. nasturtium-aquaticum* occurs throughout Europe whereas *R. microphylla* occurs mainly in the west.

102/2 × 1 Rorippa × sterilis Airy Shaw
=R. microphylla × nasturtium-aquaticum
Nasturtium microphyllum × officinale

ABD, ABS, BIRM, BM, CGE, DBN, DHM, E, GL, HAMU, HLU, K, MANCH, OXF, NMW, STA (Pre-1951)
The map is based on the work of Howard and Lyon (1950, 1951), with the addition of many field records. Though it is clearly incomplete, particularly in parts of northern England, it indicates that the hybrid may occur in the absence of either parent, particularly in the upland areas of the Welsh border and on the eastern flanks of the northern Pennines.

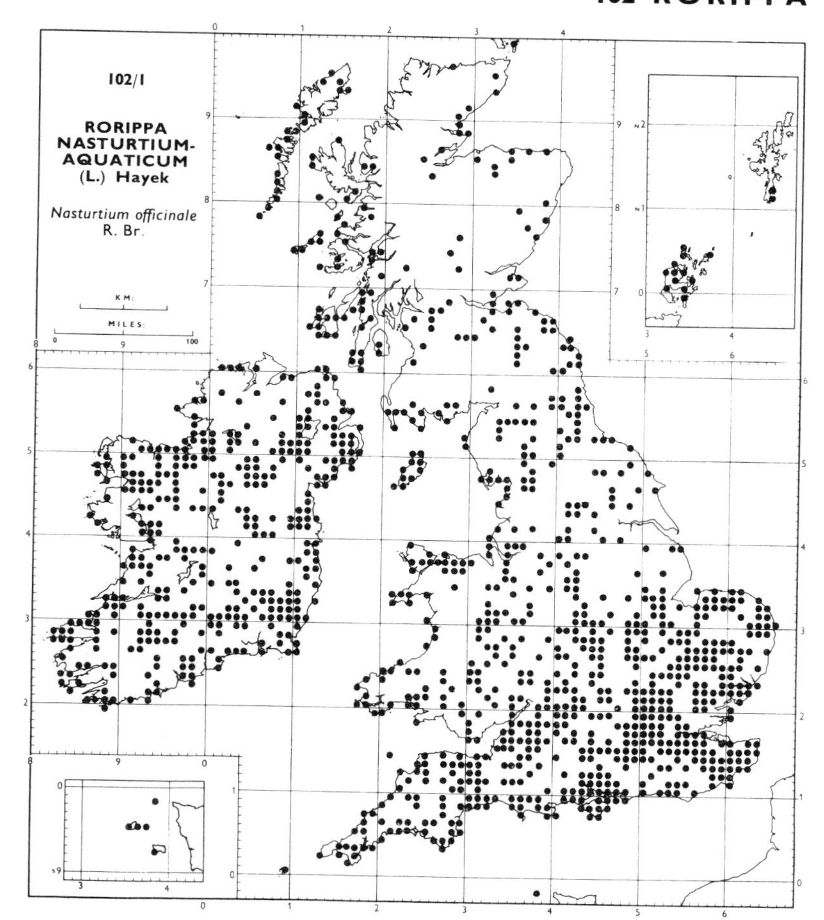

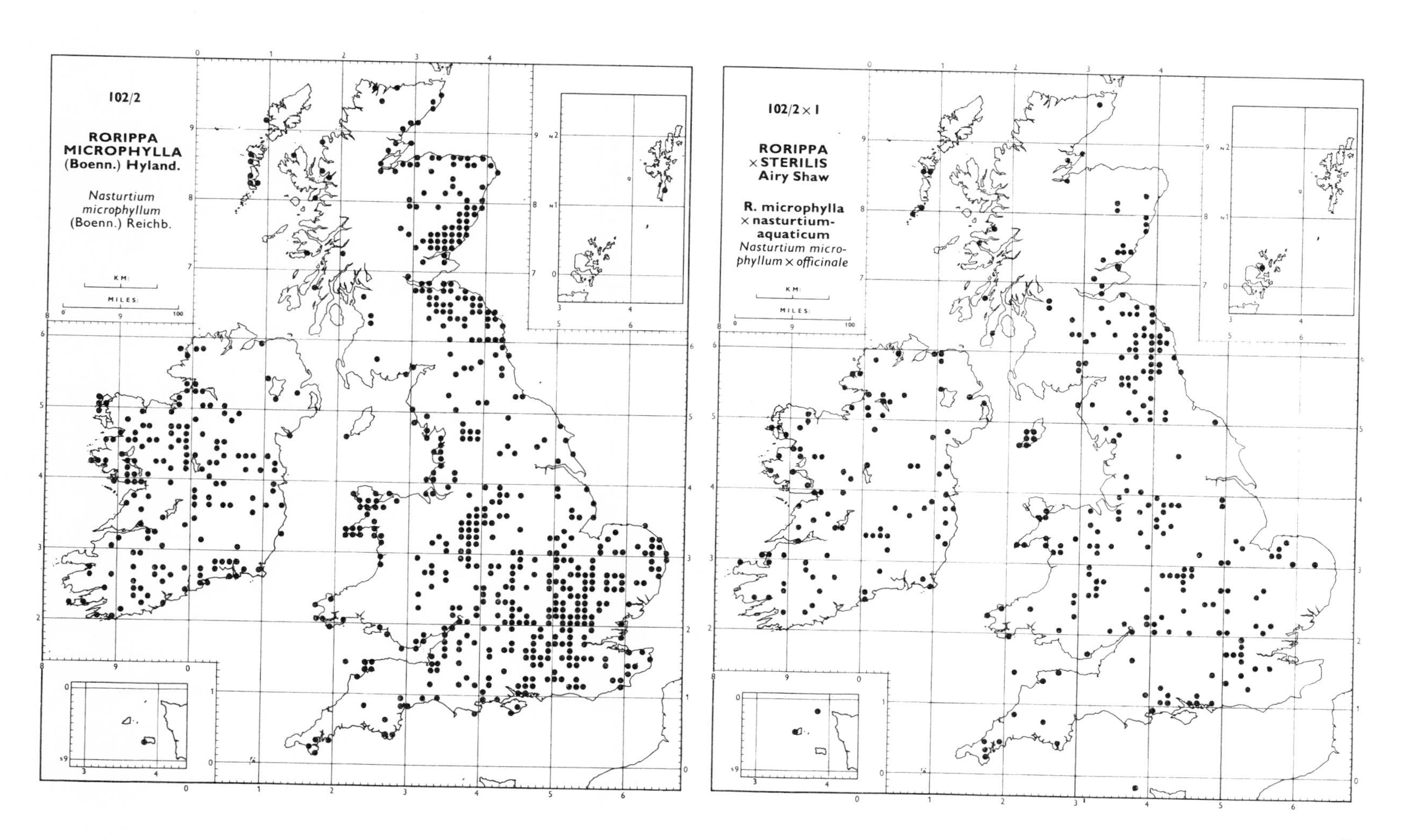

VIOLACEAE

113/9b **Viola palustris** L.
subsp. **juressi** (Neves) P. Fourn.
V. epipsila auct. brit.

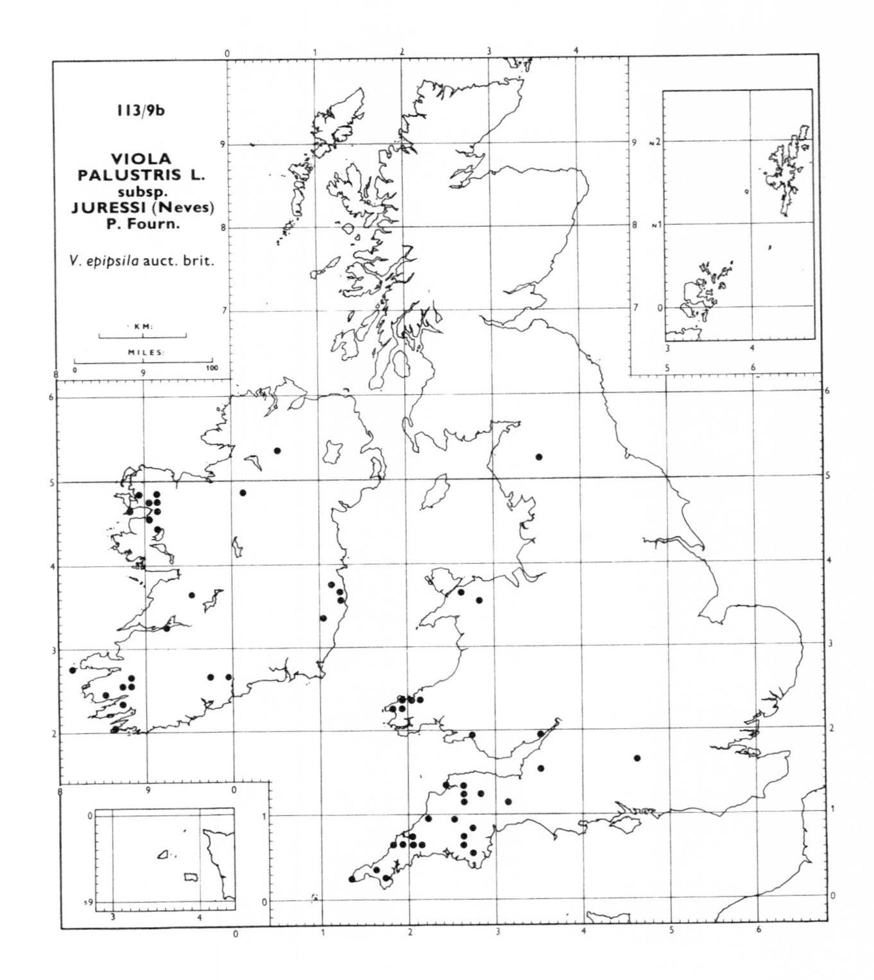

BFT, BM, CGE, DBN, NMW, OXF, SHY, TCD: Floras

Besides the records from herbaria and Floras, field records of specimens with hairy petioles have been accepted. In addition the works by Gregory (1912, 1918) were consulted and all records of *V. epipsila* were accepted with the exception of those for forma *glabrescens* Aschers. & Graebn., which is said to have the other characters of subsp. *juressi* but lacks the hairy petioles. The records made of these glabrous forms fall outside the geographical range of subsp. *juressi* as here recognized. The published map accords well with the known distribution of this taxon outside the British Isles: Portugal, Spain, and southern France (Clapham, Tutin and Warburg, 1962). Records from the following vice-counties were either not traced or have not been confirmed: 27, 37, 43, 48, 88, 90. Subsp. *palustris* appears to occur throughout the range of the species (*Atlas*, p. 54).

115/6 **Hypericum maculatum** Crantz
subsp. **maculatum**
subsp. **obtusiusculum** (Tourlet) Hayek

BFT, BM, CGE, DBN, E, GL, K, MANCH, NMW, OXF, SHY, STA, TCD

These maps have been prepared from data largely supplied by N. Robson, who established the presence of the two subspecies in Britain (Robson, 1957). There is little overlap in their geographical distribution: subsp. *obtusiusculum* is presumably the only one which occurs outside Scotland and records for the species (*Atlas*, p. 58) might be added to this map; however, because there is occasional confusion between this species and *H. perforatum*, the map has been based on herbarium material or records confirmed by Robson alone. Subsp. *obtusiusculum* is a plant of west Europe, whereas subsp. *maculatum* has a predominantly eastern distribution.

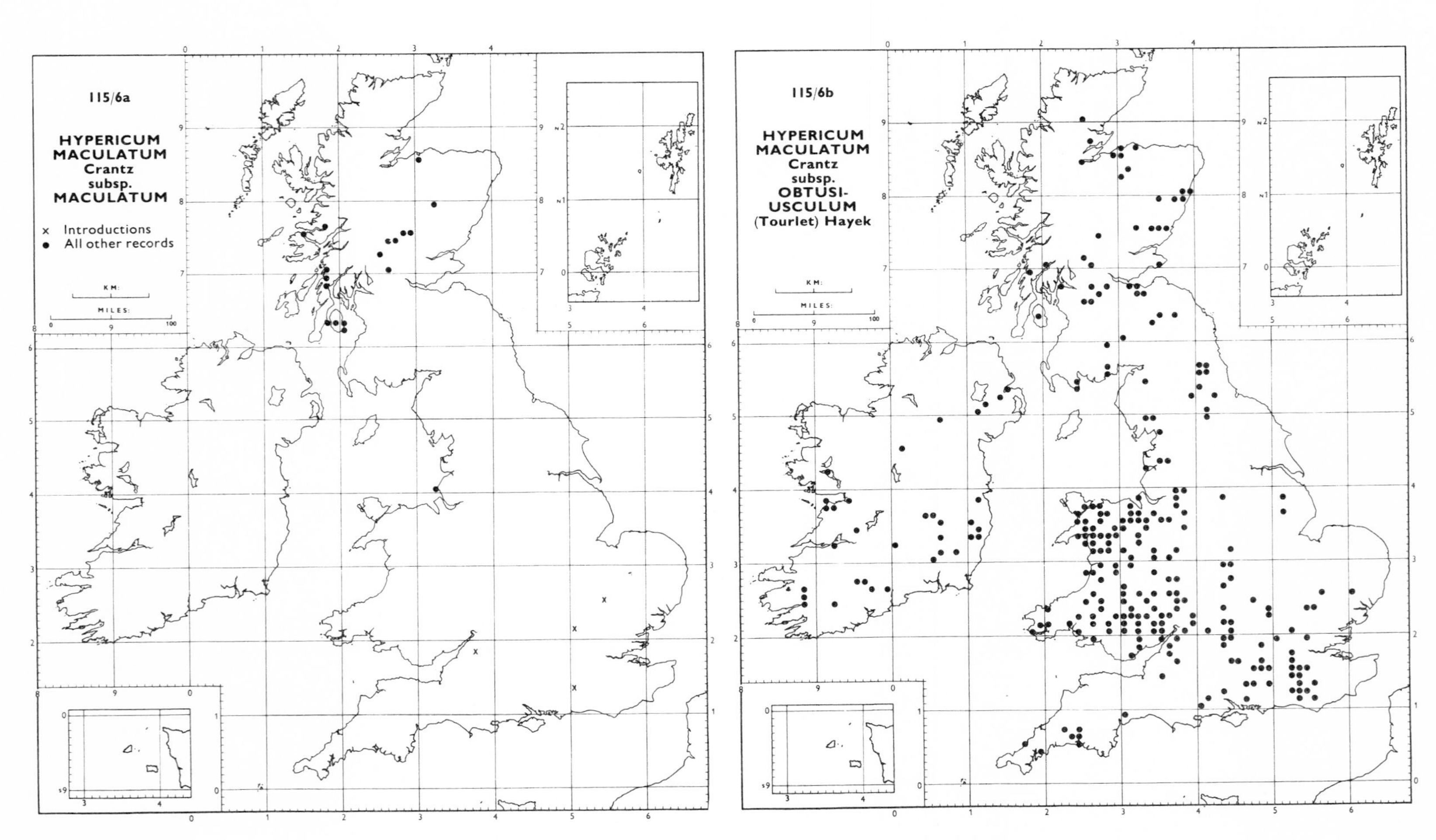

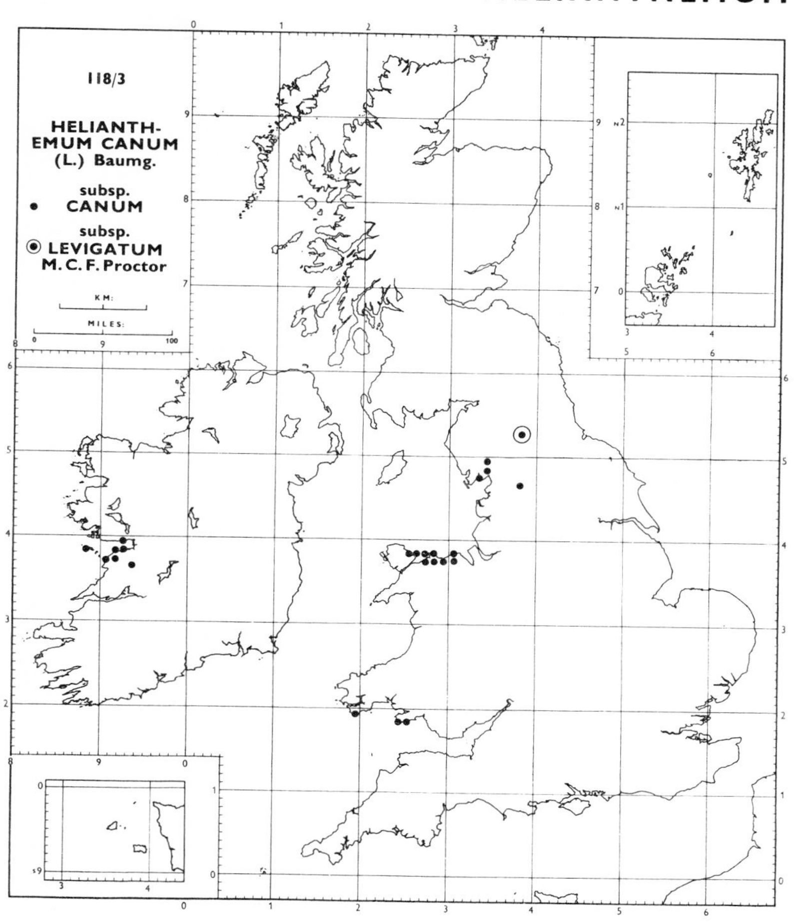

118/3 Helianthemum canum (L.) Baumg.
subsp. **canum**
subsp. **levigatum** M. C. F. Proctor

The map of these two subspecies is based on the work of Proctor (1957). Subsp. *levigatum* is an endemic confined to Cronkley Fell, Upper Teesdale (65). On the Continent *H. canum* is a very variable plant and several subspecies occur.

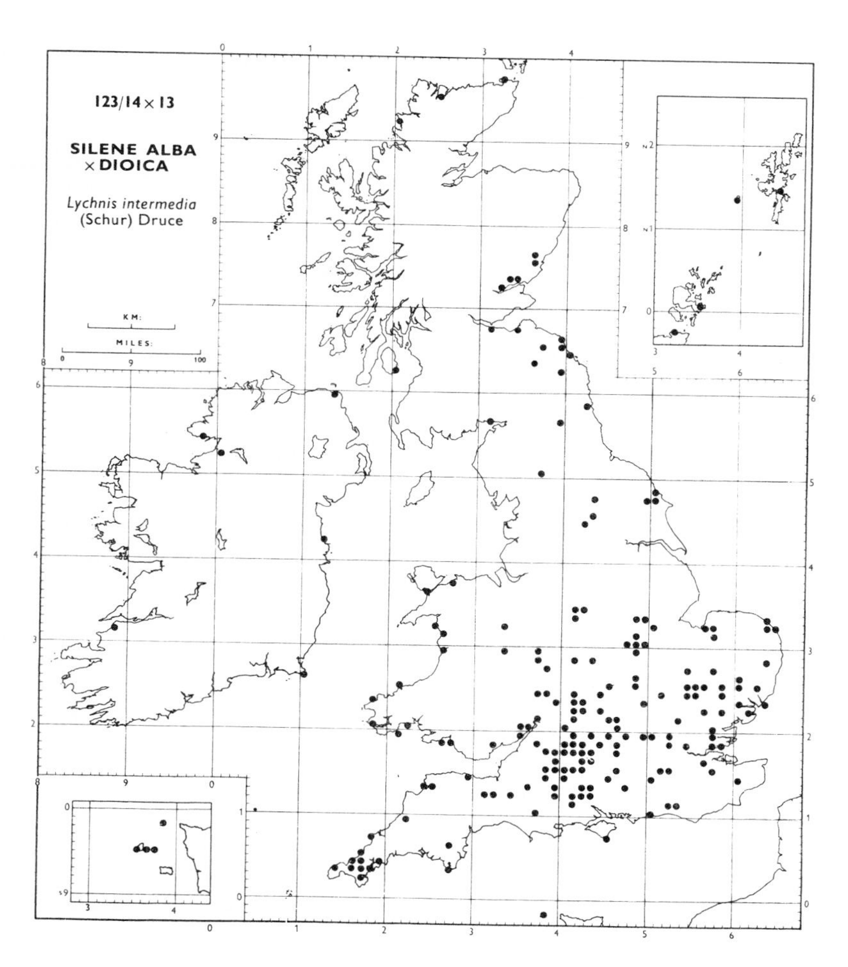

123/14 × 13 Silene alba (Mill.) E. H. L. Krause
× **S. dioica** (L.) Clairv.
Lychnis intermedia (Schur) Druce

CGE, E, NMW: Floras
This hybrid is most readily recognized by the colour of its flowers, which are varying shades of pink, although in many other characters it is also intermediate between the two parents. For this reason many field records were received. Nevertheless the map is probably very incomplete, especially in the north midlands, where it is recorded from v.-cs 56, 60, 63 and 65, but no localities have been traced. It may be expected to occur wherever the two species grow together. It is found in the absence of *S. dioica* in Cambridgeshire (29).

CARYOPHYLLACEAE

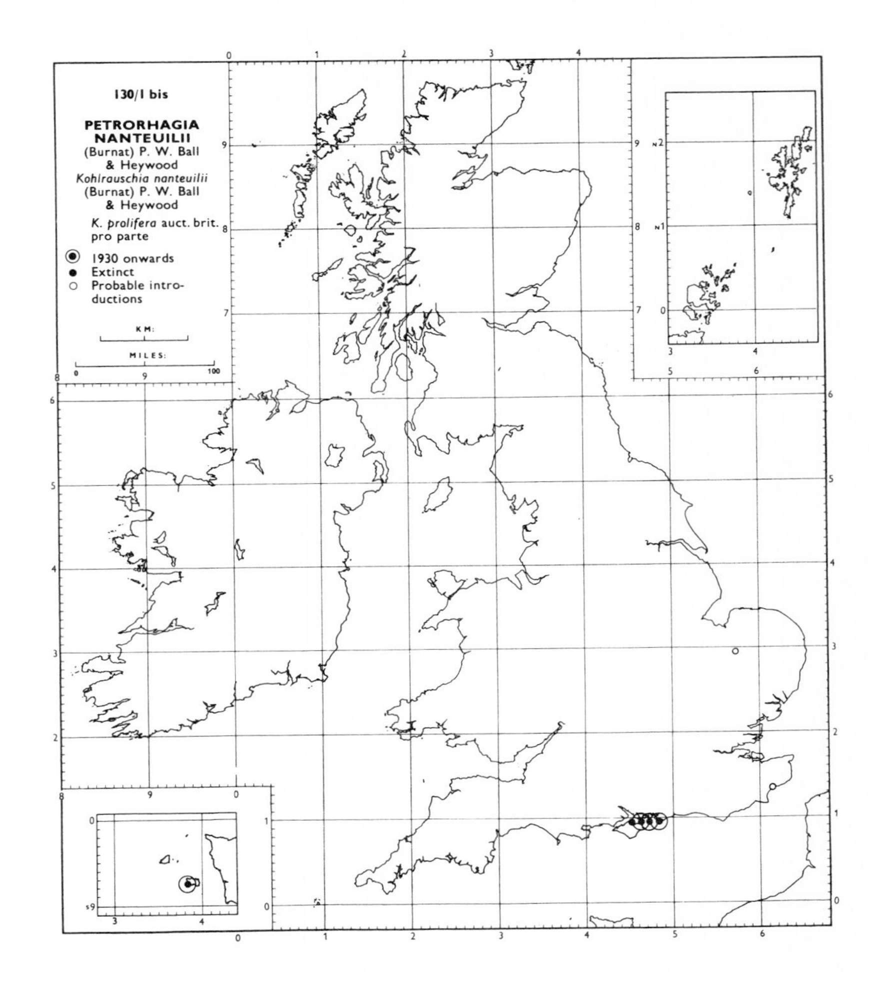

130/1 bis Petrorhagia nanteuilii (Burnat) P. W. Ball & Heywood
Kohlrauschia nanteuilii (Burnat) P. W. Ball & Heywood;
K. prolifera auct. brit. pro parte

This species was first recognized in Britain by Ball and Heywood (1962). It is native in the Canary Islands, Morocco and west Europe north to Britain. The remaining records published in the *Atlas*, p. 66, as *Kohlrauschia prolifera* are referable to **Petrorhagia prolifera** (L.) P. W. Ball & Heywood *sensu stricto*. They are all introductions, the species being native to central and southern Europe.

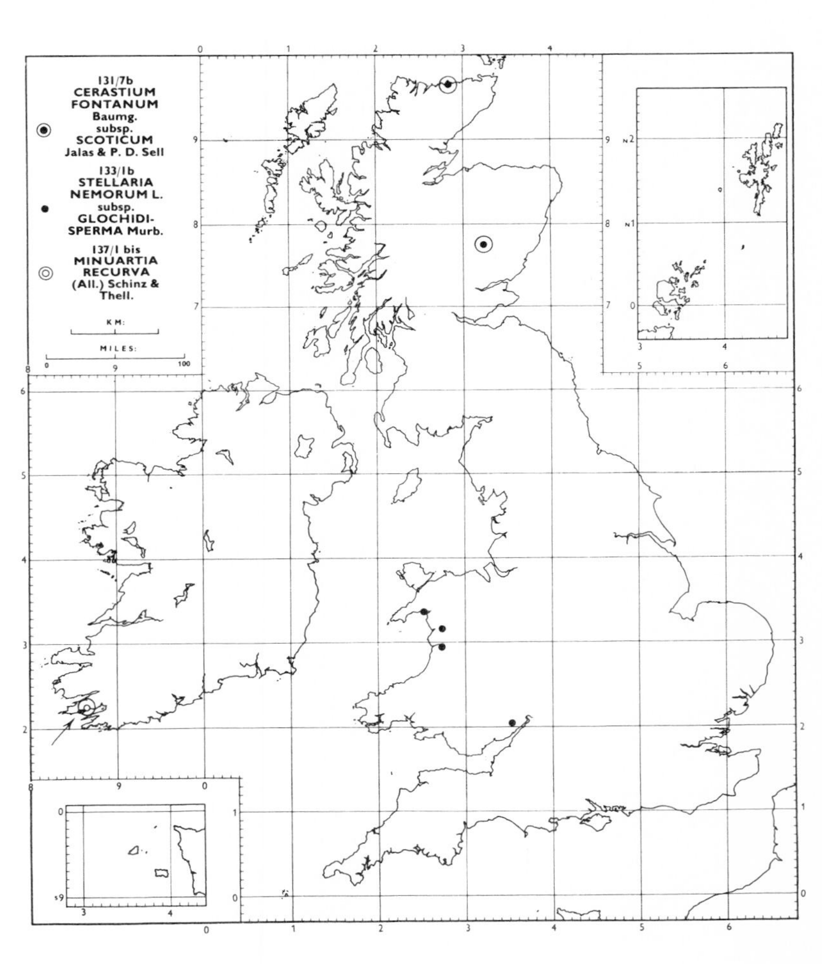

131/7b Cerastium fontanum Baumg.
subsp. **scoticum** Jalas & P. D. Sell

BM, CGE, K
This subspecies, which is endemic to the mountains of the Clova region (90) and coastal rocks at Strathy Point (108), is most nearly allied to subsp. *scandicum* Gartner from Fennoscandia and to subsp. *fontanum* from the Alps and Carpathians. It can be distinguished from all other British forms by its large seeds, 0·8–1 mm. diameter. All records in the *Atlas*, p. 68, for *C. holosteoides* refer to **C. fontanum** subsp. **triviale** (Murb.) Jalas, which is widespread throughout lowland Europe. (cf. Sell, 1967.)

133/1b Stellaria nemorum L.
subsp. **glochidisperma** Murb.

This map is based on a paper by Green (1954), with additional information supplied by P. M. Benoit. It is mainly a plant of west and central Europe.

137/1 bis Minuartia recurva (All.) Schinz & Thell.

This species was discovered in the British Isles for the first time during the summer of 1964. It is found on the mountains of southern and south-central Europe from Portugal to the south Carpathians.

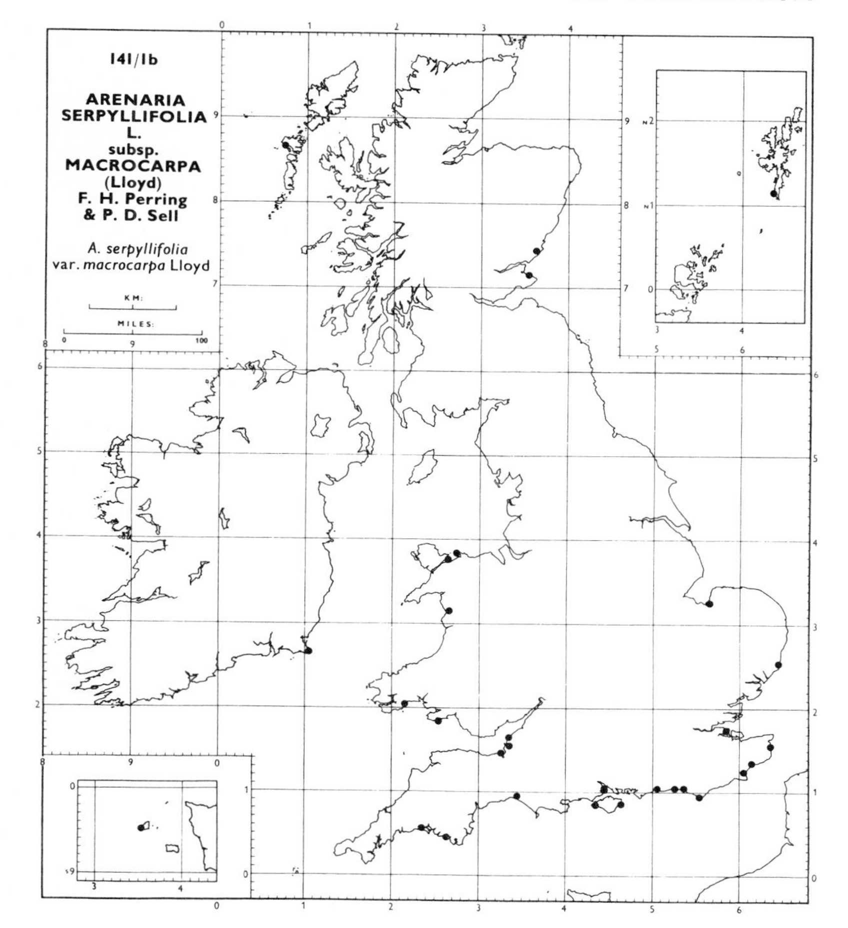

141/2 Arenaria serpyllifolia L.
subsp. **leptoclados** (Reichb.) Nyman
subsp. **macrocarpa** (Lloyd) F. H. Perring & P. D. Sell
A. serpyllifolia var. *macrocarpa* Lloyd

BFT, BM, CGE, DBN, E, NMW, NWH, SHY, SPN, TCD
All records for these two maps are based on material seen by the editor.

Subsp. *leptoclados* has a straight-sided capsule less than 3 mm. and possible to indent without fracture when mature; the pedicel is slender, about 0·3 mm. broad and upturned towards the tip; and the ripe seeds are 0·4 × 0·4 mm.

Subsp. *macrocarpa* has very dense inflorescences in which the pedicels are shorter than the sepals, and large seeds which exceed 0·6 × 0·4 mm.

Subsp. **serpyllifolia** has a capsule distinctly swollen at the base, exceeding 3 mm. and which fractures when pressed at maturity; the pedicel is stout, about 0·5 mm. broad, and straight; the ripe seeds are kidney-shaped and 0·5 × 0·4 mm.

As far as is known, the distribution of subsp. *serpyllifolia* in the British Isles is as shown in the *Atlas*, p. 76. In Europe it has a widespread distribution. Subsp. *leptoclados* is mainly in south, west and central Europe, whilst subsp. *macrocarpa* is confined to the coasts of the west. (Nomenclature, cf. Sell, 1967.)

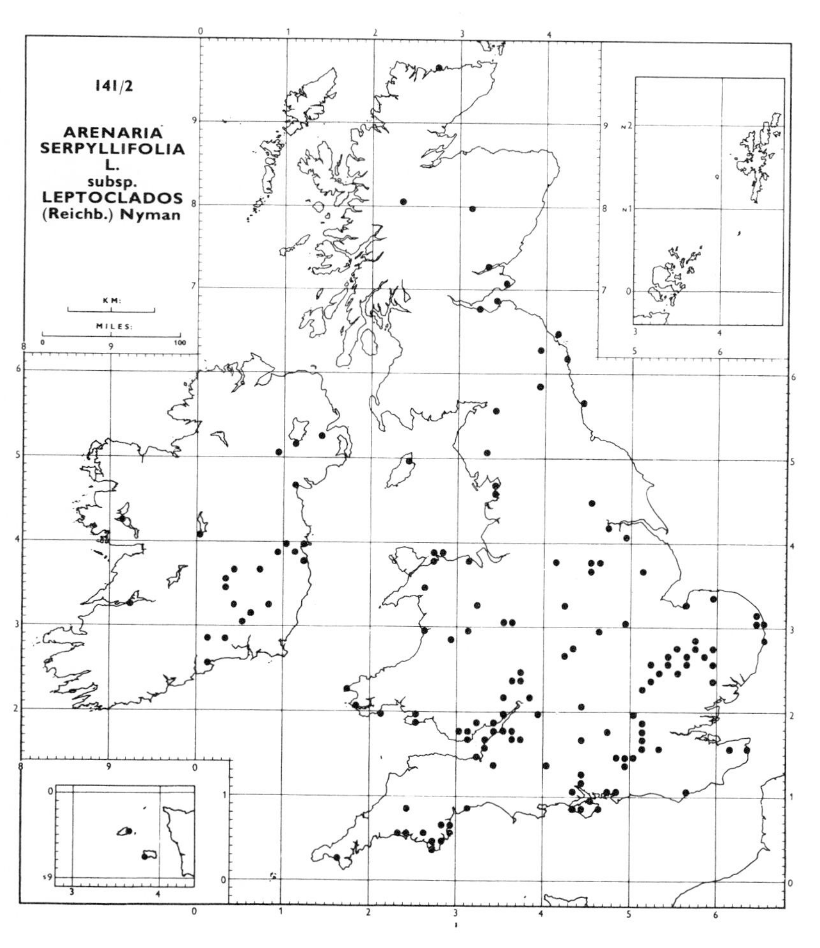

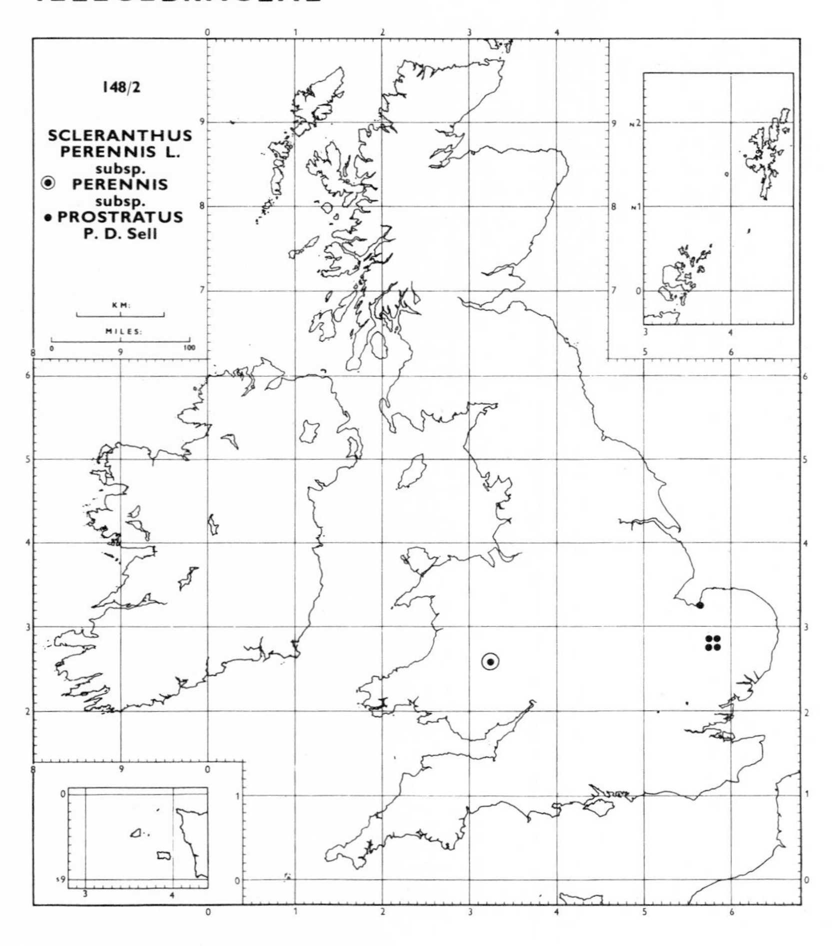

148/2 Scleranthus perennis L.
subsp. **perennis**
subsp. **prostratus** P. D. Sell

BM, CGE, NWH
The presence of two subspecies of *Scleranthus perennis* in this country has recently been recognized (Sell, 1963b). Subsp. *perennis* has the stems ascending, the internodes 6–10 mm. long, and fruits 3·5–4·5 mm. long. In the British Isles it is confined to rocky places at Stanner Rocks, Radnor (43). It is found throughout the lowlands of continental Europe. Subsp. *prostratus* has more or less procumbent stems, internodes 2–3 mm. long, and fruits 2–3 mm. long. It is a very local endemic plant of the sandy heaths of Suffolk (26) and Norfolk (28). Some of the records in the *Atlas*, p. 79, have been omitted as no specimens from these localities have been traced. A specimen recently discovered in CGE collected at Shaugh, Devon (3), in 1841 is clearly *S. perennis*, but it is inadequate to be certain to which subspecies it belongs.

149/1 Montia fontana L.
subsp. **fontana**
subsp. **variabilis** Walters
subsp. **amporitana** Sennen
subsp. *intermedia* (Beeby) Walters
subsp. **chondrosperma** (Fenzl) Walters

BM, CGE, E, GL, NMW, OXF, Lousley, Simpson, Wallace
These maps are based on specimens with ripe seed determined almost entirely by S. M. Walters, though a small amount has been determined by E. B. Bangerter, P. M. Benoit and the editor. As can be seen by comparison with the map of the species in the *Atlas* (p. 80), these taxa are under-recorded, especially in Ireland. Though four subspecies have been recognized, fertile intermediates occur. Variation in the sculpturing of the testa is related to geographical distribution: from subsp. *chondrosperma* of southern Britain with dull seeds covered with coarse tubercles to subsp. *fontana* of northern Britain in which the seeds are shining and smooth. In continental Europe also there are differences in distribution. Subsp. *fontana* is mainly in the north and centre, subsp. *variabilis* in the west and centre, and subsp. *amporitana* in the west and south-west. Subsp. *chondrosperma* occurs north to Scotland and south Sweden, but generally in drier places than the previous three subspecies, and is often on sandy ground. For a detailed revision of the species, see Walters (1953).

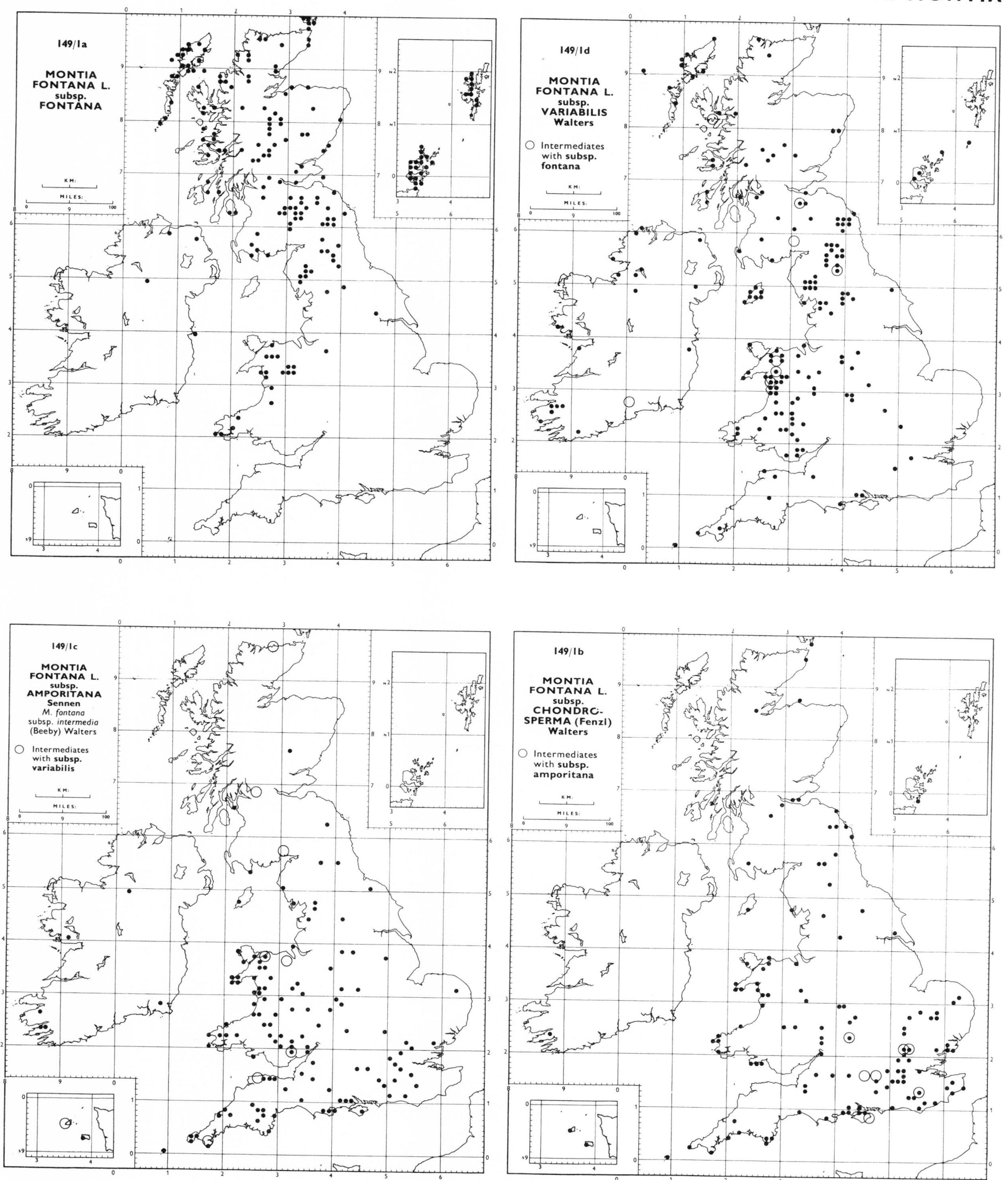
149/1a
MONTIA FONTANA L. subsp. FONTANA
KM:
MILES:
149/1d
MONTIA FONTANA L. subsp. VARIABILIS Walters
Intermediates with subsp. fontana
149/1c
MONTIA FONTANA L. subsp. AMPORITANA Sennen
M. fontana subsp. intermedia (Beeby) Walters
Intermediates with subsp. variabilis
149/1b
MONTIA FONTANA L. subsp. CHONDROSPERMA (Fenzl) Walters
Intermediates with subsp. amporitana

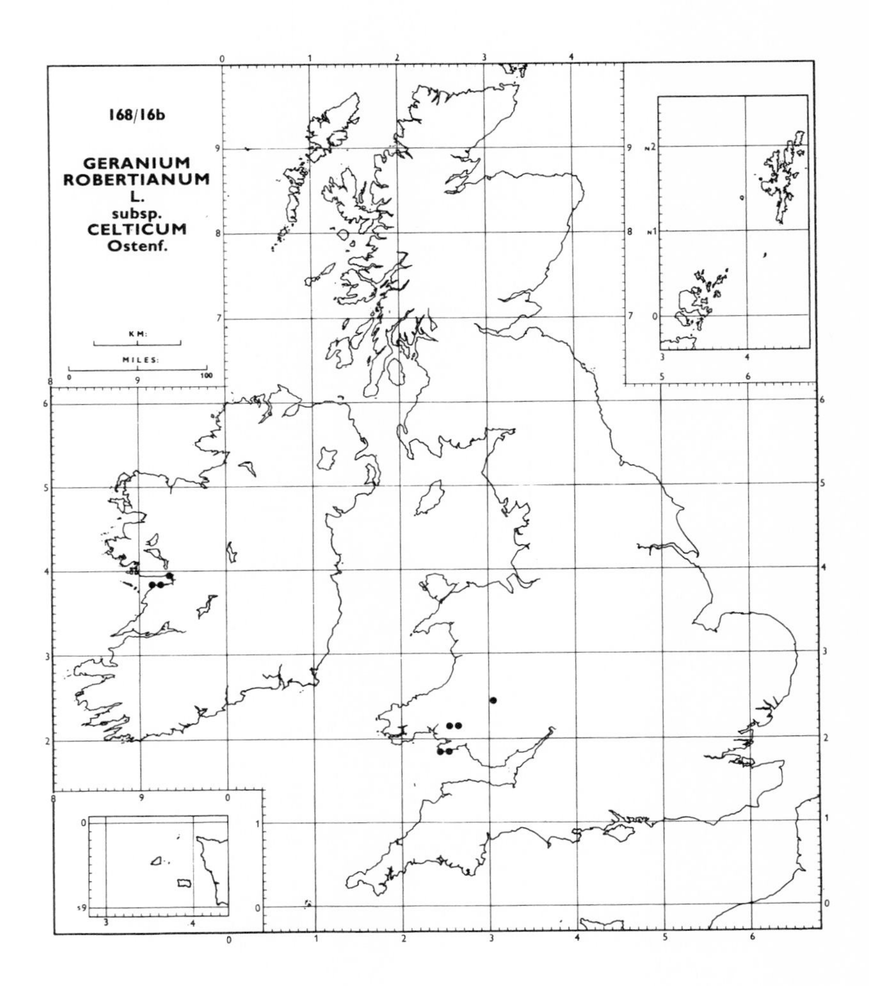

168/16 Geranium robertianum L.
subsp. **celticum** Ostenfeld
subsp. **maritimum** (Bab.) H. G. Baker

BM, CGE, K, NMW, OXF
The maps are based almost entirely on data published by Baker (1956). As far as is known, subsp. *robertianum* occurs throughout the range of the species (see *Atlas*, p. 94). Subsp. *celticum* is apparently endemic, whilst subsp. *maritimum* occurs along the Atlantic coast from Madeira to Norway.

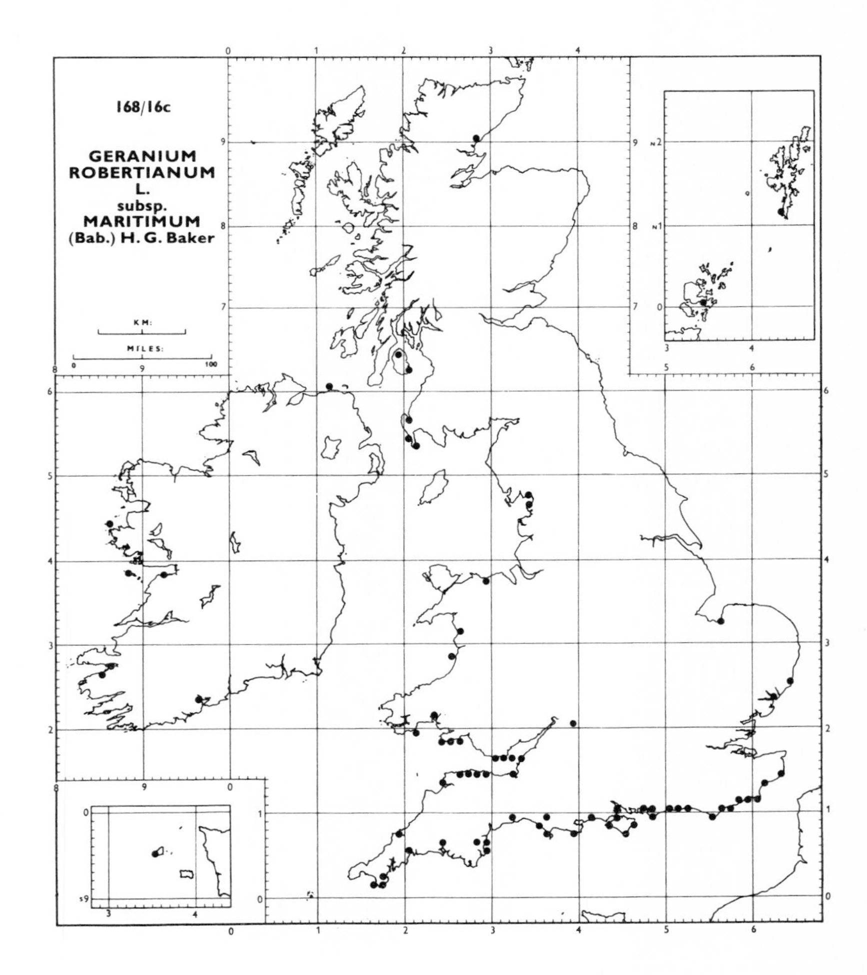

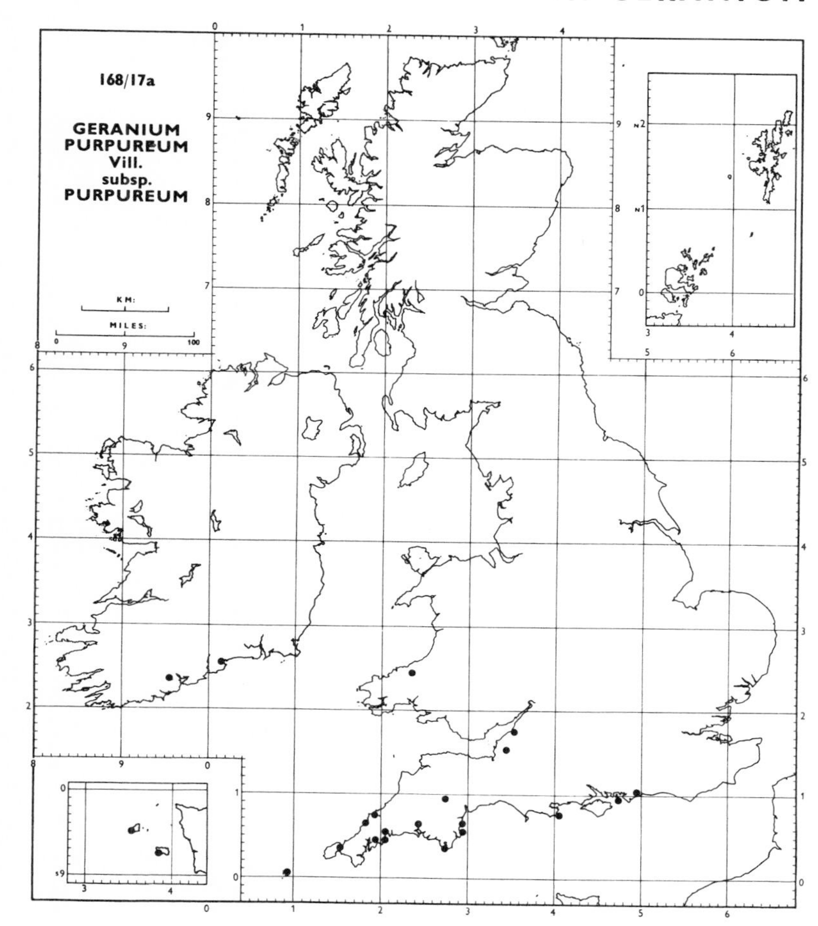

168/17 **Geranium purpureum** Vill.
subsp. **purpureum**
subsp. **forsteri** (Wilmott) H. G. Baker

BM, CGE, E, K, LDS, NMW, OXF
These maps are based almost entirely on data published by Baker (1955). It is thought possible that subsp. *forsteri* could have arisen as a result of introgression between subsp. *purpureum* and prostrate forms of *G. robertianum*. Subsp. *purpureum* is in south-west Europe and Madeira; whilst subsp. *forsteri* is endemic.

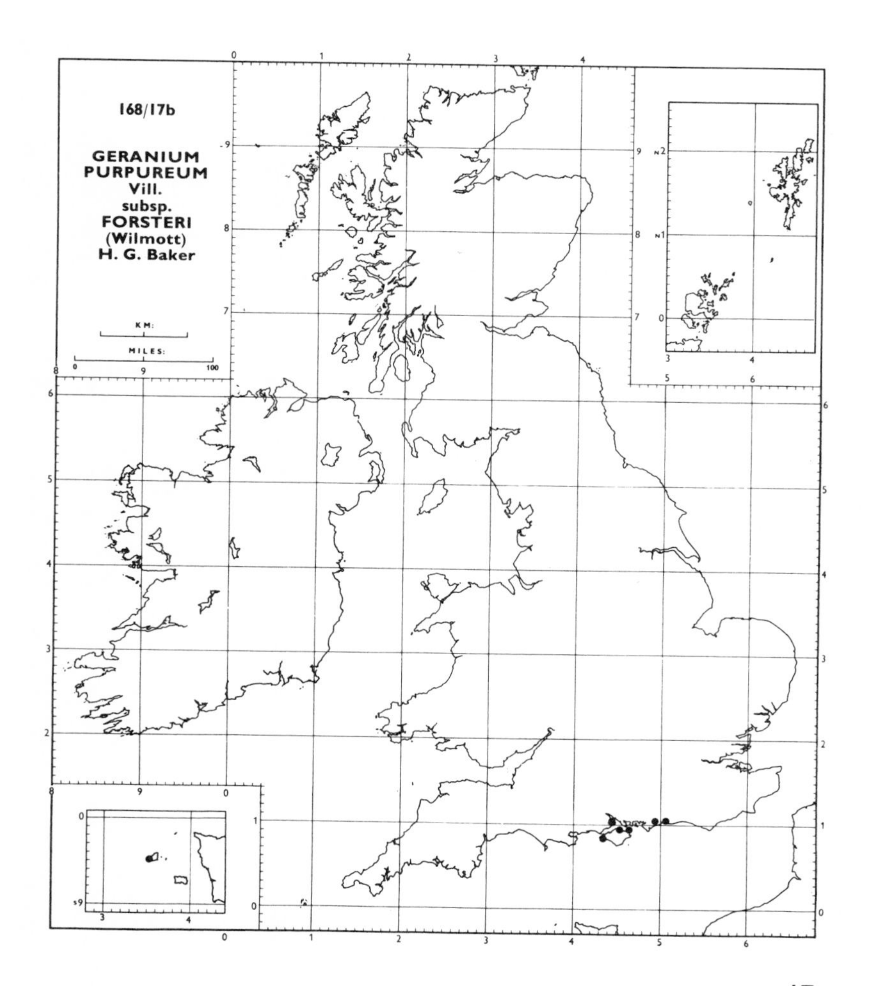

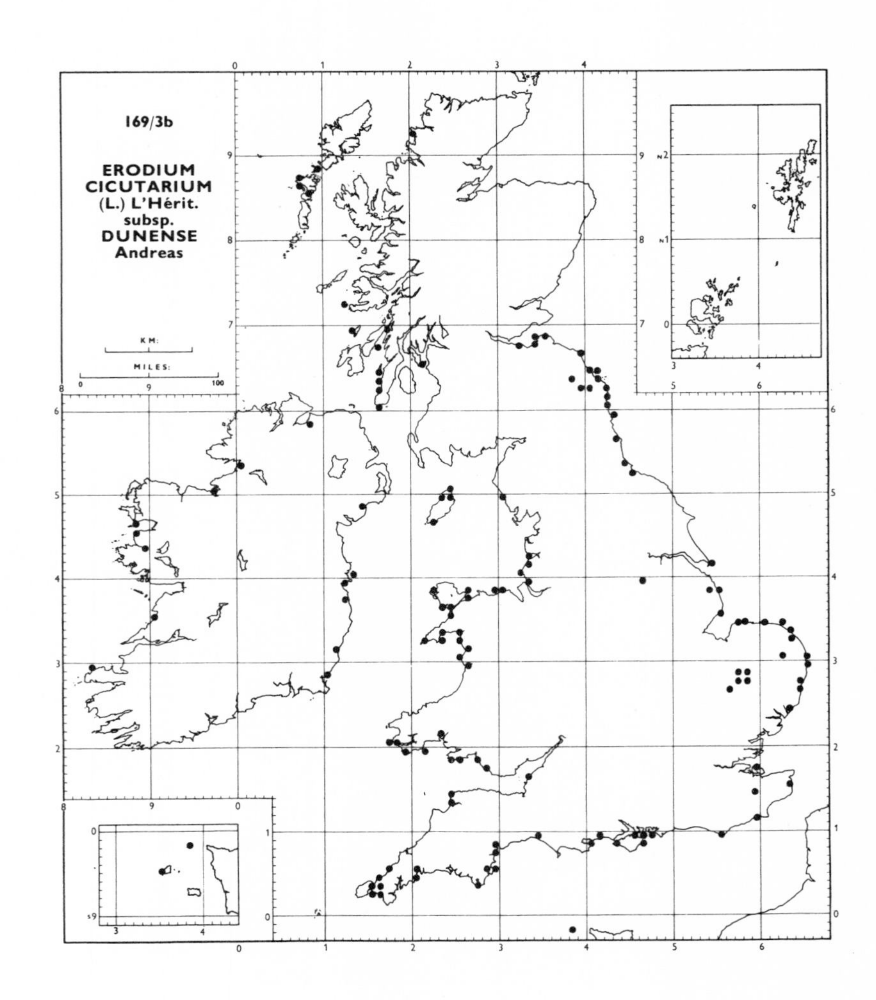

169/3b Erodium cicutarium (L.) L'Hérit. subsp. **dunense** Andreas

BFT, CGE, DBN, E, NMW, NWH, SHY, TCD
Subsp. *dunense* is not always clearly distinguishable from subsp. *cicutarium*. The map is based on herbarium material seen by the editor and by P. M. Benoit: only specimens with less than 5-flowered peduncles, fruits less than 5 mm., and pedicels less than 13 mm. have been accepted. Subsp. **cicutarium** appears to occur throughout the range of the species (see *Atlas*, p. 95). It replaces subsp. *dunense* on sand-dunes in the north of Britain. Subsp. *dunense* occasionally occurs inland on sandy heaths in the east. It is mainly a plant of the coasts of north-west Europe.

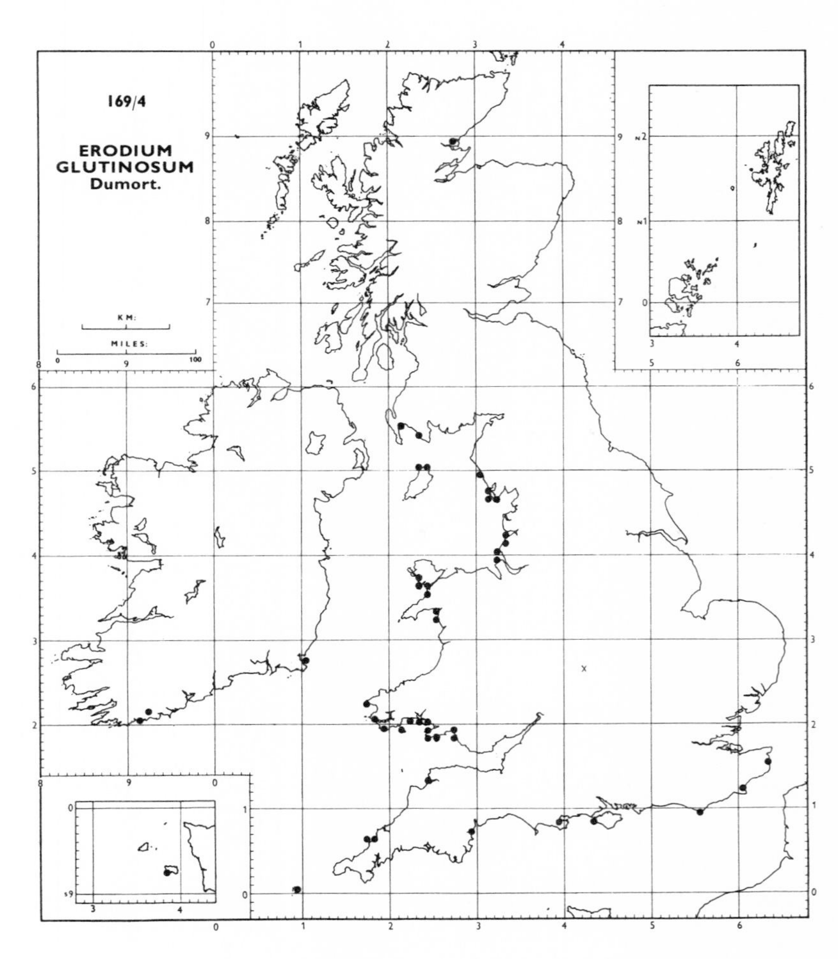

169/4 Erodium glutinosum Dumort.

BFT, CGE, DBN, E, NMW, NWH, SHY, TCD
The map is based on herbarium material seen by the editor and by P. M. Benoit. *E. glutinosum* is not always easily distinguished from *E. cicutarium* subsp. *dunense*: only specimens which were densely glandular, with flowers less than 7 mm. in diameter and with 2–3-flowered peduncles, were accepted. Perhaps more widespread in dunes along the east and south coasts of Ireland than the map suggests. It occurs along the coast of western Europe from Portugal to the Netherlands.

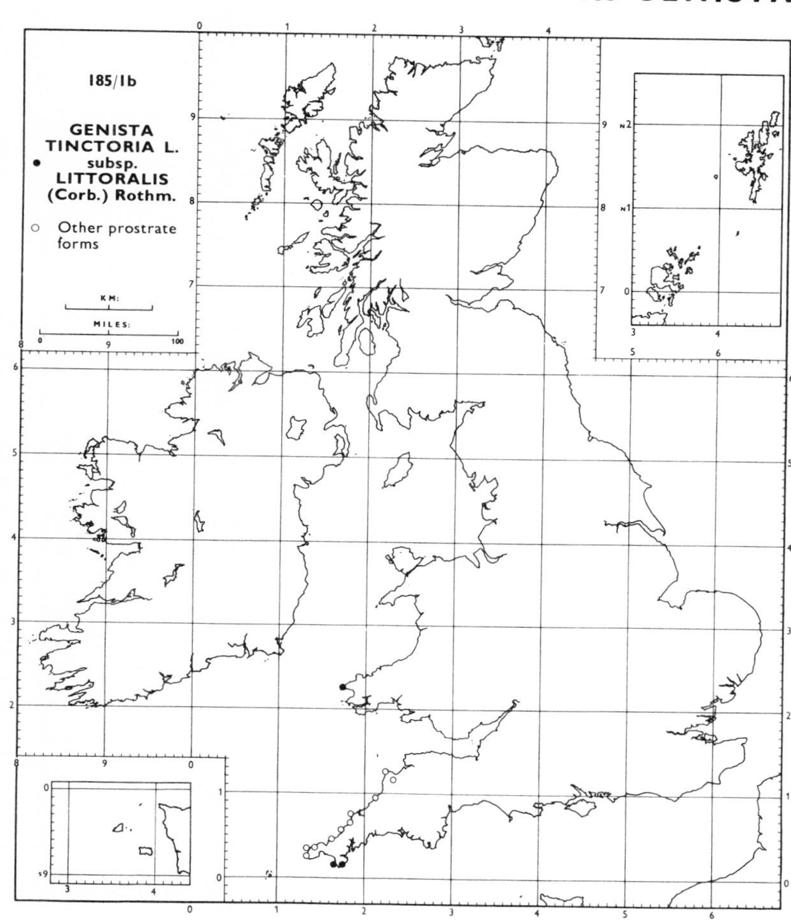

185/1b Genista tinctoria L.
subsp. **littoralis** (Corb.) Rothm.

BM, CGE, K: Floras
This prostrate subspecies has its pods hairy on the back of each valve. Prostrate forms with *glabrous* pods are recorded from the coast of Cornwall and north Devon (1, 2, 4). They resemble subsp. *littoralis* in having stem leaves 2–4 times as long as broad, whereas in subsp. **tinctoria** the leaves are 4–6 times as long as broad. This situation is similar to that in *Sarothamnus scoparius*, in which there are also two prostrate forms which occur on the coast.

All prostrate populations of *Genista tinctoria* are geographically isolated from populations of subsp. *tinctoria*, and are probably the only forms which occur in Cornwall and Pembrokeshire (1, 2, 45) (see *Atlas*, p. 99). Subsp. *littoralis* also occurs in Normandy. (cf. Sell, 1967.)

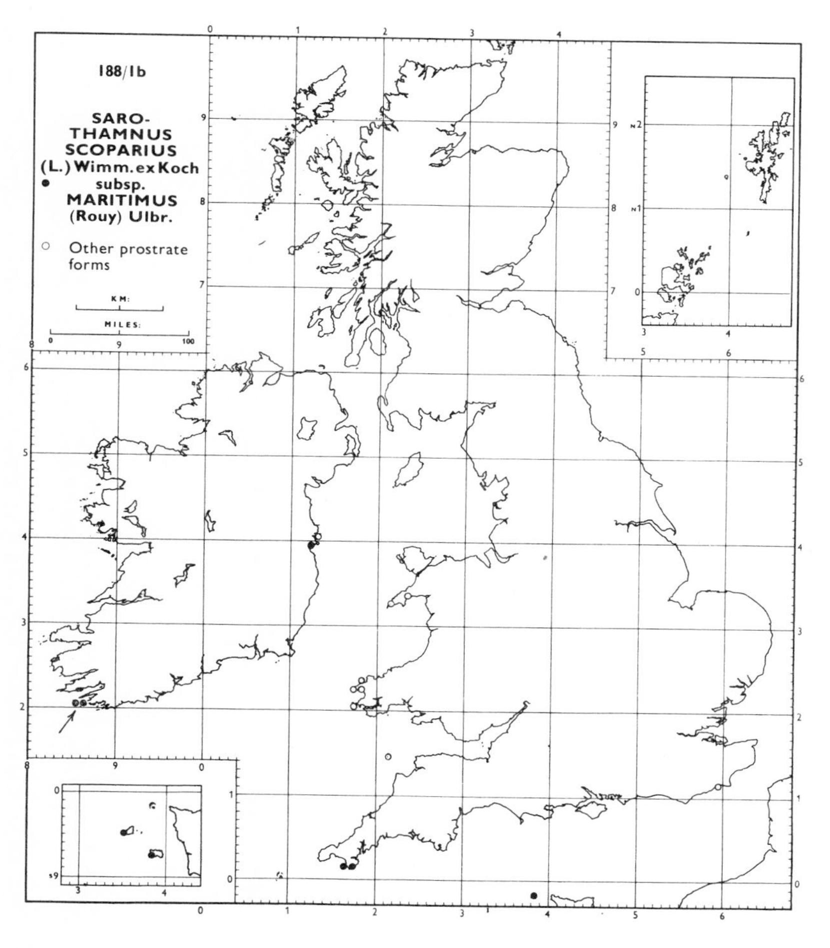

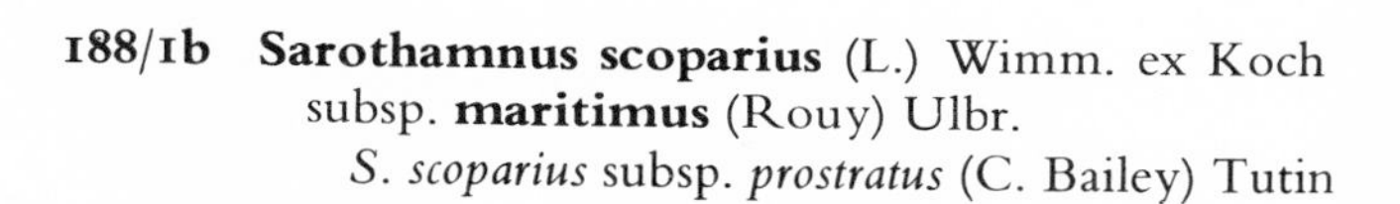
188/1b Sarothamnus scoparius (L.) Wimm. ex Koch
subsp. **maritimus** (Rouy) Ulbr.
S. scoparius subsp. *prostratus* (C. Bailey) Tutin

BFT, BM, CGE, DBN, K, NMW, TCD: Floras
This subspecies is recognized by its prostrate stems and densely woolly young twigs. It retains these characters in cultivation and breeds true. Other prostrate forms occur, but in these the young stems are only sparsely hairy. In having two prostrate forms it closely resembles *Juniperus communis* and *Genista tinctoria* (q.v.). It occurs in a few other places on the coast of western Europe.

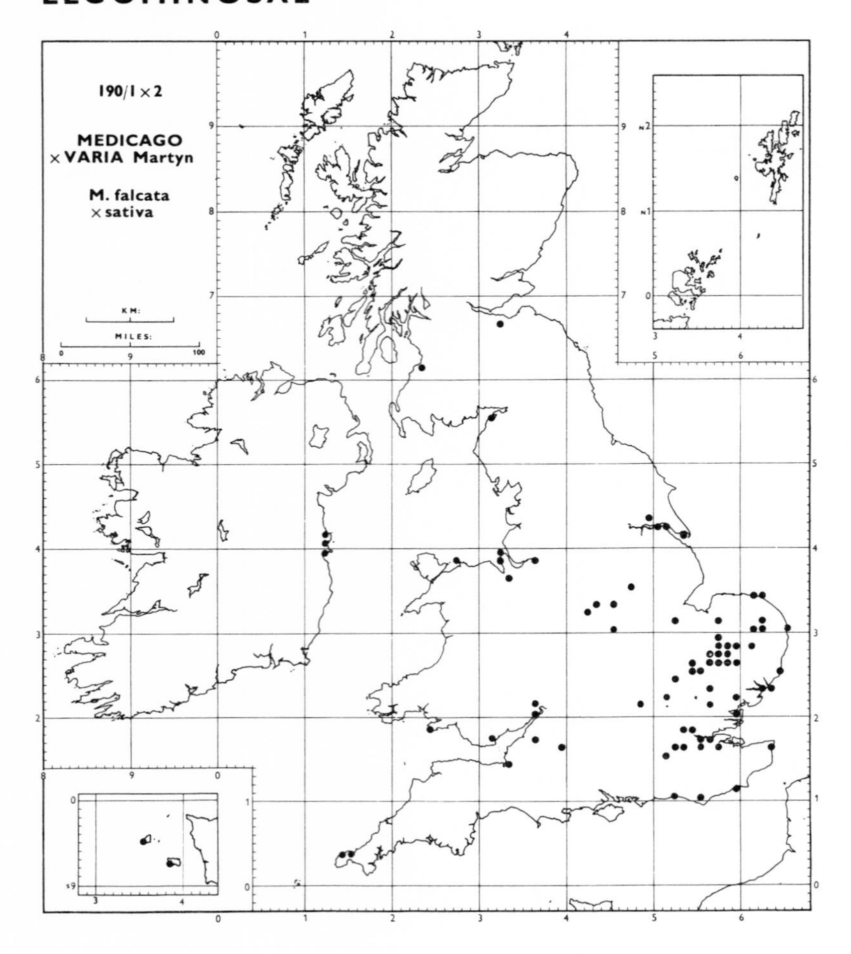

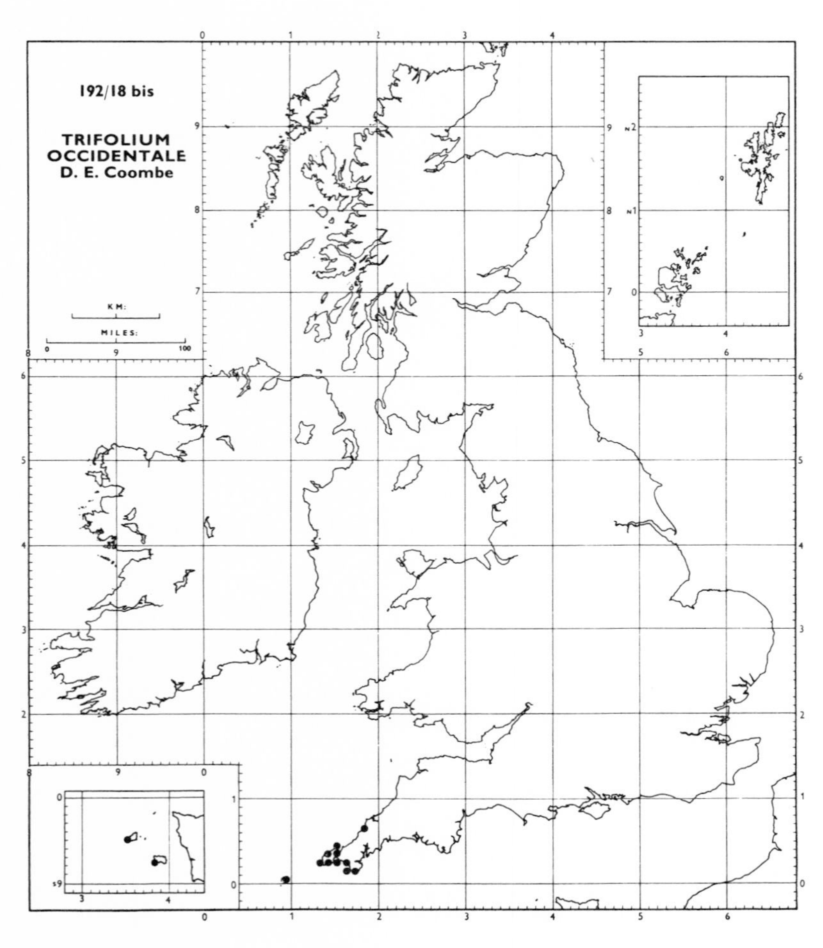

190/1 × 2 Medicago × varia Martyn
= M. falcata × sativa

BM, CGE, DBN, NMW, OXF: Floras

This fertile hybrid is found regularly in East Anglia, where the native parent, *M. falcata*, occurs frequently (*Atlas*, p. 102). Elsewhere it may be found where introduced plants of both parents are growing together by chance, and hybridization has taken place. It may also occur as a casual in the absence of one or both parents, the introduced seed itself perhaps being of hybrid origin.

193/1 Anthyllis vulneraria L.
subsp. **vulneraria**
subsp. **corbierei** (Salmon & Travis) Cullen
subsp. **polyphylla** (DC.) Arcangeli
subsp. **vulgaris** (Koch) Corb.
var. **pseudovulneraria** Sagorski
subsp. **lapponica** (Hyland.) Jalas

CGE, E, EXR, K, LIV, LIVU, MANCH

A study of this very variable species has recently been made by Cullen (1961). These maps are largely the result of specimens which he has determined. In addition a small amount of material has been determined by the editor and P. D. Sell.

Subsp. *vulneraria* is recognized by its narrow calyx, which is 2–4·5 mm. broad, with the lateral teeth obscure and appressed to the upper teeth; its stem which has appressed silky hairs over the whole length; and the leaves, which are not fleshy, and evenly distributed along the stem. Outside the British Isles it occurs only along the North Sea coast (rarely inland), in Denmark, and near the coast or on islands in the Baltic. This is the commonest subspecies in the British flora, and it is undoubtedly severely under-recorded (see *Atlas*, p. 109).

Subsp. *corbierei* has a uniformly spreading stem pubescence, and fleshy leaves which occur in the lower part of the stem only. It is endemic to the coasts of Anglesey (52), Cornwall (1) and the Channel Islands (cf. Sell, 1967).

Subsp. *polyphylla* resembles subsp. *vulneraria*, but the pubescence of the stem is spreading in the lower part, and the stem itself is very robust and erect. As a native it is a plant of central Europe, being most abundant in Czechoslovakia, Austria and Hungary. It is probably introduced with crops in Britain.

Subsp. *vulgaris* var. *pseudovulneraria* has the calyx 5–7 mm. broad which is covered with appressed silky hairs and has the obvious lateral teeth, not appressed to the upper teeth. The stems are usually numerous (10–30). Subsp. *vulgaris* is a central European subspecies which reaches its northern limit (as var. *pseudovulneraria*) in Europe in Northumberland (68).

Subsp. *lapponica* is fairly closely related to subsp. *vulgaris*, but the calyx is covered with shaggy hairs, and the stems are usually few in number (1–5 (–10)). Outside the British Isles it is recorded from Sweden, Norway and Finland.

192/18 bis Trifolium occidentale D. E. Coombe

BM, CGE, K

This species has recently been described (Coombe, 1961). It is closely related to *T. repens*, but is morphologically, genetically, cytologically and ecologically distinct. In the British Isles it occurs in relatively dry coastal habitats where it flowers about a month earlier than *T. repens*. It occurs also on the coast of north-east Spain and is probably widespread in west and north-west France. It is adapted to withstand midsummer drought.

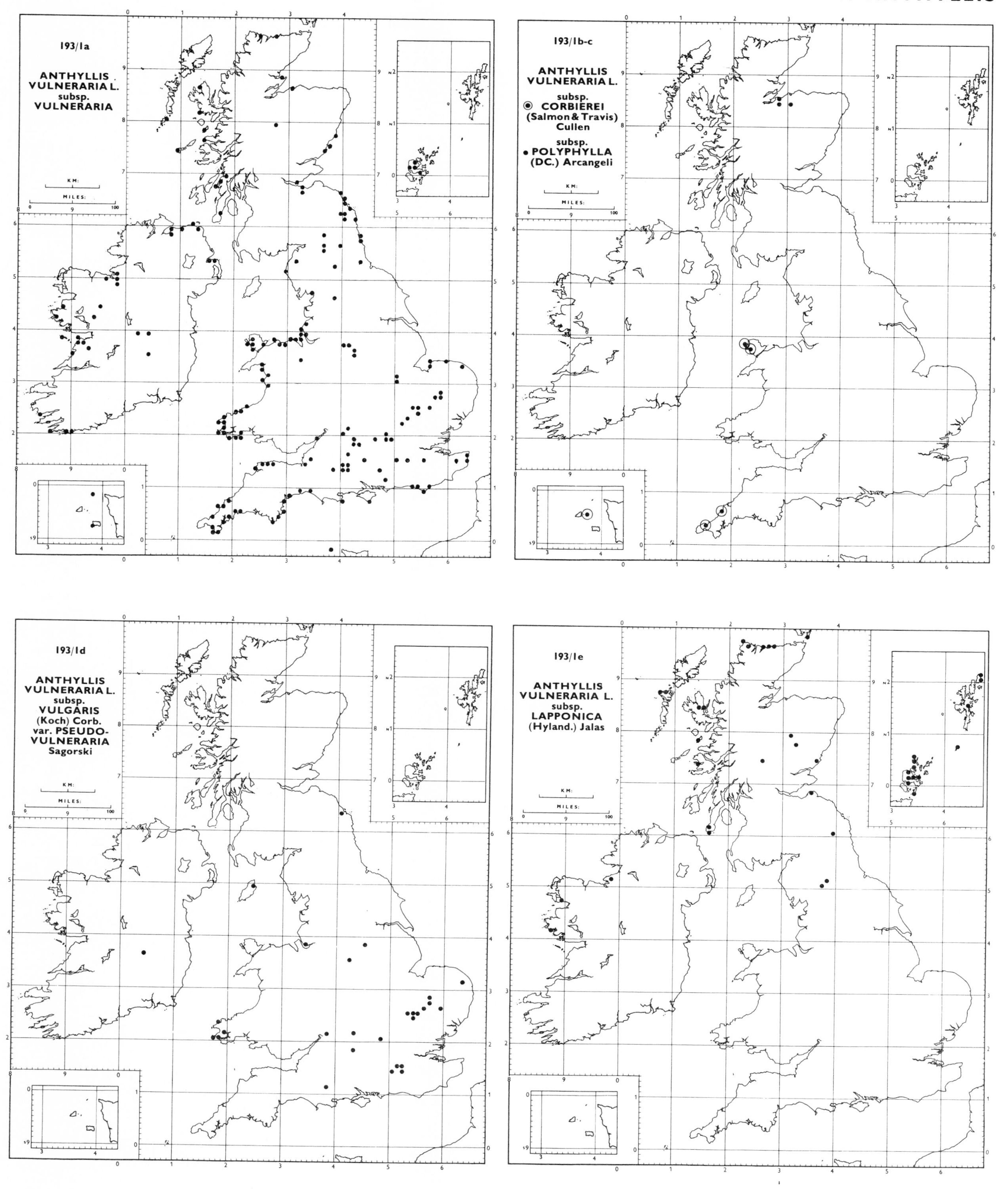
193/1a
ANTHYLLIS VULNERARIA L. subsp. VULNERARIA
KM:
MILES:
193/1b-c
ANTHYLLIS VULNERARIA L.
subsp. CORBIEREI (Salmon & Travis) Cullen
subsp. POLYPHYLLA (DC.) Arcangeli
193/1d
ANTHYLLIS VULNERARIA L. subsp. VULGARIS (Koch) Corb. var. PSEUDO-VULNERARIA Sagorski
193/1e
ANTHYLLIS VULNERARIA L. subsp. LAPPONICA (Hyland.) Jalas

211/11 RUBUS FRUTICOSUS L. *sensu lato*

BIRA, BIRM, BM, CGE, LCN, MANCH, NMW, NOT: Floras

E. S. Edees has selected the following sixteen taxa to demonstrate the main types of distribution of *Rubus* species occurring in the British Isles. They were chosen because they are, in his opinion, well-defined species about which there has been little or no dispute amongst *Rubus* specialists past or present. Because of this he has felt that it was reasonable to base the maps largely on literature references, though he added to these data his own widespread knowledge, and records from a number of herbaria.

The preparation of the maps has been identical with the exception of that of *R. ulmifolius*. Mr Edees prepared maps after consulting the Floras of Great Britain and the herbaria listed above. Further material in three herbaria (CGE, DBN, NMW) has recently been determined by B. A. Miles, and these data and his own personal data have been added to the maps by the editor, who has also consulted the Irish local Floras. The Irish records for all the species must be regarded as inadequate: they do little more than indicate whether or not a species occurs in that country.

The map of *R. ulmifolius* was prepared in a similar way to those of the other fifteen, but in addition a special attempt was made to collect field records for this very widespread and easily identifiable species.

The notes on the species which follow, are based on observations which E. S. Edees made whilst preparing the maps. The note on *R. ulmifolius* is based on the editor's experience.

All the British *Rubus* species are pseudogamous polyploids, with the exception of *R. ulmifolius*. The polyploids are not necessarily derived only from *R. ulmifolius*; they might have been derived from other European diploids which may or may not now exist. The only primary diploid sexual species now widespread in Europe are *R. ulmifolius* and *R. tomentosus* Borkh. For a summary of the genetics and distribution of British *Rubi*, see Haskell (1961).

Section **SUBERECTI** P. J. Muell.

211/11.2 Rubus scissus W. C. R. Wats.
R. fissus auct. mult., non Lindl.

This is a plant of acid heaths and moors which may be expected wherever these habitats occur throughout the British Isles. It is a widespread plant in northern Europe.

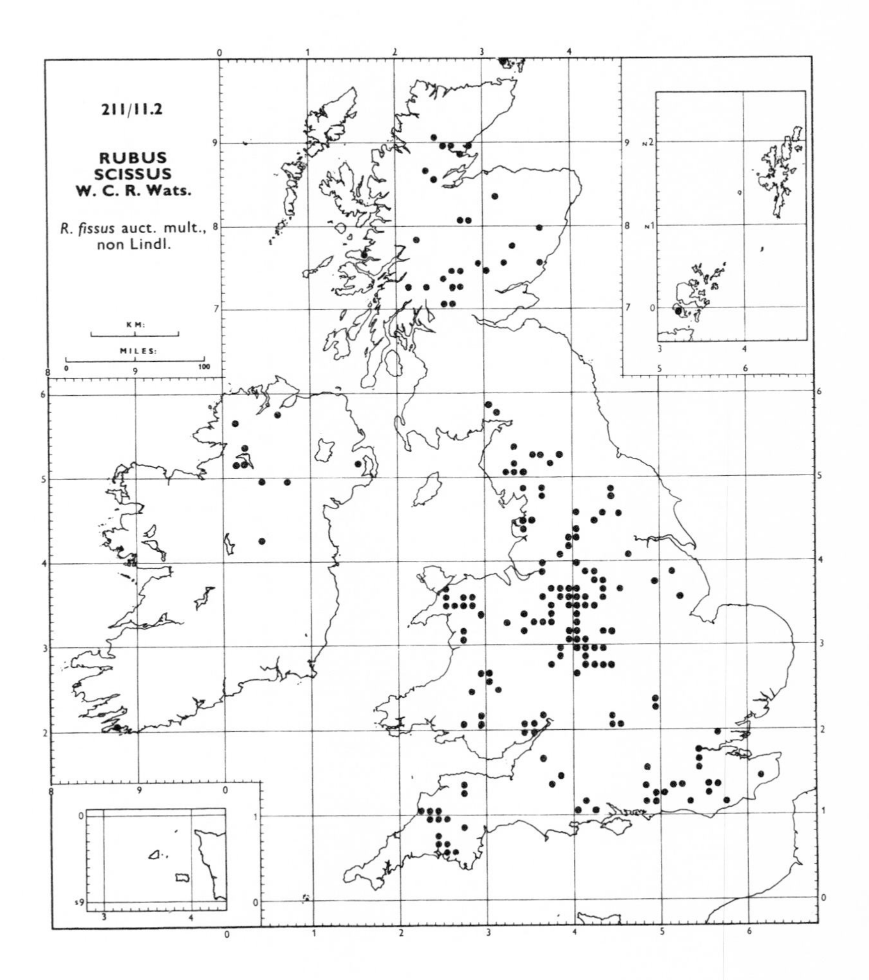

Section **TRIVIALES** P. J. Muell.

211/11.17 Rubus sublustris Lees

R. corylifolius var. *sublustris* (Lees) Leighton;
R. corylifolius subsp. *sublustris* (Lees) Riddelsdell;
R. semialterniflorus Sudre forma *sublustris* (Lees) Hruby

The map includes all literature records where *R. sublustris* is distinguished from *R. conjungens*, except for *The Flora of Perthshire* (White, 1898), though these records may be quite reliable. There should be more records for Kent (15, 16), the *Flora of Kent* (Hanbury and Marshall, 1899) considering it too generally distributed to make detailed locations necessary. *R. sublustris* is certainly commoner in the north than the map shows, but it evidently is less common in Wales. Perhaps the distribution south of the 100 km. northings "4" line is fairly accurately known.

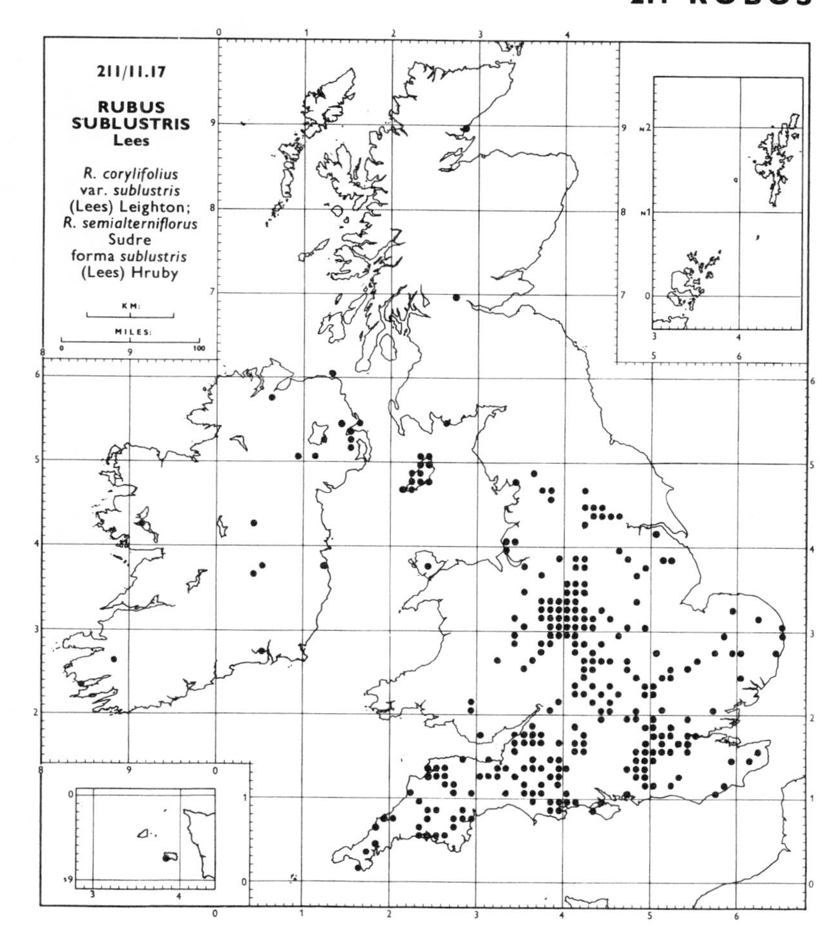

Section **SYLVATICI** P. J. Muell.

211/11.52 Rubus nemoralis P. J. Muell.

R. selmeri Lindeb.; *R. pistoris* Barton & Riddelsdell

This is the *R. nemoralis* of Watson (1958), but not of Rogers (1900). It is one of the commonest and most widely distributed British species, being particularly common in Scotland. The map ought to show a much denser concentration of records in the north, though of course there are extensive areas of high ground where no brambles grow (see *Atlas*, p. 120). Outside the British Isles it is recorded from north-west Europe.

211/11.59 Rubus lindleianus Lees

R. vulgaris var. *lindleianus* (Lees) Focke

This is one of the most widely distributed and one of the best-known of British and Irish brambles. There ought to be a few more records for the north of England and the south of Scotland, but on the whole (Ireland excepted) the map shows the distribution satisfactorily. It is a widespread plant of western Europe.

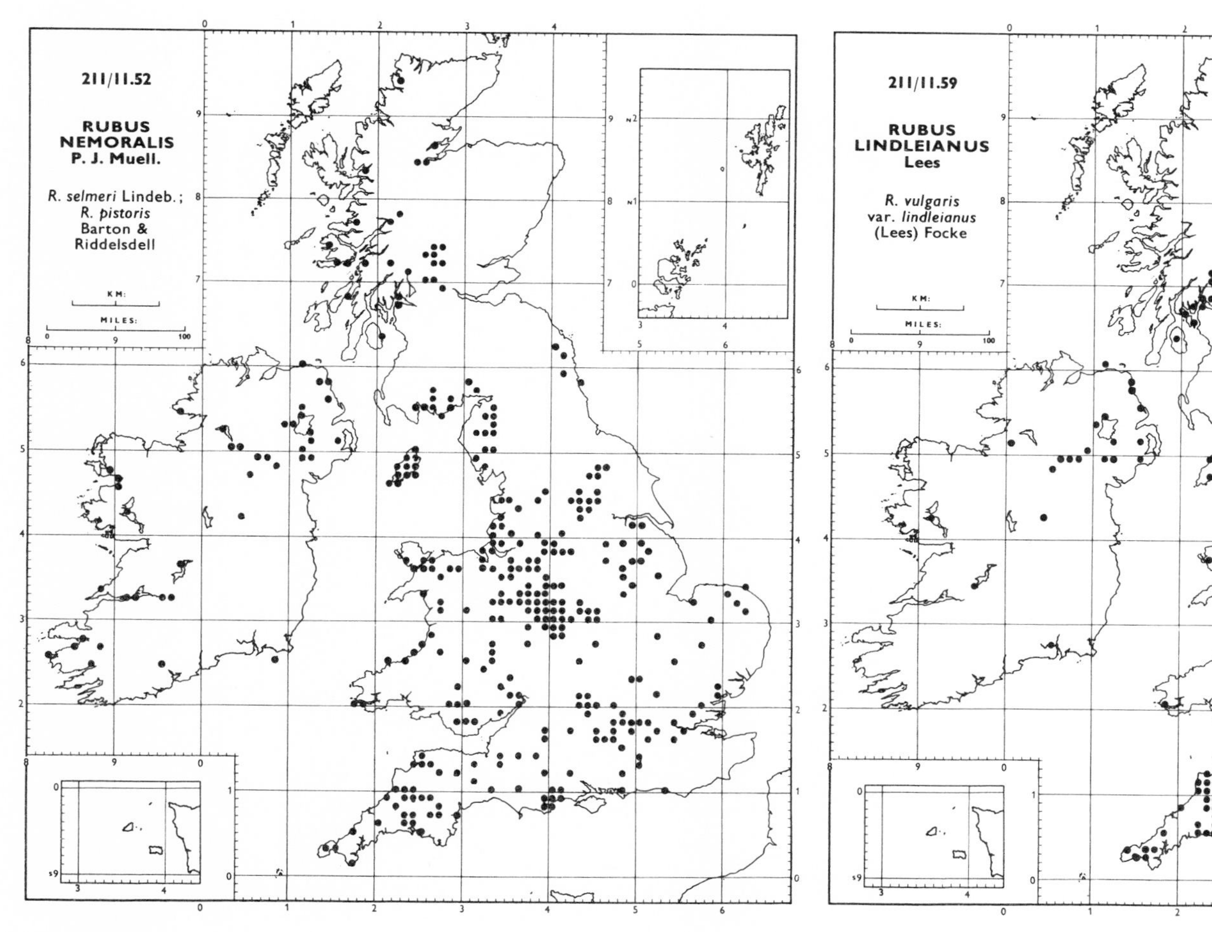

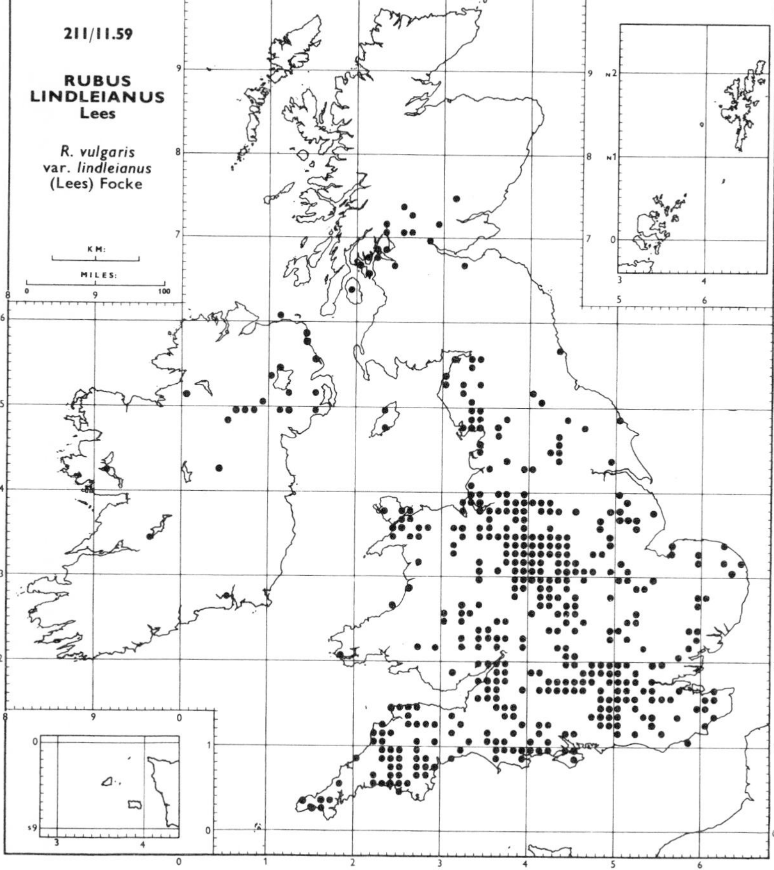

ROSACEAE

Section **SYLVATICI** P. J. Muell.

211/11.114 Rubus rubritinctus W. C. R. Wats.
R. cryptadenes Sudre, non Dumort.; *R. erythrinus* auct. brit., non Genev.; *R. argenteus* auct. brit., non Weihe & Nees

The map, though doubtless far from complete, particularly for Ireland, shows well enough the south-westerly distribution of this bramble. Records given by Watson (1958) for v.-cs 12, 15, 23–26, 67 and H. 33 have not been confirmed. Outside the British Isles it is known only from France.

211/11.125 Rubus lindebergii P. J. Muell.
R. lacustris Rogers

This is a local species which is probably under-recorded in northern England and southern Scotland. It also occurs in Denmark and south Sweden.

211/11.126 Rubus errabundus W. C. R. Wats.
R. scheutzii sensu Rogers, non Lindeb.

This endemic species has its centre of distribution around the northern part of the Irish Sea, including the Isle of Man (71). It also occurs in central Scotland and is probably under-recorded in the country in between. There are several outlying localities recorded in the literature, but only the two in Devon and Sussex (3, 13) from which herbarium material has been seen have been accepted.

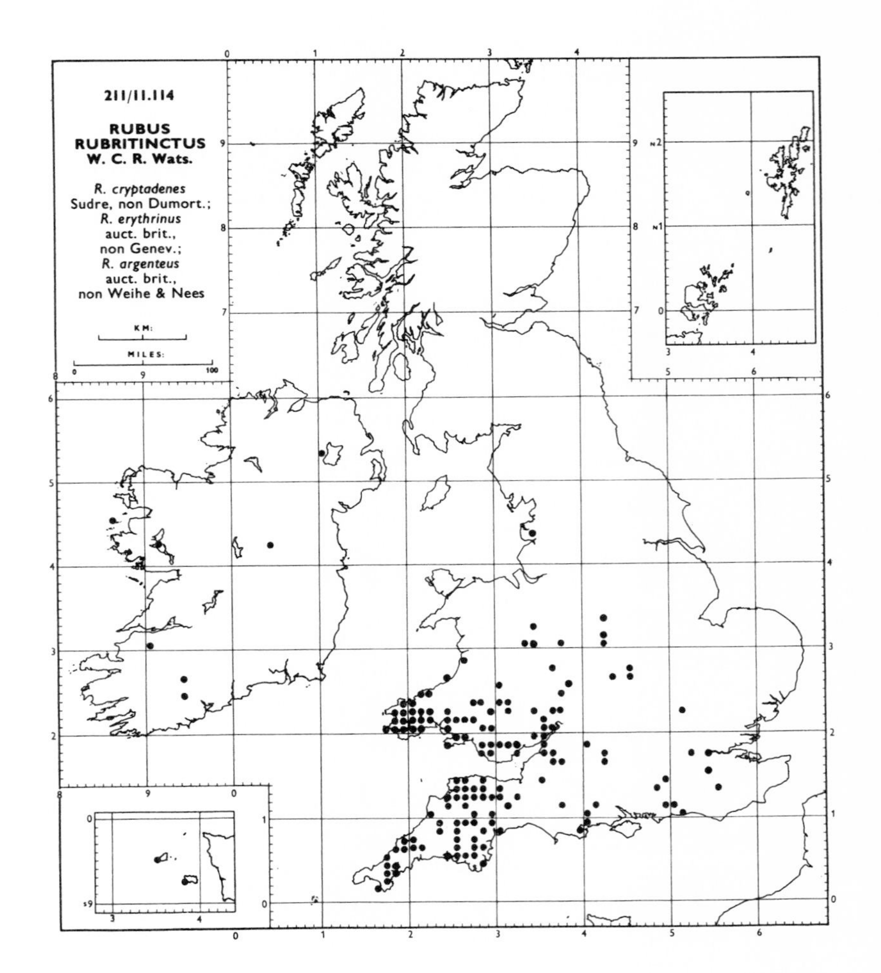

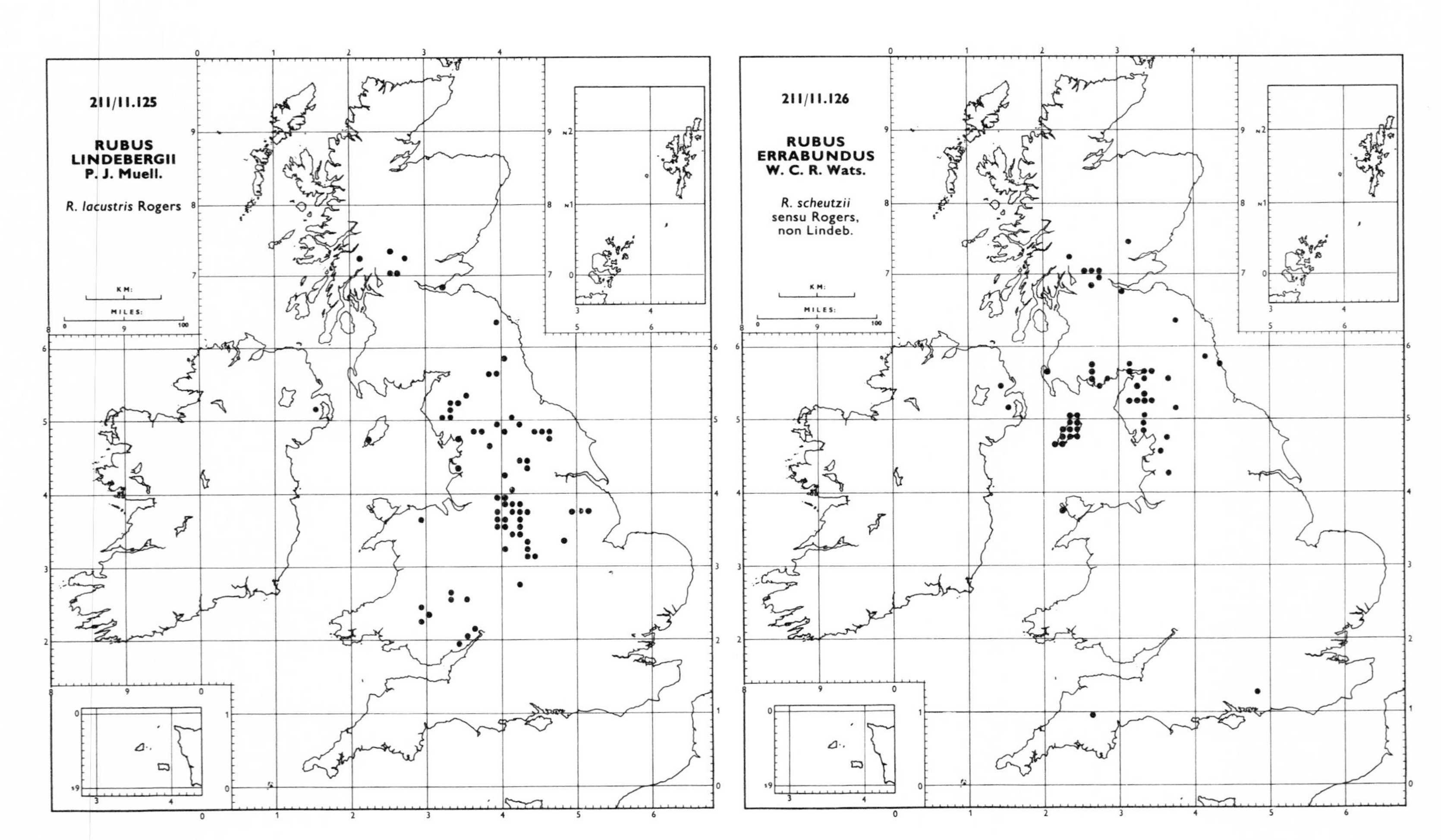

Section **SYLVATICI** P. J. Muell.

211/11.128 Rubus silurum (A. Ley) Druce
R. nemoralis var. *silurum* A. Ley

The map illustrates very well an endemic species with a typically Welsh distribution. Herbarium material has been seen from nearly all the stations plotted on the map, including the outliers in Oxfordshire, Gloucestershire and Staffordshire (23, 34, 39). Records from v.-cs 6, 8, 60 and 104 have not been confirmed.

Section **DISCOLORES** P. J. Muell.

211/11.129 Rubus ulmifolius Schott
R. rusticanus Merc.; *R. discolor* Weihe & Nees

This is by far the commonest hedgerow bramble of England and the only one with normal sexuality. Over large areas of the chalk it is the only species to be found; it is also frequent on heavy clays. The only areas from which it seems to be absent in the south are on acid soils, particularly the dry sands. The limit of the species northwards is not accurately known, though it is clearly of very rare occurrence in Scotland. It has been found in every part of Ireland where it has been looked for, the map showing clearly the recent journeys of the editor in that country. It probably occurs throughout Ireland except on the wet acid soils of the west. Common in western Europe, and also occurring in north-west Africa. It hybridizes with many species, particularly *R. caesius* L.

Section **SPRENGELIANI** (Focke) W. C. R. Wats.

211/11.146 Rubus sprengelii Weihe

The map probably gives a reasonable impression of the actual distribution of this species. Its apparent rarity in Scotland, Ireland and Wales is a real phenomenon. This well-known species is widespread in western Europe from northern France to Denmark.

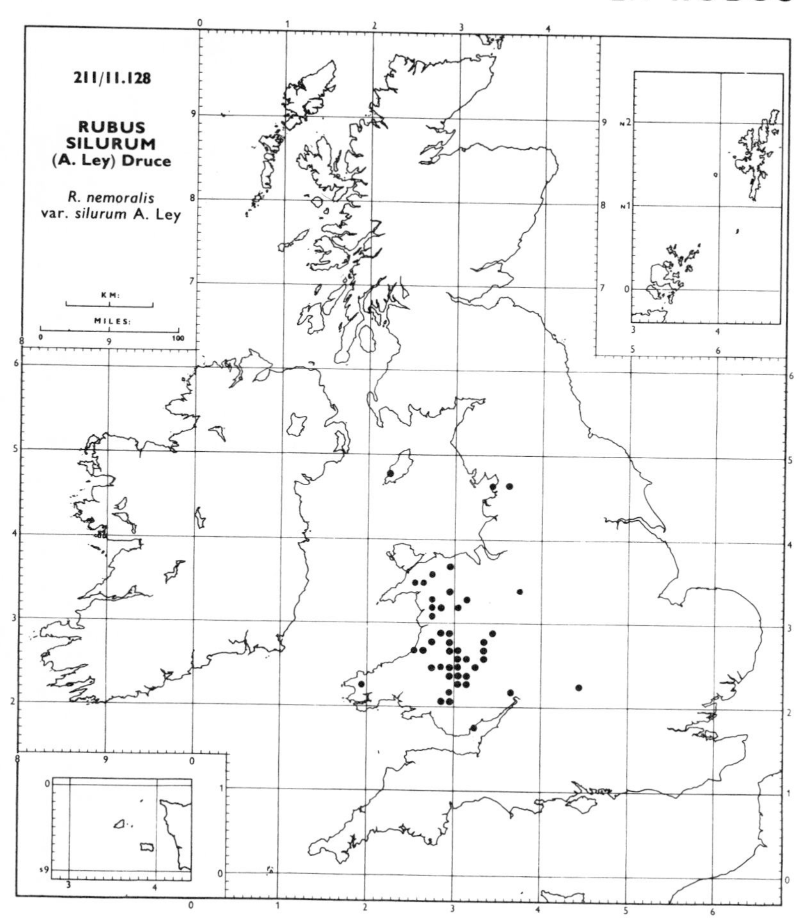

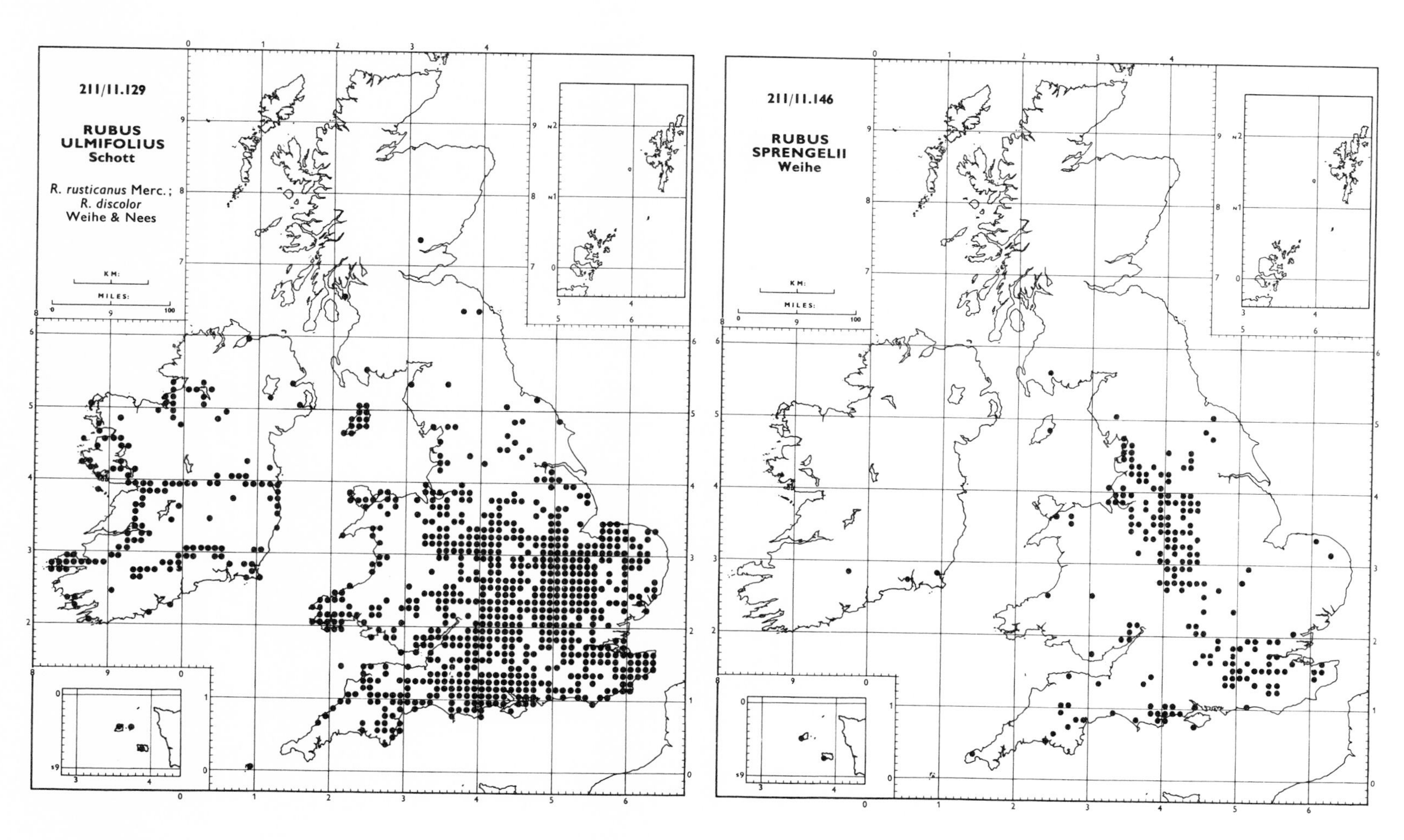

ROSACEAE

Section **APPENDICULATI** (Genev.) Sudre

211/11.165 Rubus vestitus Weihe & Nees

R. diversifolius Lindl.; *R. leucostachys* auct., non Sm.

R. leightonianus Bab.

The map is probably very incomplete, but accurate as far as it goes. About half the records are based on herbarium material or on plants examined in the field. The others are taken from recent Floras. Unfortunately several of these, e.g. Kent, Berkshire, Bedfordshire, Gloucestershire and south Lancashire (15, 16, 22, 30, 33, 34 and 59) consider the plant too common for localities to be given, hence there are incomplete records for these areas on the map.

Published records for Devonshire, Somerset, Dorset and Hampshire (3–6, 9–12) have been omitted because Rilstone (1952) says that *R. vestitus* is absent from Cornwall, and Riddelsdell (1939) that the Devonshire records are seldom of the typical form.

The Welsh distribution is fairly accurate, though more coastal records might be expected to occur. There should be many more records from the north of England, but as far as can be judged the species is absent from Scotland. It is generally distributed in central Europe from Belgium to Silesia.

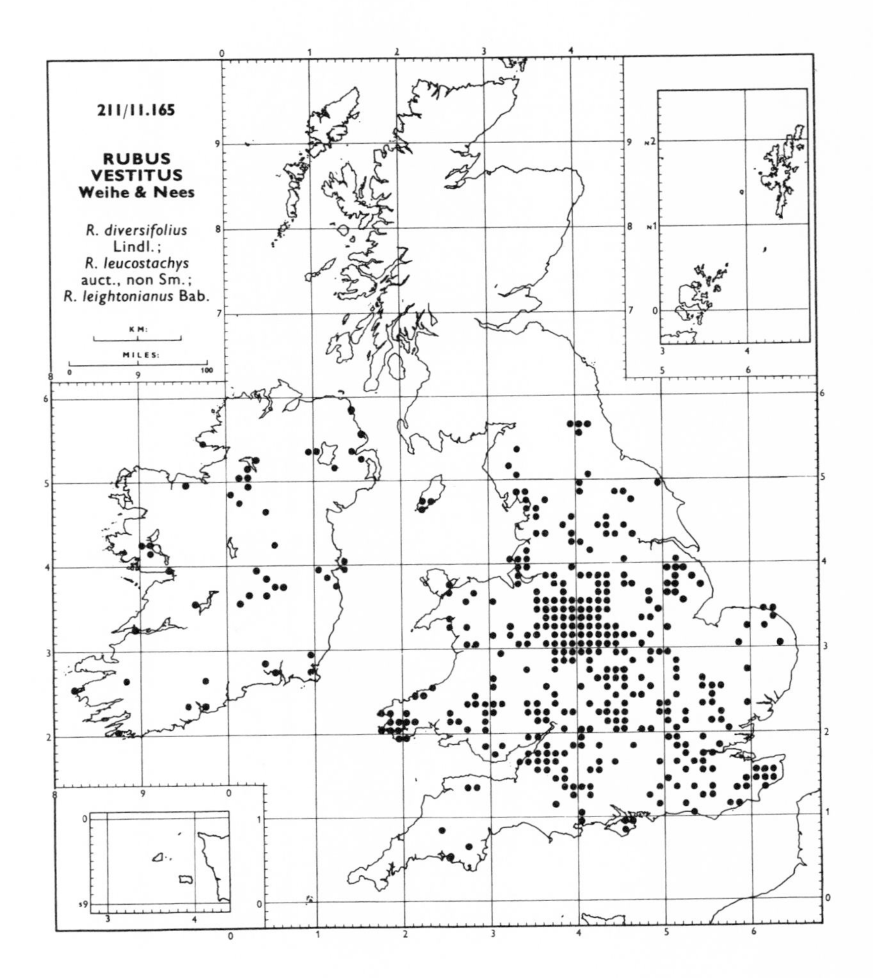

211/11.212 R. echinatus Lindl.

R. discerptus P. J. Muell.

R. fuscoater var. *echinatus* (Lindl.) Bab.

R. rudis auct. ante 1890, non Weihe & Nees

A common species of the clays, as well as on sand and gravel. Outlying records from v.-cs 25, 42 and 99 given by Watson (1958) have not been confirmed. Recorded from France, Portugal, Switzerland and Bavaria. (Nomenclature, cf. Sell, 1967.)

211/11.284 Rubus rufescens Muell. & Lefev.

R. rosaceus var. *infecundus* Rogers

R. hystrix var. *infecundus* (Rogers) Riddelsdell

A plant of woods in shade or sun. Although stated by Watson (1958) to occur in England, Scotland, Wales and Ireland, this species appears to be almost confined to central and southern England. Records given by Watson for v.-cs 10–12, 49, 58, 67, 70, 99 and H. 38 have not been confirmed. Also known from north France and Belgium.

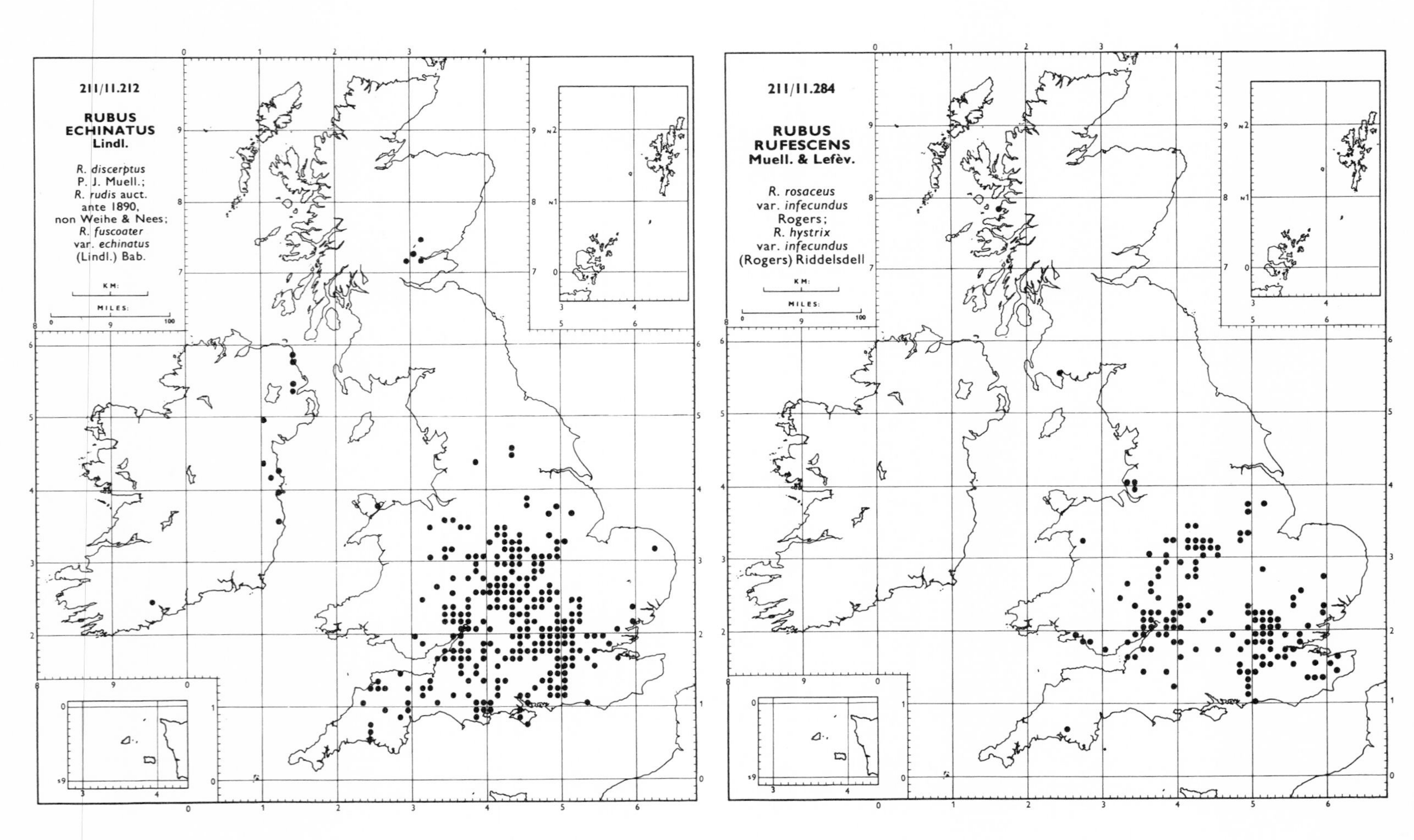

Section **APPENDICULATI** (Genev.) Sudre

211/11.310 Rubus leightonii Lees ex Leighton
R. radula var. *anglicanus* Rogers
R. anglicanus (Rogers) J. W. White

This is a very distinct species. The map is accurate as far as present knowledge goes, though there ought perhaps to be a few more records on the south coast. It is extremely rare in Wales and almost certainly does not occur in Scotland or Ireland. The density of records in the midlands is natural and it does not reflect observer bias, though E. S. Edees does live in that area: many of the records are in fact from Leicestershire (55), which he has not yet explored. Records for Scotland are doubtful, and those from Cornwall (1, 2) have been omitted because, although the species is recorded from that county by Rilstone (1952), it is omitted from the list given by Watson (1958). The following v.-c. records of Watson's have not been confirmed: 35, 36, 90. It occurs also in north-west France.

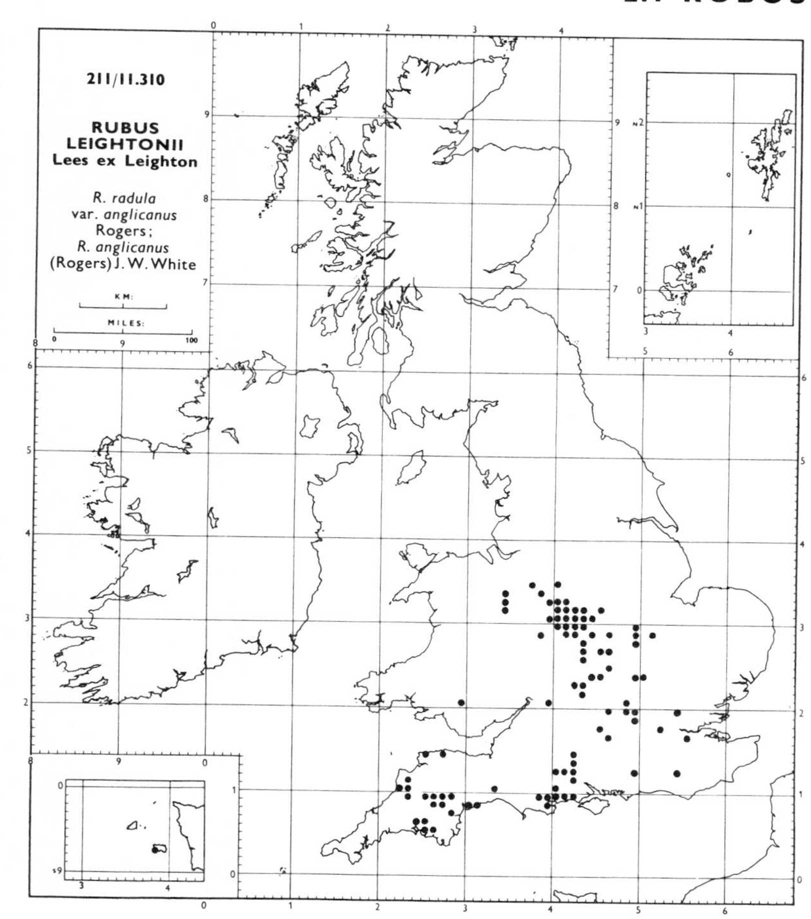

Section **GLANDULOSI** P. J. Muell.

211/11.330 Rubus murrayi Sudre
R. rosaceus subsp. *adornatus* sensu Rogers
R. adornatus sensu Rogers, non P. J. Muell. ex Wirtg.

The map of this endemic species is fairly complete according to present knowledge. The record from v.-c. 26 given by Watson (1958) has not been confirmed.

211/11.356 Rubus dasyphyllus (Rogers) E. S. Marshall
R. koehleri var. *dasyphyllus* Rogers

This is probably the commonest bramble of the Section *Glandulosi* in the north of England. North of the 100 km. northings "4" line the species is probably under-recorded. South of the "4" line the map is reasonably accurate, though the concentration of records in Staffordshire and Derbyshire (39, 57) is probably in part determined by the fact that E. S. Edees lives in the former county. Only known elsewhere in Europe from northern France and Denmark.

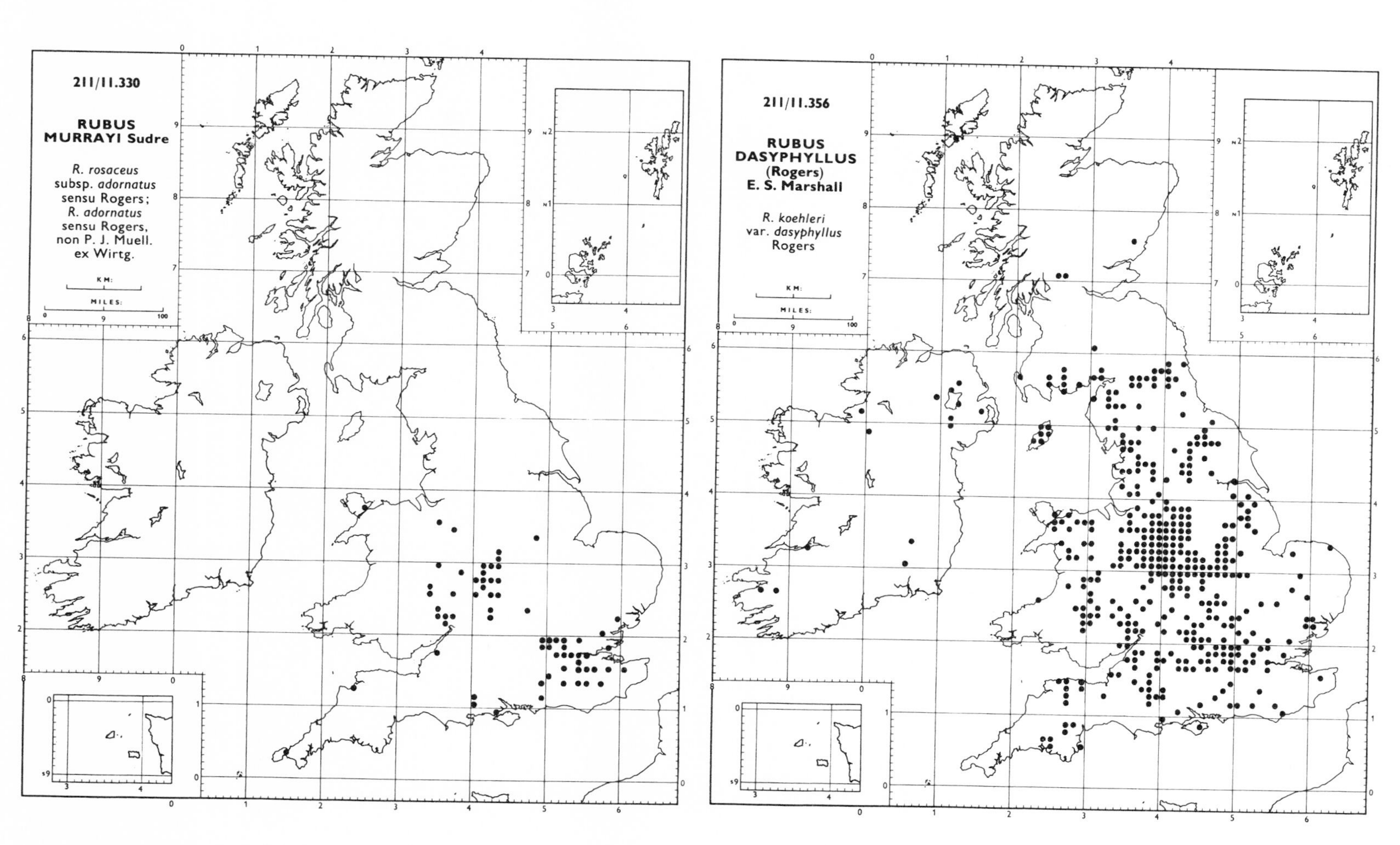

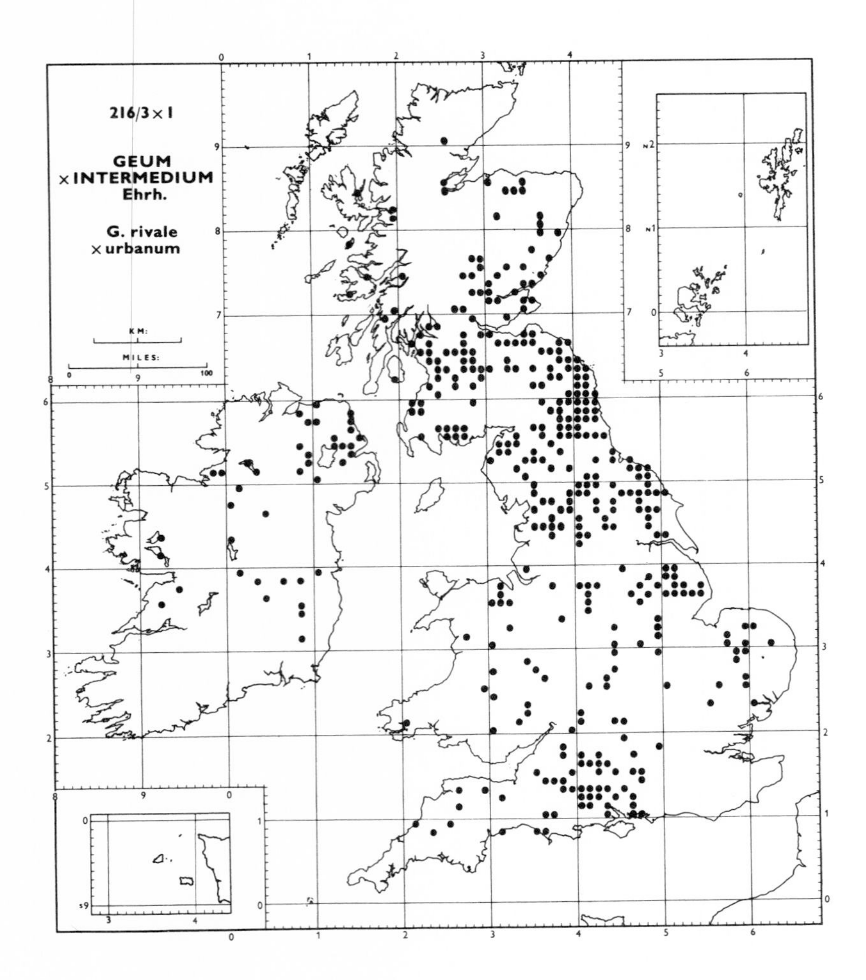

216/2 x 1 Geum × intermedium Ehrh.
= G. rivale × urbanum

Floras

G. urbanum is a species of hedges, roadsides, woodland margins and clearings, whereas *G. rivale* is found in marshy meadows and wet woods and on stream banks and wet rock ledges in mountains. Wherever the species meet a hybrid swarm results. Rarely, if ever, does the hybrid occur in the absence of the parents. It occurs throughout those parts of the British Isles in which the geographical distributions of the parents overlap (*Atlas*, p. 124), except in southern Ireland. In the latter area *G. rivale* is almost entirely a mountain plant, and it is unlikely that the habitats of the two parents meet. Records for v.-cs 79, 93, 106, 111, H. 25, 28 and 34 have not been traced. Those for v.-cs H. 17 and 38 were errors.

This hybrid is widespread in central Europe.

220 ALCHEMILLA L.

ABD, BFT, BIRM, BM, CGE, CLE, DBN, E, EXR, K, KLE, LCN, LTN, LTR, MANCH, MNE, NMW, OXF, QMC, SLBI, TCD

The methods used in the preparation of the maps of this genus have varied according to the difficulty with which the species can be identified. Well-developed specimens of the three widespread taxa *A. filicaulis* subsp. *vestita*, *A. glabra* and *A. xanthochlora* are easily distinguished, and all records of these from lowland Britain have been accepted. Records of the rarer and more local taxa, particularly those from the mountains, cannot be so readily accepted. Preparation of the maps of these species has depended largely upon the work of M. E. Bradshaw and S. M. Walters, who have seen material from all the main collections in the British Isles, determined many specimens sent for identification, and made field records of their own. Sometimes selected material which they have not personally verified has been included, but in these cases the records are differentiated from those of Bradshaw and Walters. For further details see papers by Walters (1949a, 1952) and Bradshaw (1962, 1963a and b, 1964).

All the native British *Alchemilla* species are high-polyploid apomicts, and with the exception of the endemic *A. minima* are widespread on the European Continent, where the genus as a whole is somewhat northern-montane in general distribution. This feature is reflected in the northern pattern of distribution for the total records of *Alchemilla vulgaris* L. sensu lato in the British Isles (see *Atlas*, p. 125).

220/1 Alchemilla alpina L.

For the map of this species, in which there is no taxonomic difficulty in the British Isles, see *Atlas*, p. 125.

220/2 Alchemilla conjuncta Bab.

Native in the Jura and west Alps, and once regarded as a native in Clova, Angus (90), and Glen Sannox, Arran (100), but now thought to have been introduced, though well established in Clova among rocks by streams. It occurs occasionally as an escape from gardens elsewhere. The record from v.-c. 104 has not been traced.

220/3 Alchemilla vulgaris L. *sensu lato*

220/3.1 Alchemilla glaucescens Wallr.

A. minor auct. brit., non Huds.

This is a local species of limestone grassland in north-west Scotland and the Ingleborough area (64) with an isolated locality in Leitrim (H. 29). A continental species, widespread in central and east Europe. The record for v.-c. 66 was an error.

220/3.4 Alchemilla subcrenata Buser

One of the rarest species of *Alchemilla* in the British Isles, first discovered in 1951. It is one of the group of three northern montane species of *Alchemilla* confined to the Teesdale region in Britain, which is its most westerly locality in Europe. It occurs throughout the Alps, Scandinavia and much of central and eastern Europe.

220/3.5 Alchemilla minima Walters

This is the only endemic species of *Alchemilla* in Britain, known from closely grazed *Festuca-Agrostis* grassland on two mountains in the Craven district of the north Pennines. Dwarf variants of *A. filicaulis* subsp. *vestita* also occur in this area, but transplant experiments carried out by Bradshaw (1964) have shown that they differ from *A. minima* in morphological characters, habit and ecological requirements.

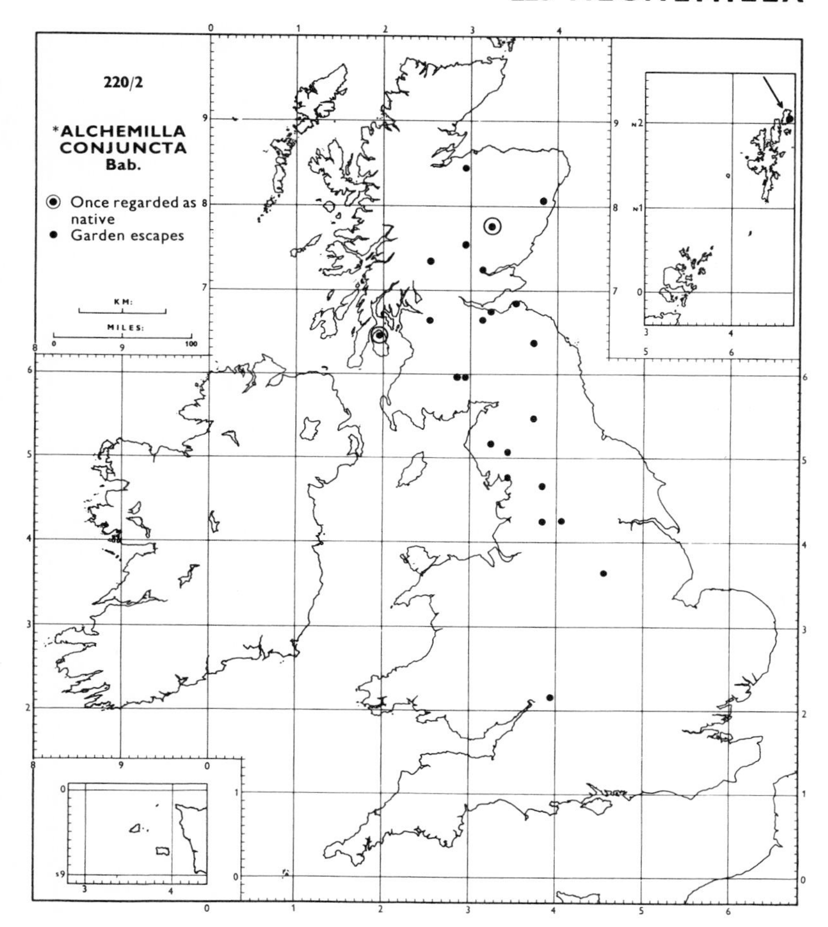

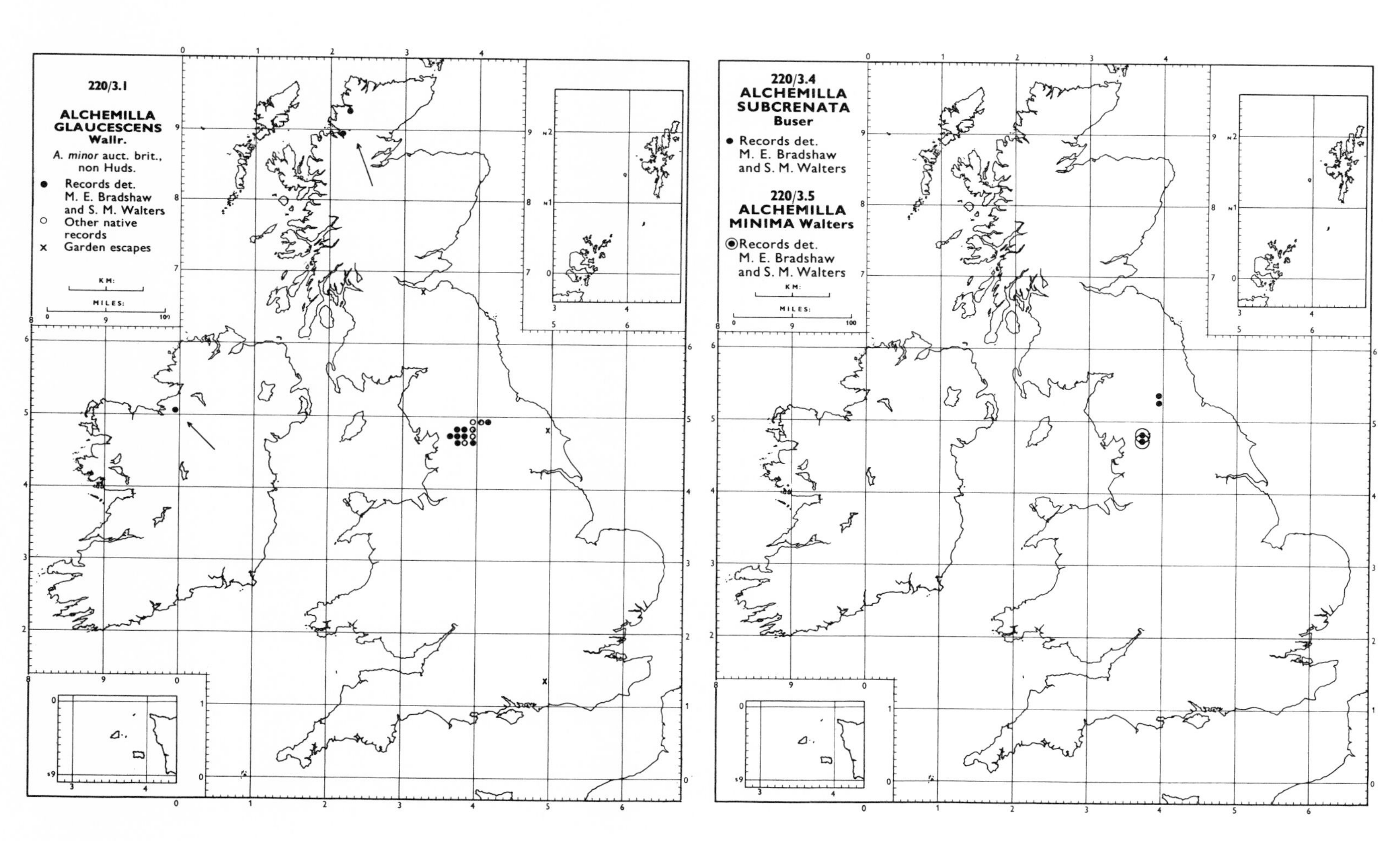

220/3.6 **Alchemilla monticola** Opiz

This species is frequent or abundant on roadsides and in hay-meadows in Teesdale and Weardale (65, 66) where it is undoubtedly native. It has occurred in Surrey (17) and Buckingham (24), but only in small quantity as a casual introduction. A record for v.-c. 37 has not been traced, and the records from v.-cs 64 and 96 are known to have been errors.

This is another of the group of three northern montane species of *Alchemilla* which occur in the Teesdale region. It is widespread in Europe from north Scandinavia to the southern Alps, and throughout much of central and eastern Europe extending to west Siberia.

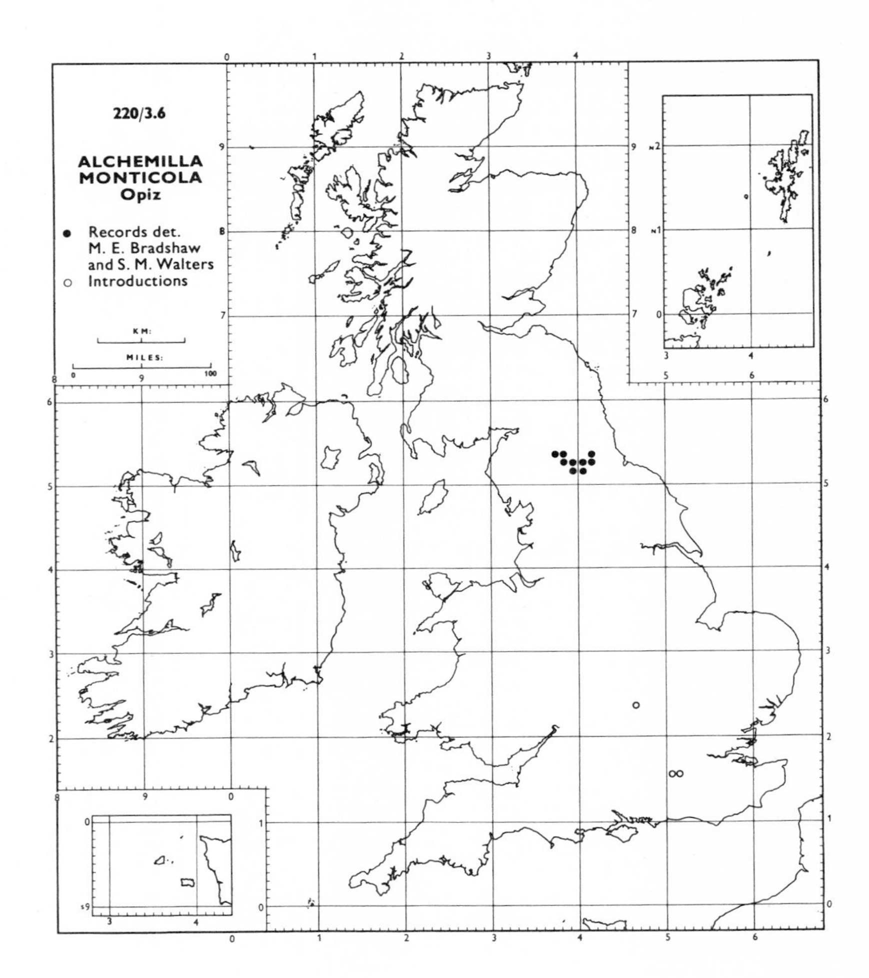

220/3.2–3 **Alchemilla filicaulis** Buser
subsp. **vestita** (Buser) M. E. Bradshaw
A. vestita (Buser) Raunk.
subsp. **filicaulis**

Until recently these taxa were regarded as species, but work by Bradshaw (1963b) has shown that no clear-cut morphological distinction can be made between them, as intermediates occur. She has therefore suggested that in view of the general difference in geographical distribution which they show, the two taxa be ranked as subspecies of *A. filicaulis*. Subsp. *vestita* is the most widely distributed *Alchemilla* in the British Isles and almost the only one in the south of England.

Subsp. *filicaulis* is not uncommon above 1,500 ft (450 m.) on Scottish mountains and ascends to over 3,000 ft (900 m.). It occurs less frequently on the mountains of England and Wales and is very rare in Ireland. It descends to sea-level in Orkney (111) and Shetland (112), where subsp. *vestita* is absent. Both subspecies have an amphi-Atlantic distribution, occurring in Iceland, Greenland and North America as well as in north and west Europe, but subsp. *filicaulis* is constantly more northern or alpine.

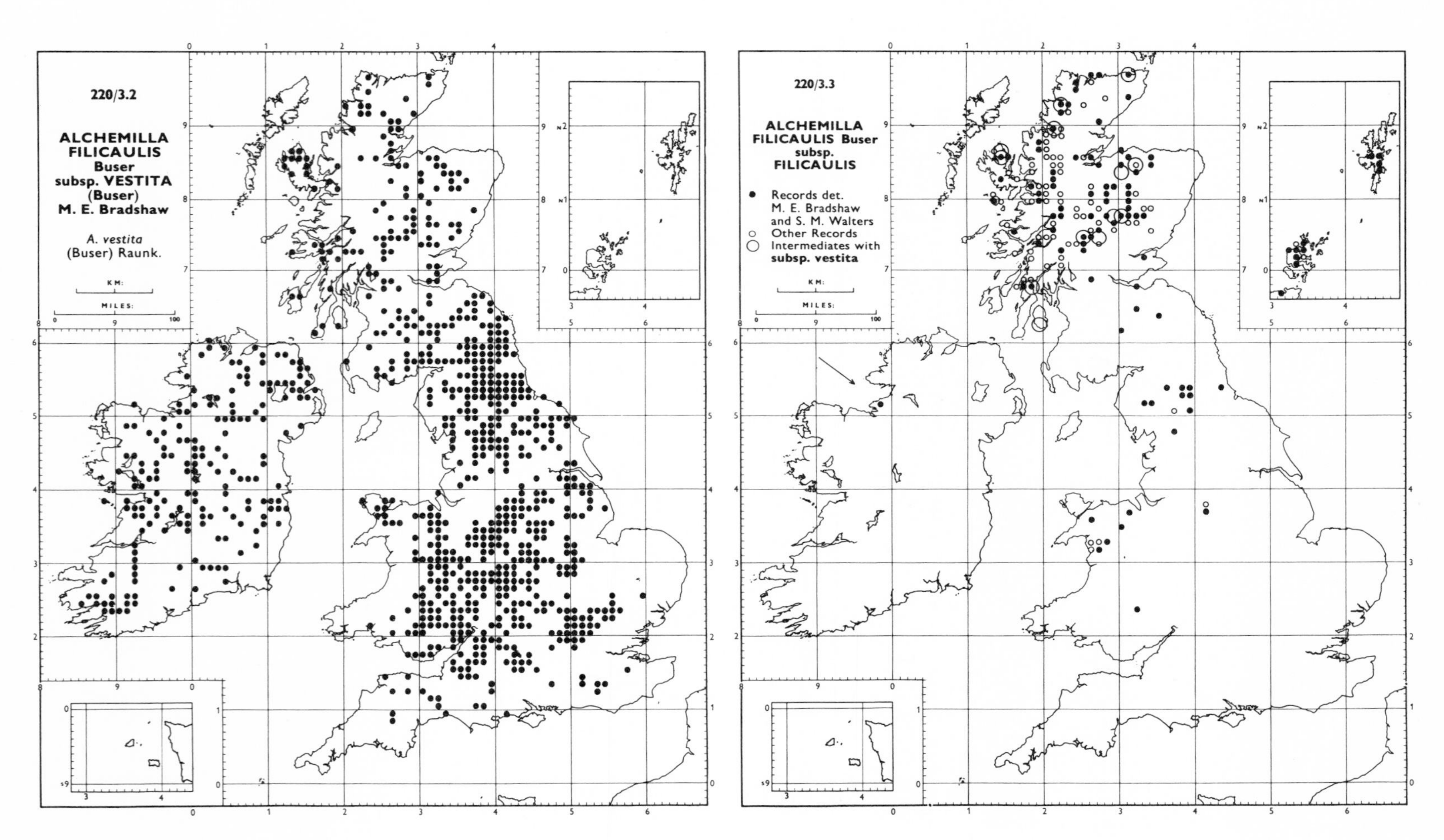

220/3.7 **Alchemilla acutiloba** Opiz

This species is occasional on roadsides and in hay-meadows in Upper Teesdale and Weardale (66), where it grows with *A. monticola* and other, common, species.

This is the third of the group of three northern-montane species of *Alchemilla* confined to the Teesdale region in Britain. Elsewhere it is recorded from Scandinavia and north-west Germany, eastwards to the Ob region of west Siberia and south to the Alps and the Volga-Don region.

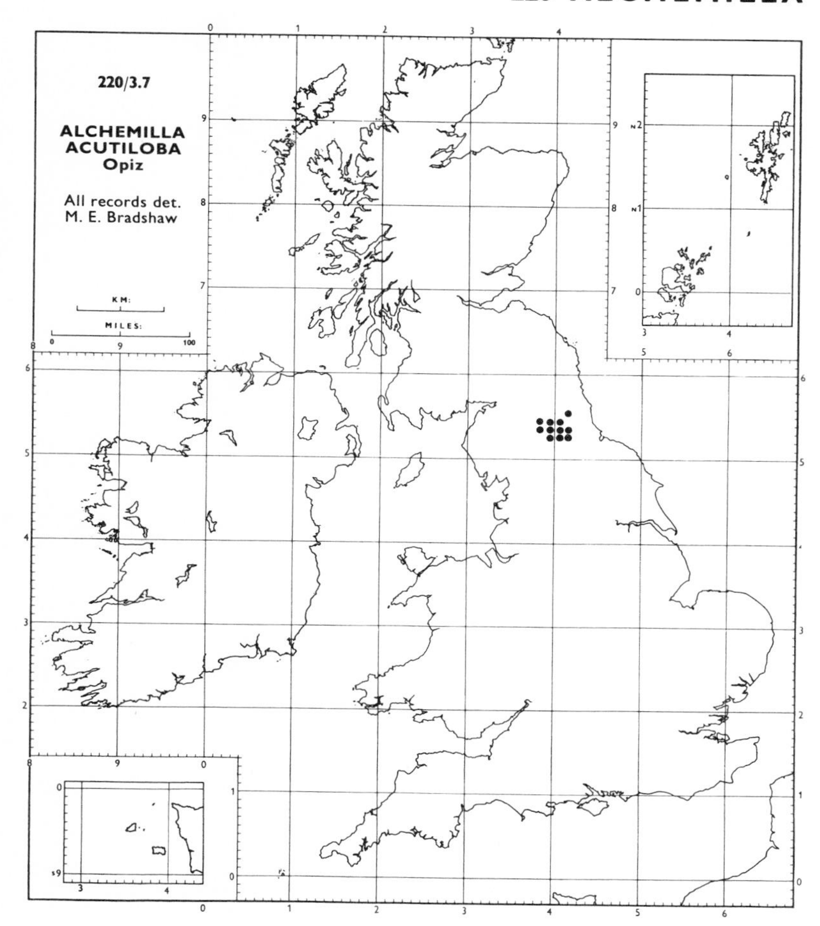

220/3.8 **Alchemilla xanthochlora** Rothm.

One of the three common species of *Alchemilla*, abundant in upland Britain on roadsides and in pastures and meadows, but not on high mountains and rare in north Scotland. It has a suboceanic distribution in Europe closely paralleling that of Beech, *Fagus sylvatica*.

220/3.9 **Alchemilla glomerulans** Buser

This is a species of damp, usually somewhat acid, rock ledges in the mountains, mainly at altitudes above 2,000 ft (600 m.), but descending to 600 ft (180 m.) in Teesdale (66), where it occurs on more basic strata. It is arctic-alpine in its distribution, being widespread in the Arctic and on Scandinavian mountains, and occurring rarely in the Jura, Alps and Pyrenees.

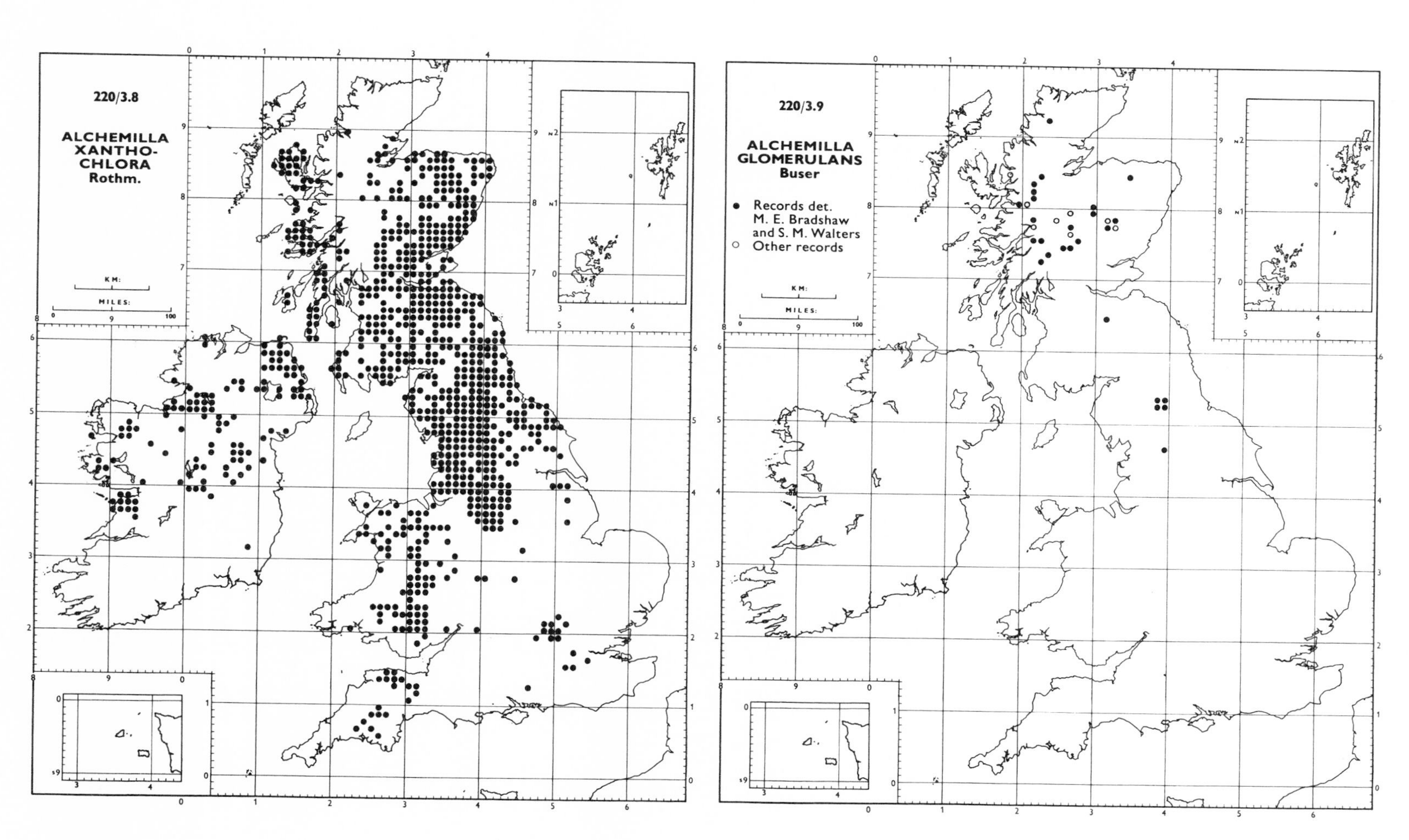

ROSACEAE

229/3.10 **Alchemilla glabra** Neygenf.

One of the three commonest species of *Alchemilla* and the most frequent in the mountains of northern England, and Scotland where it ascends to over 4,000 ft (1,250 m.). Widespread in western and central Europe, from north Norway to north Italy, extending eastwards to central Ural.

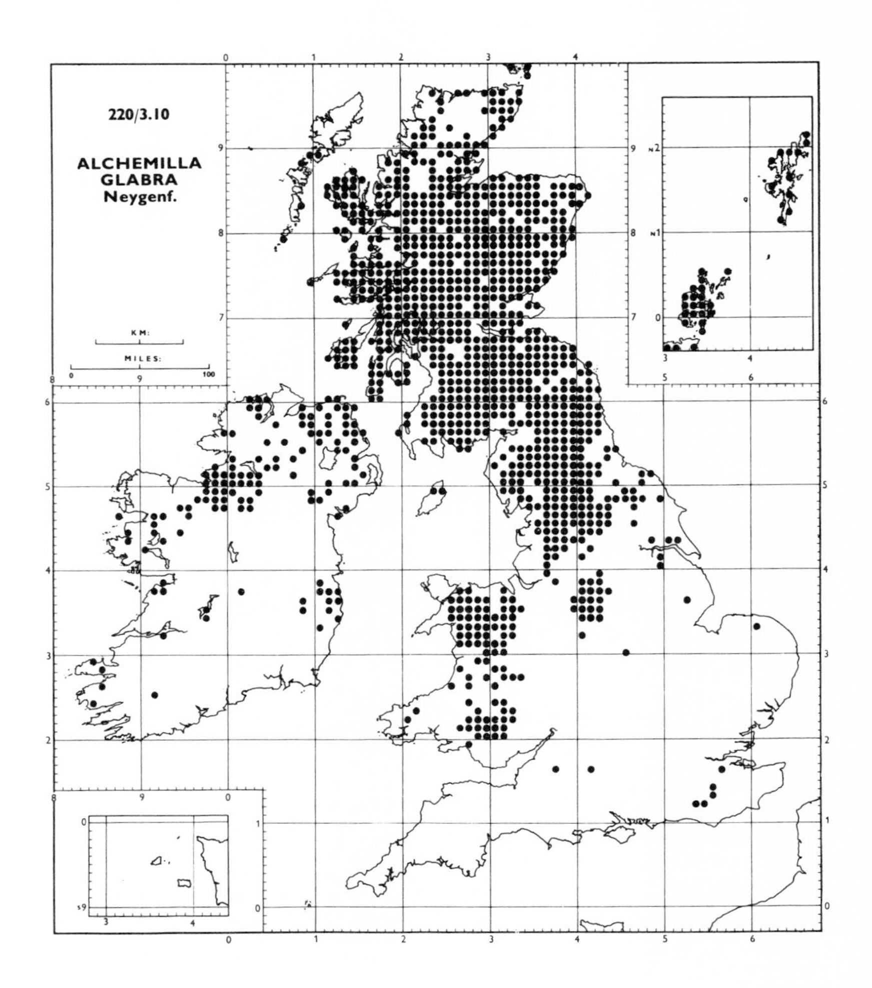

220/3.11 **Alchemilla wichurae** (Buser) Stéfansson

This species is restricted to basic soils in the mountains of north England and Scotland, ascending to 3,000 ft (900 m.). It is essentially an Arctic species extending from east Greenland to Finland. Recorded from v.-cs 94 and 110 in error.

220/3.12 **Alchemilla mollis** (Buser) Rothm.

The commonest, and showiest, of the *Alchemilla* species grown in gardens, and occasionally well established some distance away from cultivation. A native of Romania and Asia Minor.

220/3.12 bis **Alchemilla tytthantha** Juz.

This native of the Crimea, with a history of cultivation in Botanic Gardens, has recently been found established in Selkirk (79), Berwick (81) and Stirling (86). (See Bradshaw and Walters, 1961.)

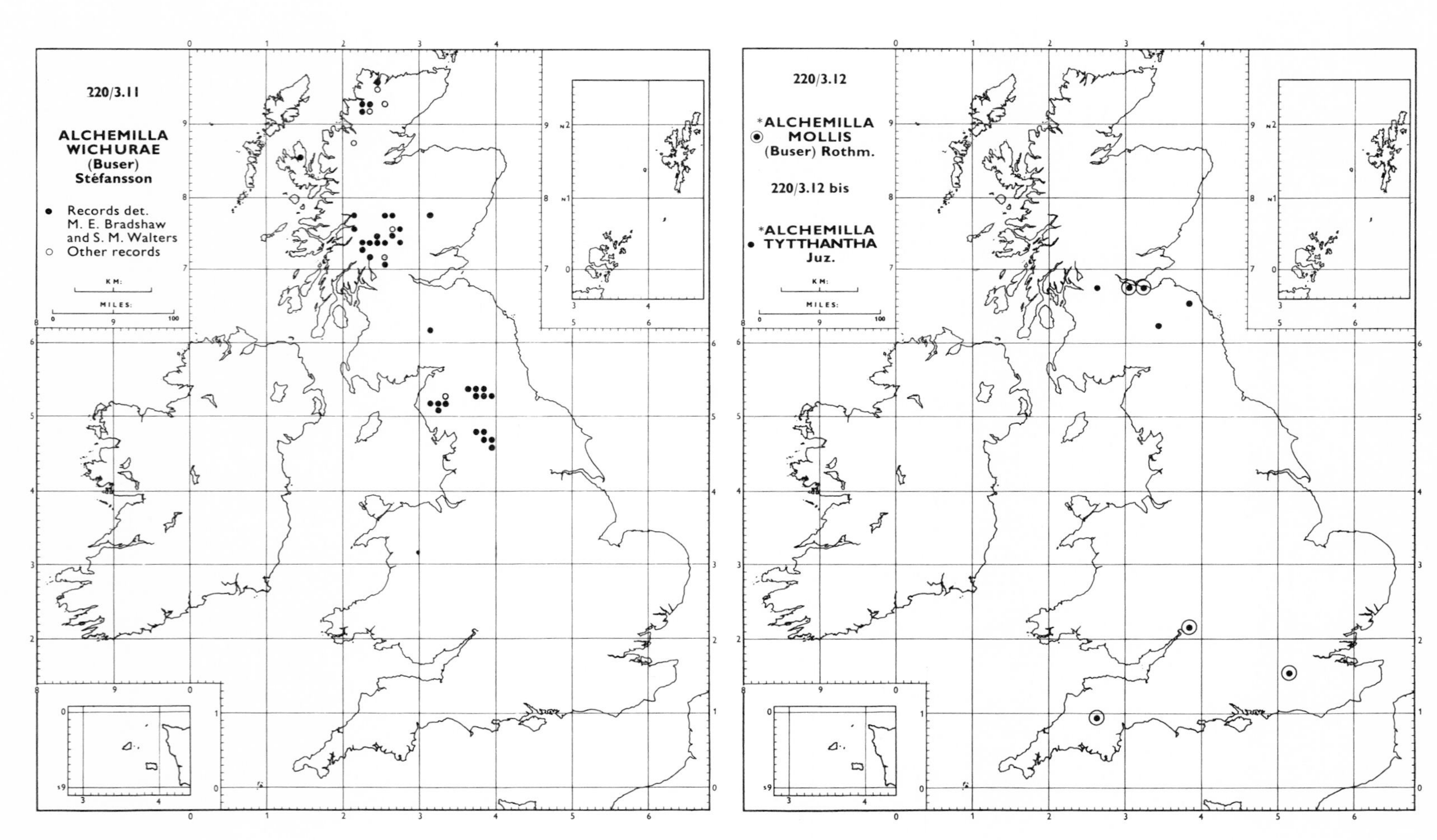

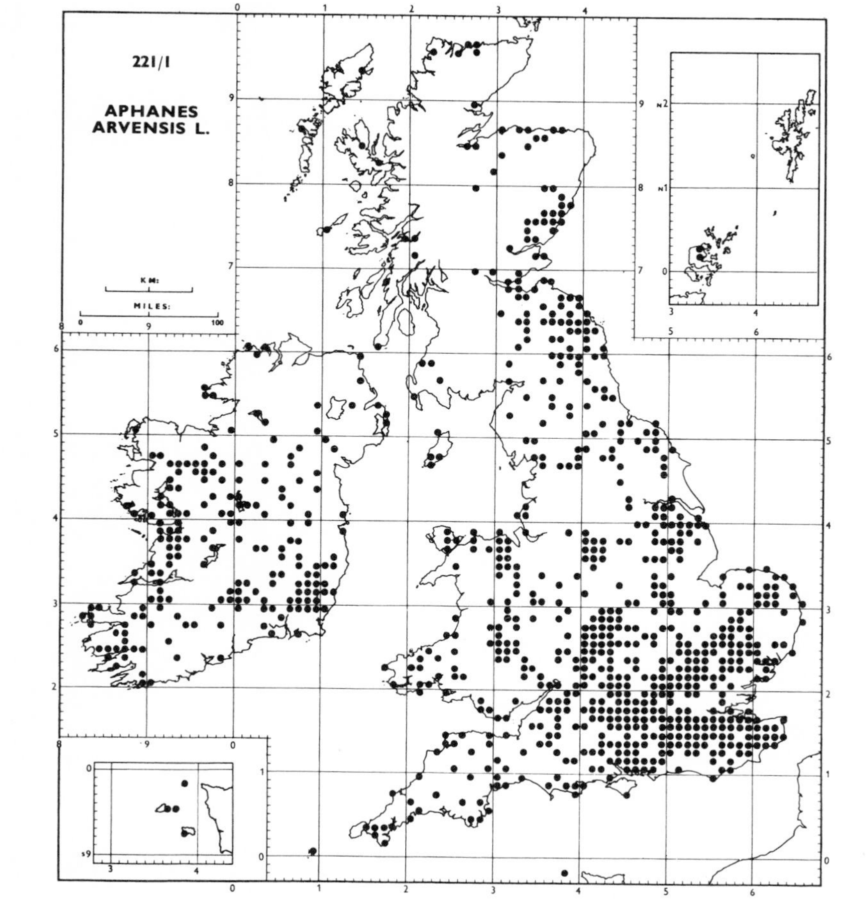

221/1 **Aphanes arvensis** L.

221/2 **Aphanes microcarpa** (Boiss. & Reut.) Rothm.

BM, CGE, K, OXF, TCD

These maps are based on fresh material sent in for identification, and herbarium material, determined by S. M. Walters and the editor. Field records made by reliable recorders following the paper by Walters (1949b) are also included. The maps reflect well the different soil tolerances of the two species: whereas *A. arvensis* is indifferent to soil acidity, *A. microcarpa* is more or less restricted to acid soils.

The maps are reasonably complete, but clearly the two have not been separated in northern Ireland and some parts of the midlands of England (see the map of *A. arvensis* agg., *Atlas*, p. 26).

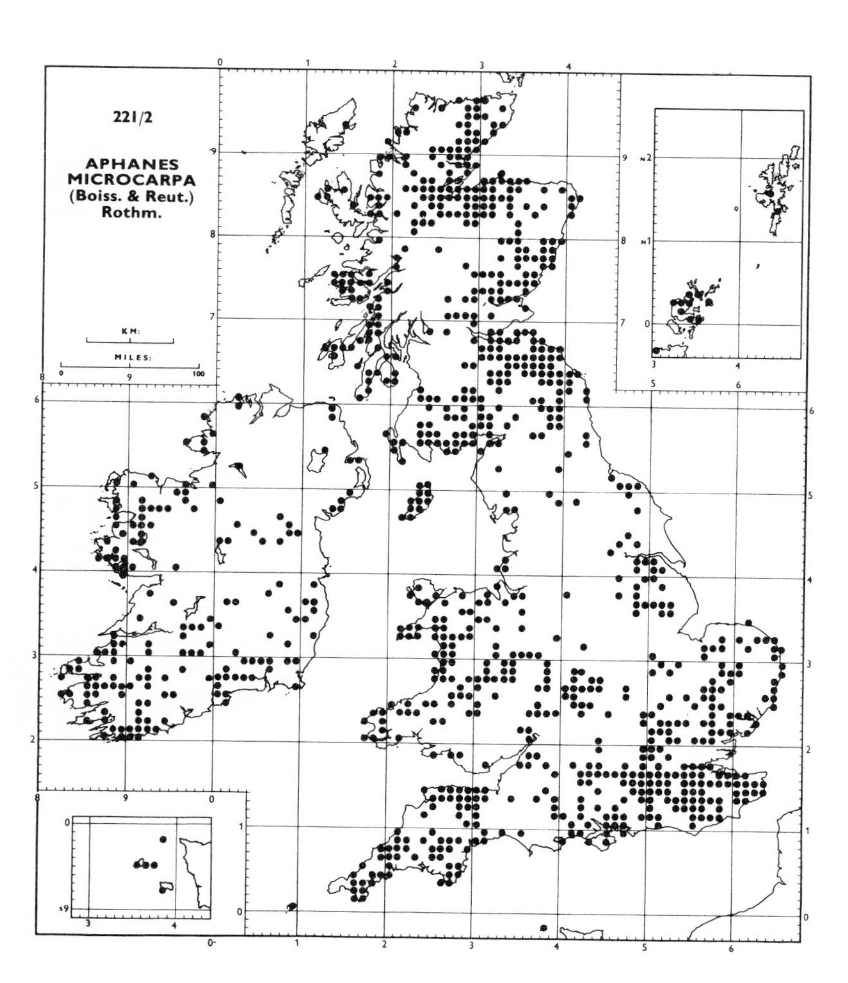

3

ROSACEAE

232 SORBUS L.

The maps of this difficult genus were prepared with the guidance of the late E. F. Warburg, who identified material in all the main national herbaria, and saw many of the plants in the field.

The majority of the maps are probably up to date as far as records go. Many of the polyploid species are apomictic and confined to very small areas of cliff or open habitats. For further details Warburg's account in Clapham, Tutin and Warburg (1962) should be consulted.

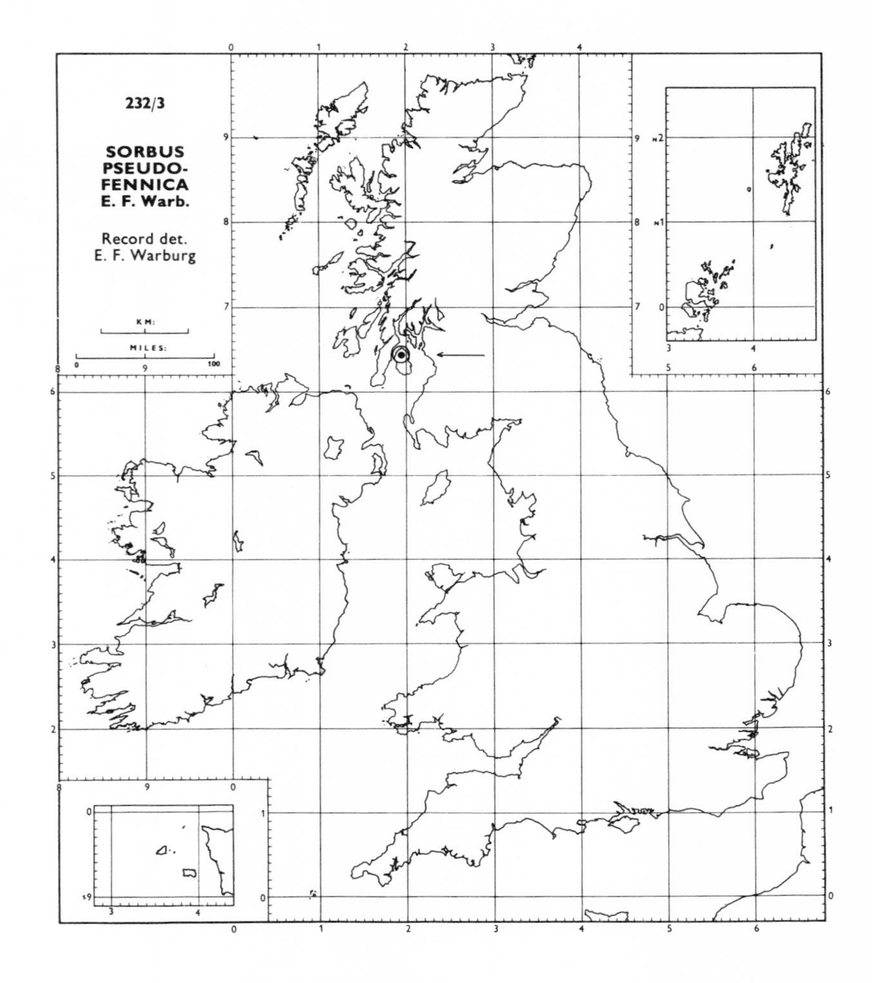

232/3 Sorbus pseudofennica E. F. Warb.

Endemic to a steep granite stream bank in Glen Catacol, Arran (100). The closely allied *S. hybrida* L. is widespread in Fennoscandia.

232/4 Sorbus intermedia (Ehrh.) Pers. *sensu lato*

Species of this aggregate probably arose as a result of hybridization between *S. aucuparia* L. and *S. aria* (L.) Crantz *sensu lato*. All the native species occur almost entirely outside the native range of *S. aria* L. *sensu stricto* as now recognized.

232/4.1 Sorbus intermedia (Ehrh.) Pers. *sensu stricto*

A commonly planted tree, sometimes bird-sown and fully naturalized in a few places. Records from v.-cs 21, 34, 87, 92, 98, 100 and 104 have not been traced. The record for v.-c. 35 was an error. Native of south Sweden, Bornholm, the Baltic States and north-east Germany.

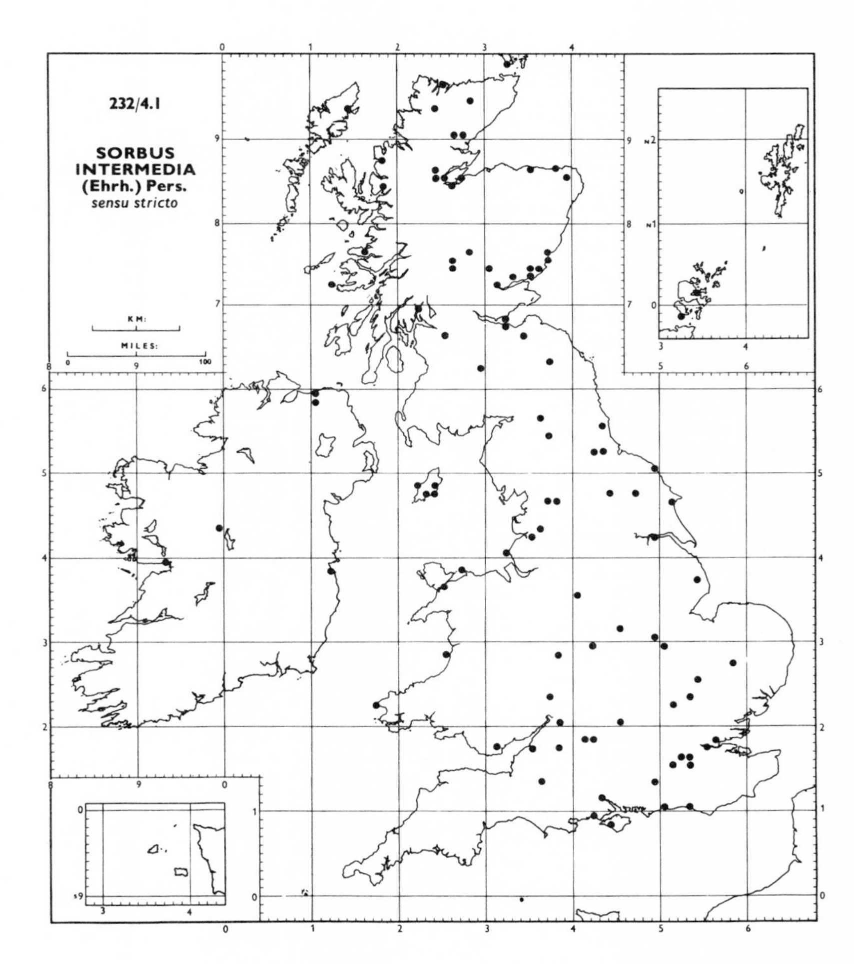

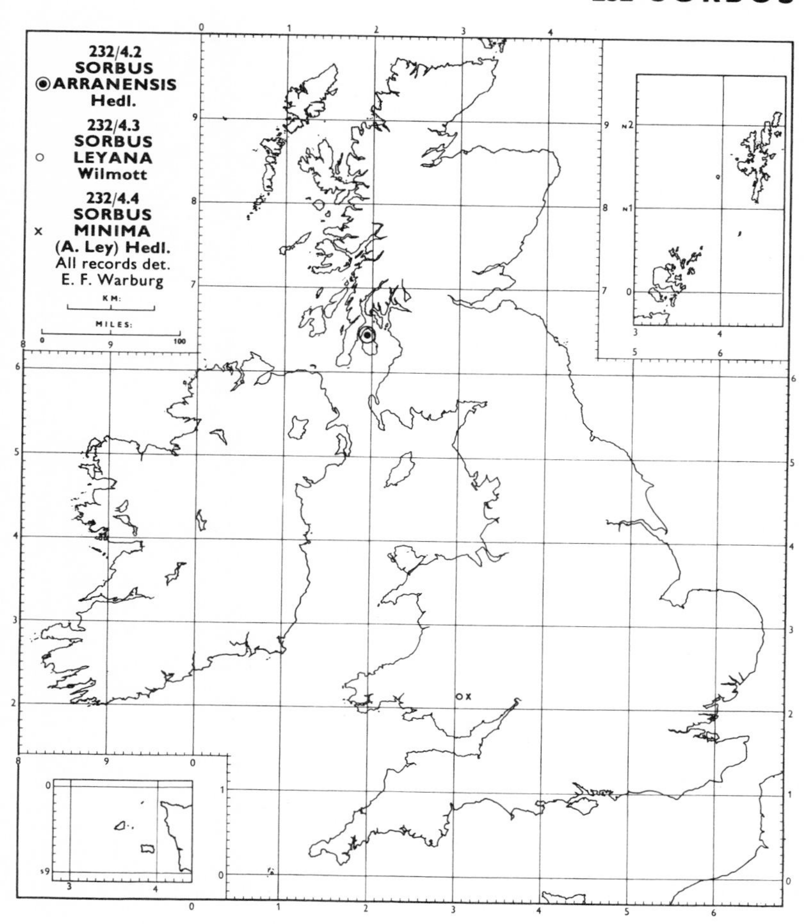

232/4.2 Sorbus arranensis Hedl.

Endemic to steep granite stream banks in Glen Easan Biorach and Glen Catacol, Arran (100). Norwegian plants referred to this species are not identical with it.

232/4.3 Sorbus leyana Wilmott

Endemic to two carboniferous limestone crags near Dan-y-Graig north of Merthyr Tydfil, Brecon (42).

232/4.4 Sorbus minima (A. Ley) Hedl.

Endemic to carboniferous limestone crags near Crickhowell, Brecon (42).

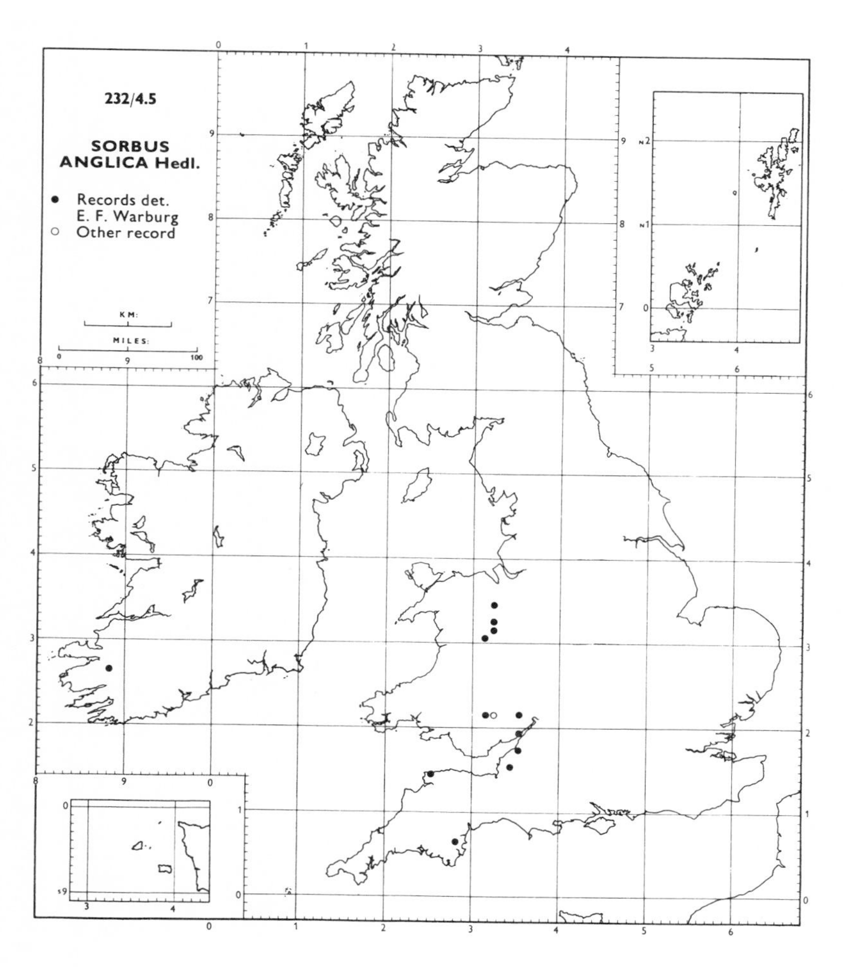

232/4.5 Sorbus anglica Hedl.

An endemic species found on crags and in rocky woods in a number of disjunct localities nearly always on carboniferous limestone. Populations tend to differ slightly from one another. Closely allied species occur in Norway, the Alps and the Pyrenees.

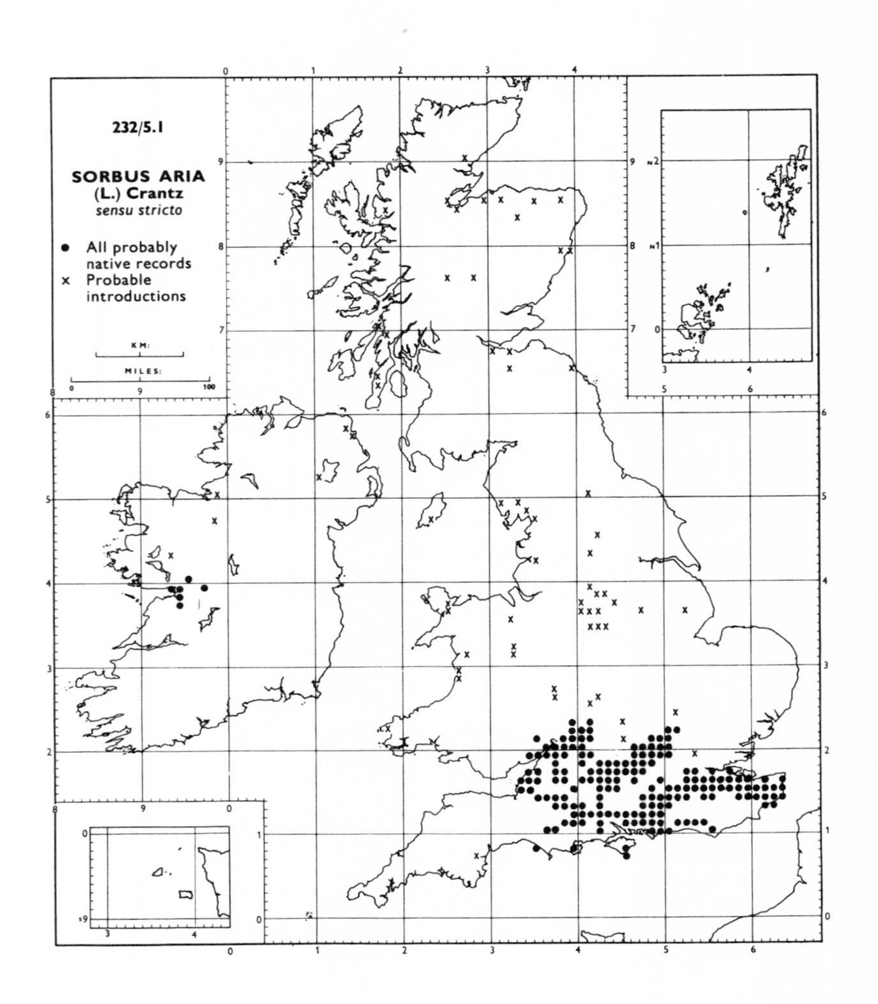

232/5 **Sorbus aria** (L.) Crantz *sensu lato*

This group of nine species consists of a diploid *S. aria* sensu stricto (2n=34) and a number of triploids and tetraploids. The tetraploid species occur very widely outside the native range of *S. aria* sensu stricto.

232/5.1 **Sorbus aria** (L.) Crantz *sensu stricto*

In marked contrast to the other British species of *Sorbus*, *S. aria* is very variable. It is found in woods and scrub, usually on chalk or limestone, but locally on sandstone. It is much planted and its native distribution is not easy to define. Trees from Galway (H. 17) with smaller leaves are perhaps referable to a distinct species.

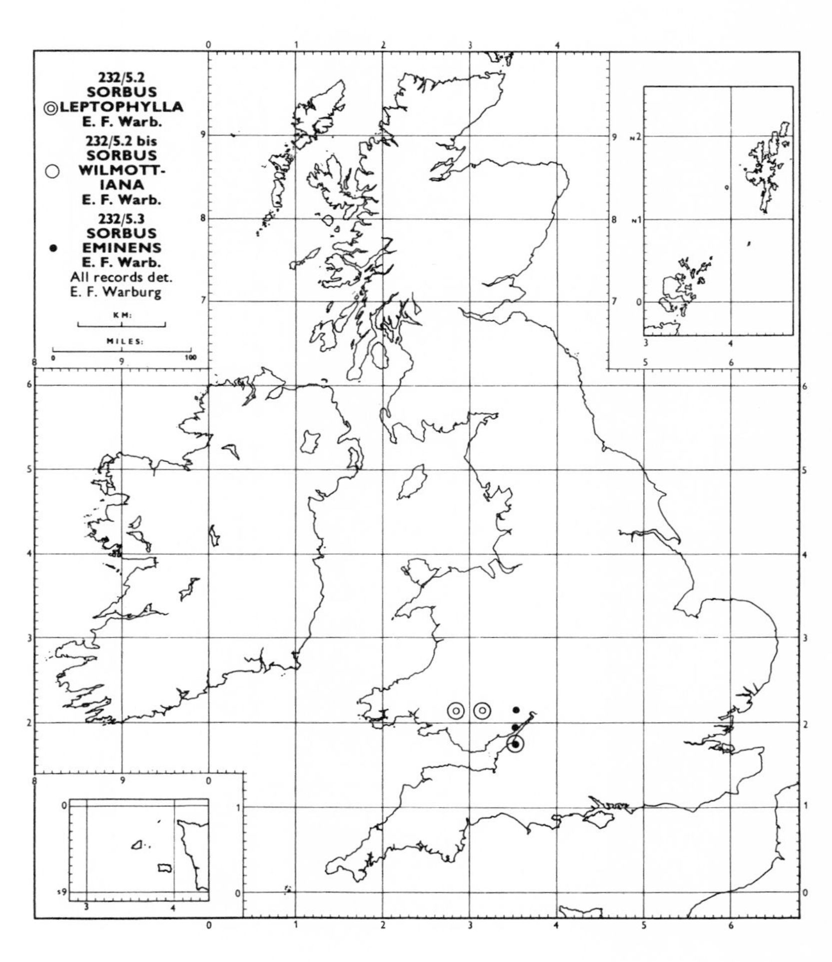

232/5.2 **Sorbus leptophylla** E. F. Warb.

Only known with certainty from shady limestone crags in two localities in Brecon (42), where it replaces *S. aria*. Probably also occurs in Montgomery (47), but fruiting specimens have not been seen.

232/5.2 bis **Sorbus wilmottiana** E. F. Warb.

Endemic to rocky limestone woodland and scrub in the Avon Gorge (6, 34). (cf. Sell, 1967.)

232/5.3 **Sorbus eminens** E. F. Warb.

Endemic to woods on carboniferous limestone in the Wye Valley (34–36) and the Avon Gorge (6, 34). Trees from around Symonds Yat (34) have subrhombic leaves and deeper teeth.

232/5.4 **Sorbus hibernica** E. F. Warb.

Endemic to rocky open woods and scrub on carboniferous limestone across central Ireland, with an isolated record in Antrim (H. 39).

232/5.5 **Sorbus porrigentiformis** E. F. Warb.

Endemic to crags and rocky woods on limestone, usually carboniferous, in south Devon (3), north Somerset (6), the Wye Valley and Wales. The very closely allied *S. porrigens* Hedl. is a tree of Asia Minor.

232/5.8 **Sorbus vexans** E. F. Warb.

Endemic to the coastal area between Lynmouth, Devon (4), and Culbone, Somerset (5). Unlike related species, it does not grow on limestone.

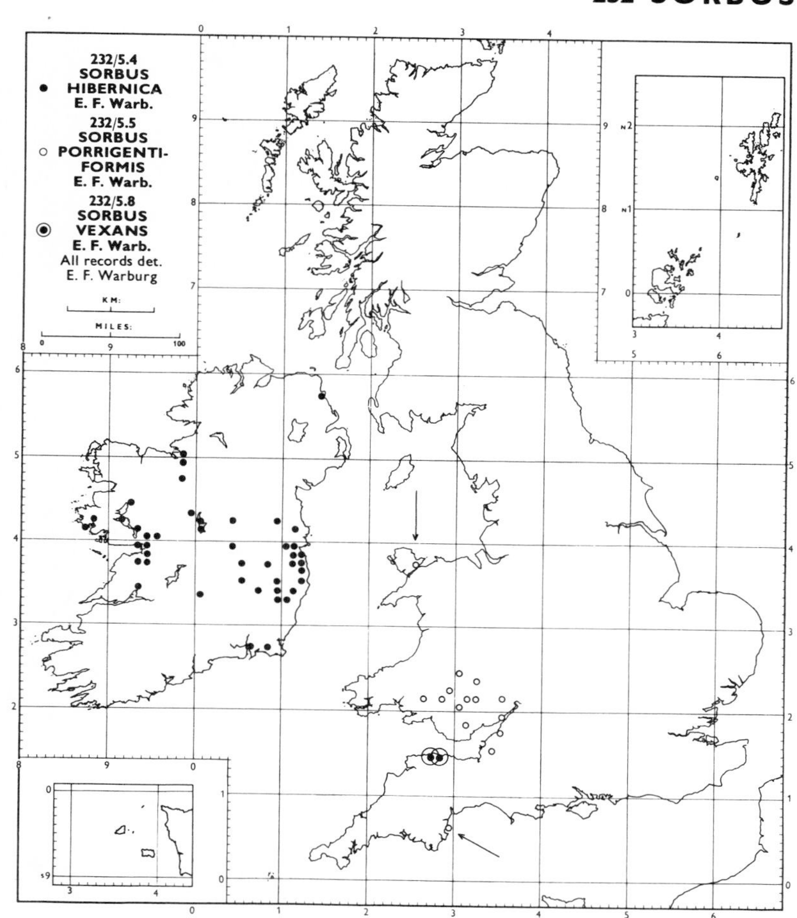

232/5.6 **Sorbus lancastriensis** E. F. Warb.

Endemic to several localities on carboniferous limestone round Morecambe Bay in Lancashire and Westmorland (60, 69).

232/5.7 **Sorbus rupicola** (Syme) Hedl.

Local on crags and among rocks, usually limestone and nearly always basic, from south Devon (3) through Wales and central England to north Scotland, and in north and west Ireland. Also occurs in Scandinavia.

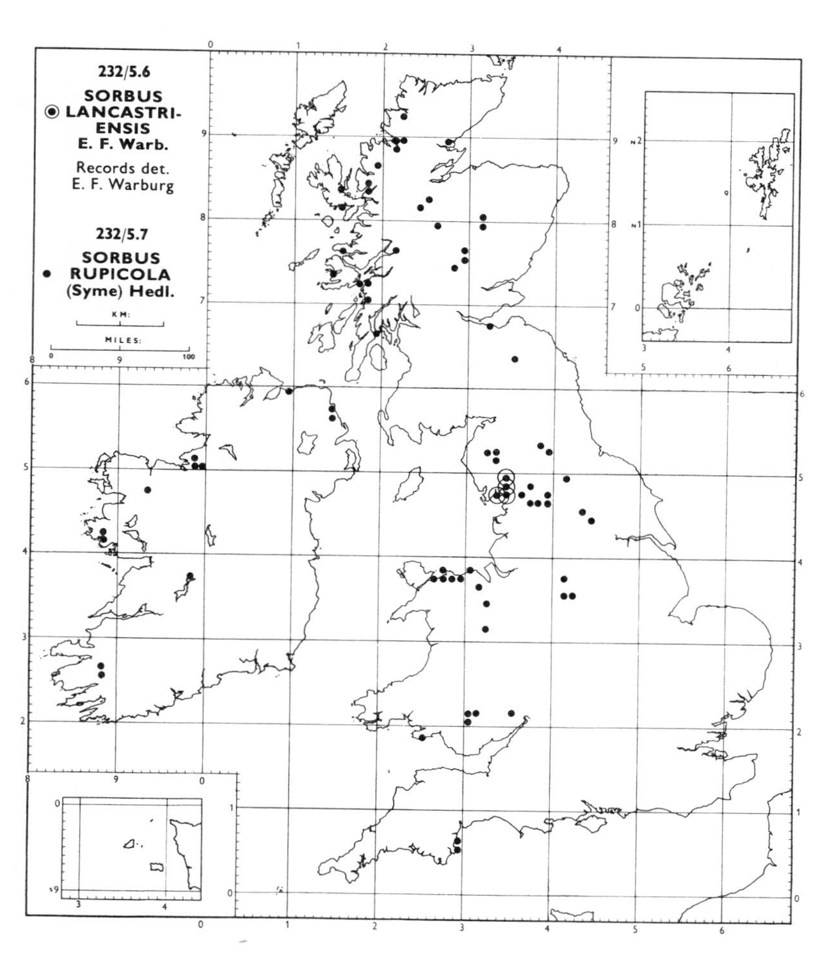

232/6 **Sorbus latifolia** (Lam.) Pers. *sensu lato*

The three species in this aggregate which occur in Britain almost certainly arose as a result of hybridization between *S. aria* and *S. torminalis* (L.) Crantz.

232/6.1 **Sorbus bristoliensis** Wilmott

Endemic to rocky woods and scrub on carboniferous limestone in the Avon Gorge (6, 34).

232/6.2 **Sorbus subcuneata** Wilmott

Endemic to open woods of *Quercus petraea* between Minehead, Somerset (5), and Watersmeet, Devon (4). A tree growing at Watersmeet and similar to other specimens collected in the area, possibly from the same tree, differs from *S. subcuneata* in its broader leaves which are more sharply and deeply lobed.

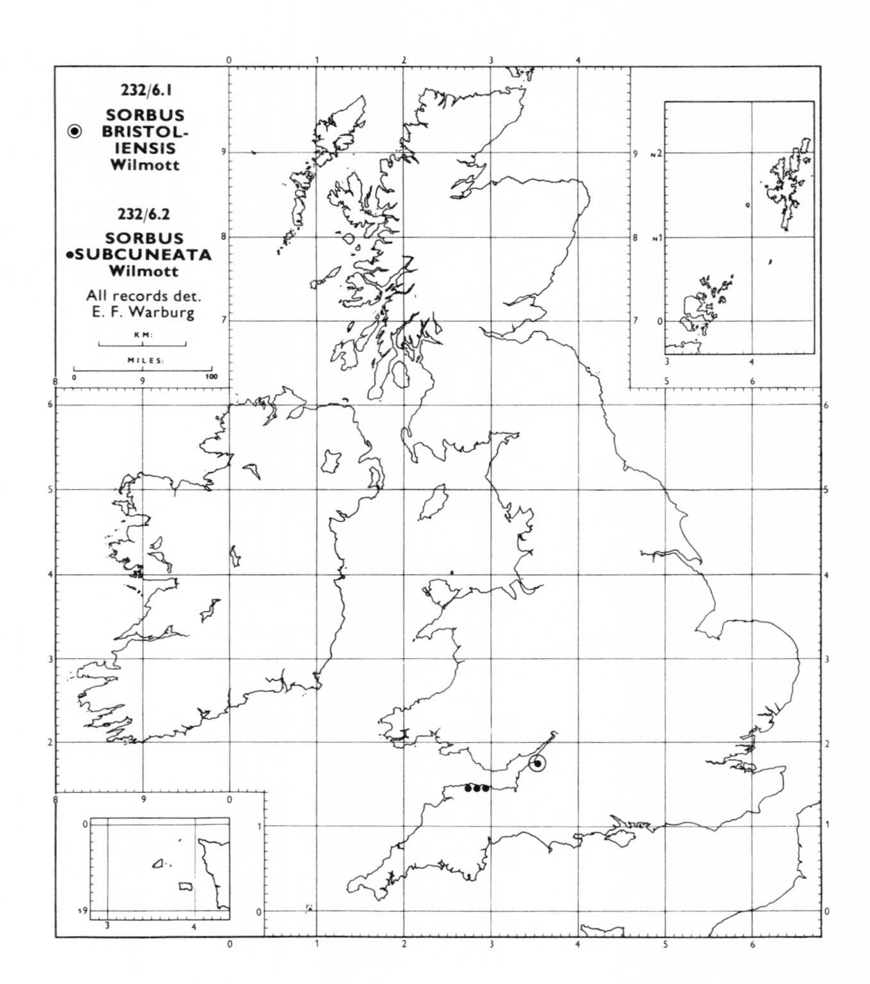

232/6.3 **Sorbus devoniensis** E. F. Warb.

Widespread in Devon (3, 4) where it is the commonest species of the genus, and just extending into Cornwall (2). It occurs also in Kilkenny, Wexford and Carlow (H. 11–13), where it is apparently native.

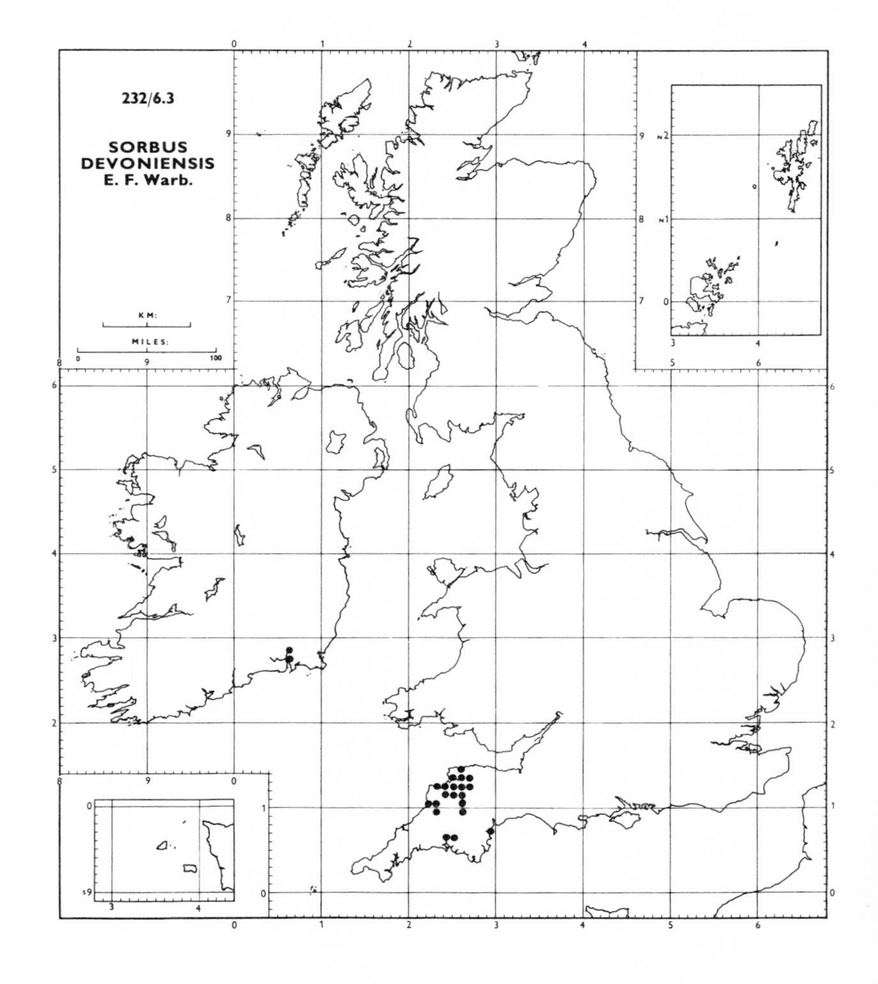

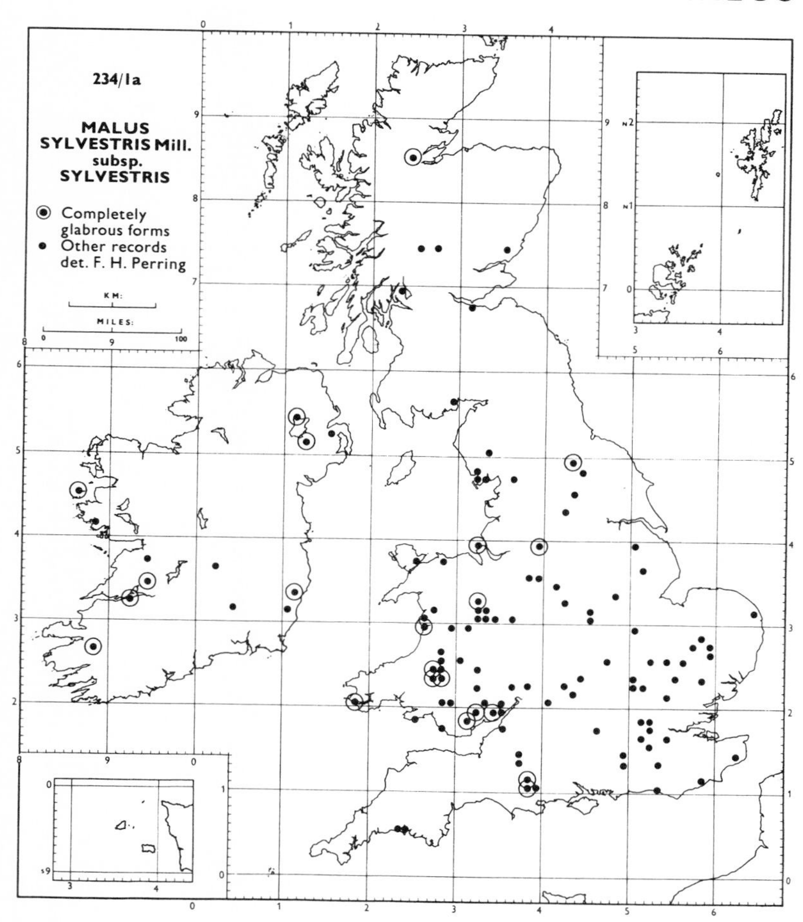

234/1 Malus sylvestris Mill.
subsp. **sylvestris**
subsp. **mitis** (Wallr.) Mansf.

BFT, BM, CGE, DBN, E, NMW, NWH, SHY, TCD

These maps are based entirely upon herbarium material seen by the editor, and his own field records. Intermediates in degree of hairiness between the two subspecies are widespread; these have not been included on the maps. Only specimens in which hairs were confined to the calyx, and with few scattered hairs on the under-surface of the leaves have been allowed as subsp. *sylvestris*. Material included as subsp. *mitis* had to have the leaves densely pubescent beneath, and the pedicels, receptacle and the outside of the calyx tomentose.

Completely glabrous forms occur occasionally, especially as woodland trees in the west and north. This suggests that subsp. *sylvestris* is a native woodland tree which is glabrous. Since the introduction of subsp. *mitis* introgression has taken place, particularly in the more highly populated areas where pure subsp. *sylvestris* is now becoming difficult to find. The native range may be almost that indicated by the map of the species (*Atlas*, p. 132). Subsp. *sylvestris* is native throughout temperate Europe, whereas subsp. *mitis* is only native in south-east Europe and Asia.

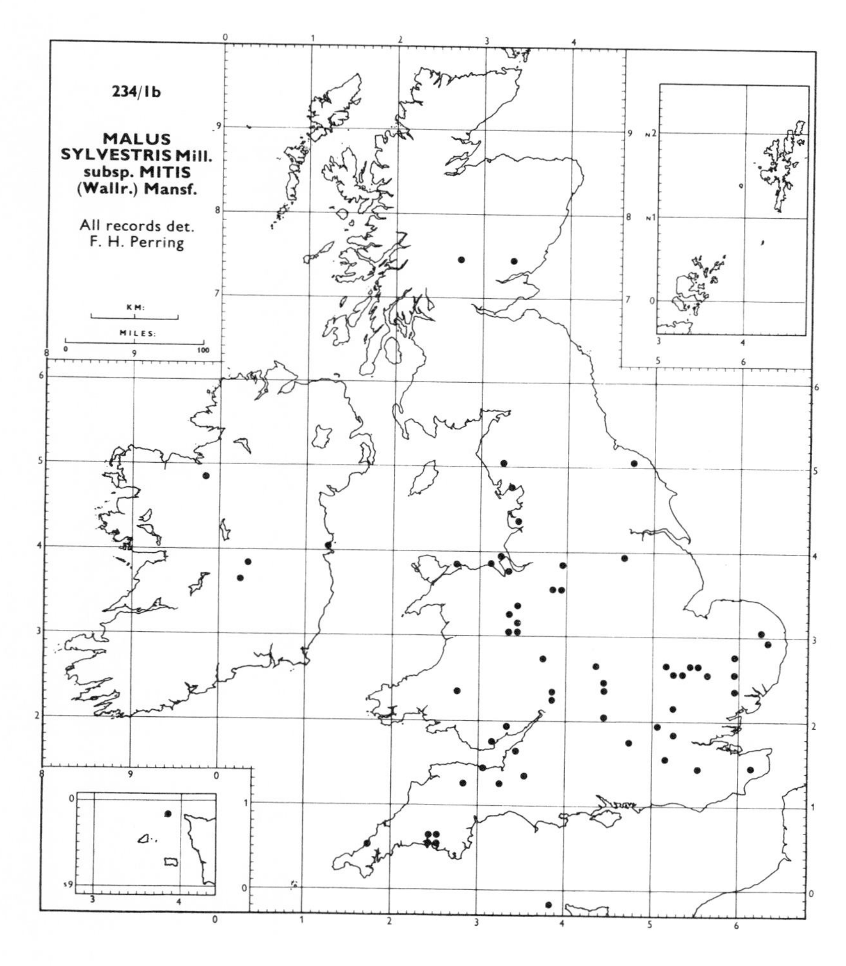

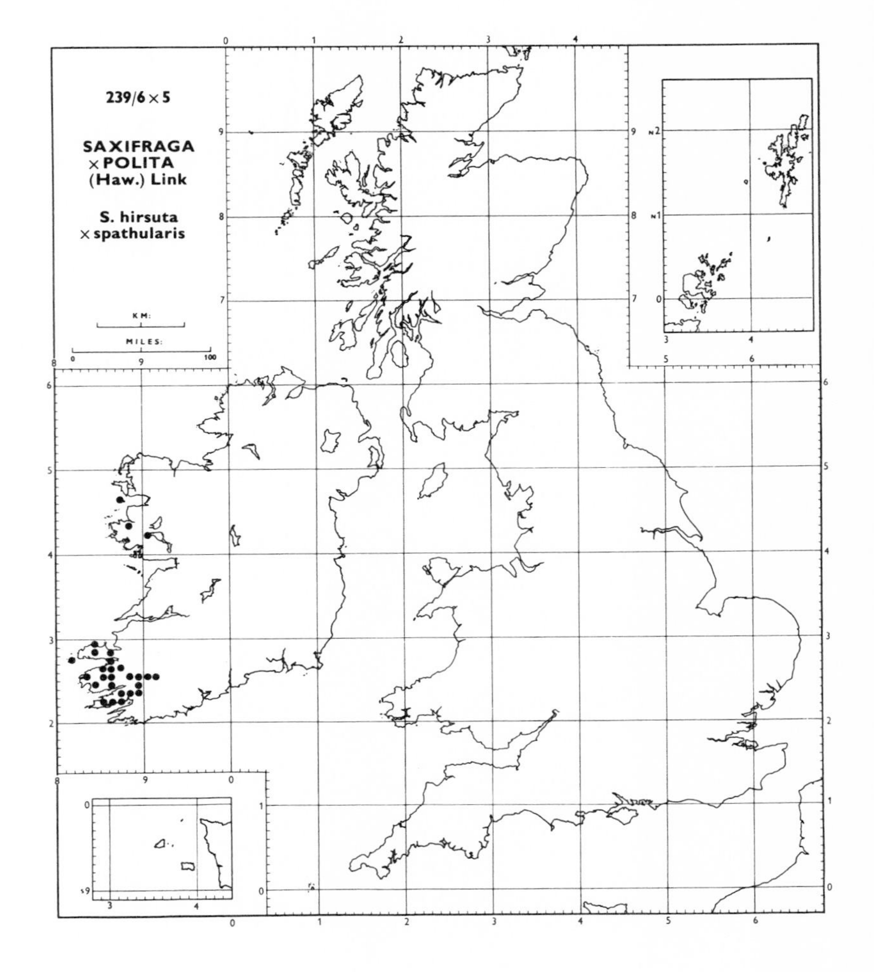

239/6×5 Saxifraga ×polita (Haw.) Link
=S. hirsuta × spathularis

TCD: Floras
It is interesting that this hybrid occurs in Galway and Mayo (H. 16, 27) in the absence of one of the parents, *S. hirsuta*, which is now confined to Kerry and W. Cork (H. 1–3) (see *Atlas*, pp. 136–7). It also occurs in northern Spain.

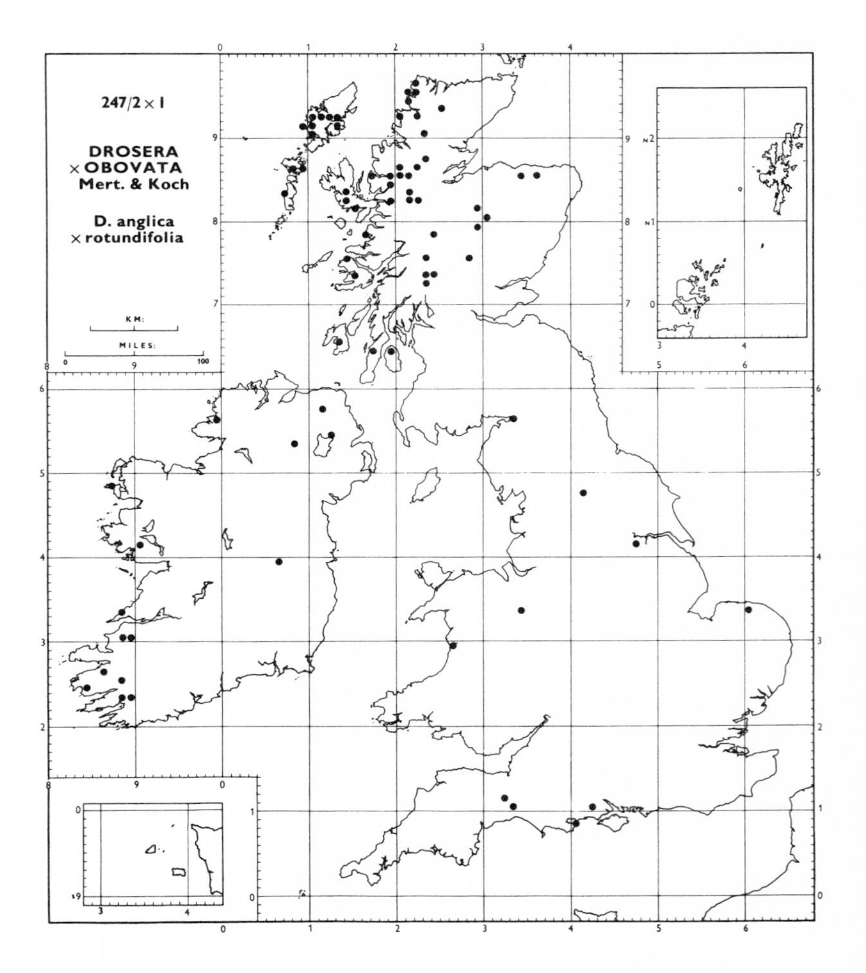

247/2×1 Drosera ×obovata Mert. & Koch
=D. anglica × rotundifolia

BM, CGE, DBN, E, NMW, OXF, TCD: Floras
This map suggests that the hybrid is infrequent, though it has been recorded throughout the range of the two parents (see *Atlas*, pp. 141–2). The reason for the high concentration of records in north-west Scotland and the Outer Hebrides (110) is not readily apparent.

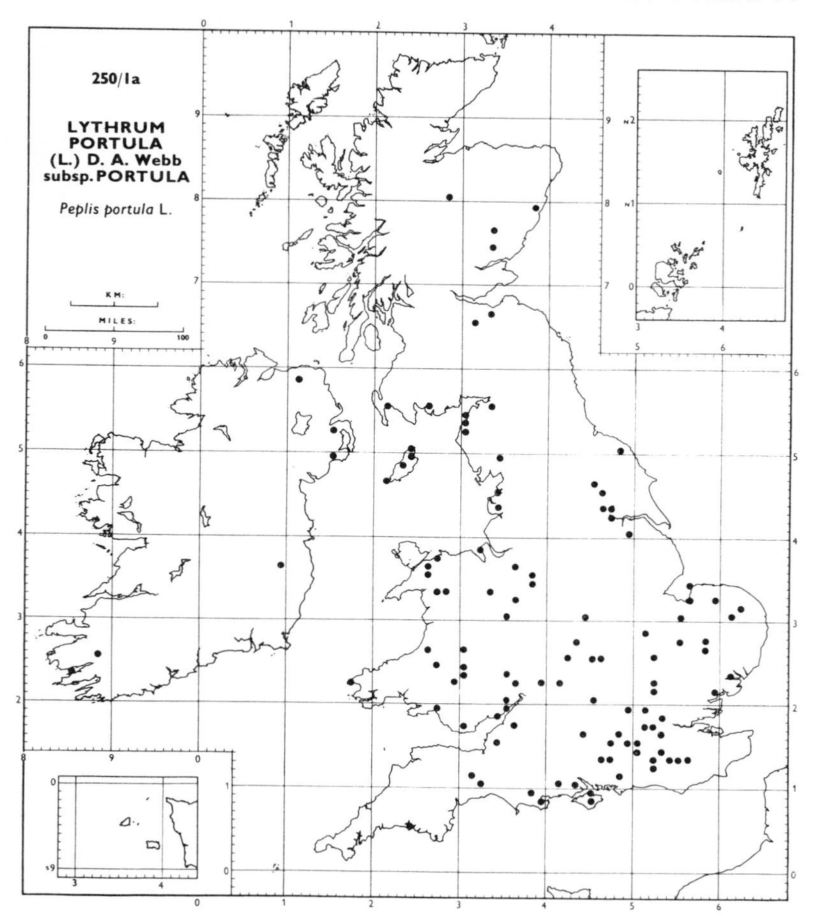

250/1 **Lythrum portula** (L.) D. A. Webb
Peplis portula L.
subsp. **portula**
subsp. **longidentata** (Gay) P. D. Sell

DBN, TCD, Simpson
In addition to the above herbaria consulted by the editor data were extracted from the following by D. E. Allen and published in *Watsonia* (Allen, 1954): BIRM, BM, CGE, CLE, K, MANCH, NMW, OXF. He gave the records for subsp. *longidentata* in full, but those for subsp. *portula* in map form only.

As Allen pointed out, these taxa represent extremes connected by intermediates. In the type material of subsp. *longidentata* the epicalyx segments are 1·5–2 mm. long, whereas in subsp. *portula* the segments attain only 0·5 mm. in length at the most. Intermediates have been omitted from these maps.

It is clear that the differences in distribution of the two subspecies within the British Isles are reflected elsewhere: subsp. *longidentata* is known only from France, Spain, Portugal, Algeria and the Azores, whereas subsp. *portula* is recorded from most parts of Europe. (Nomenclature, cf. Sell, 1967.)

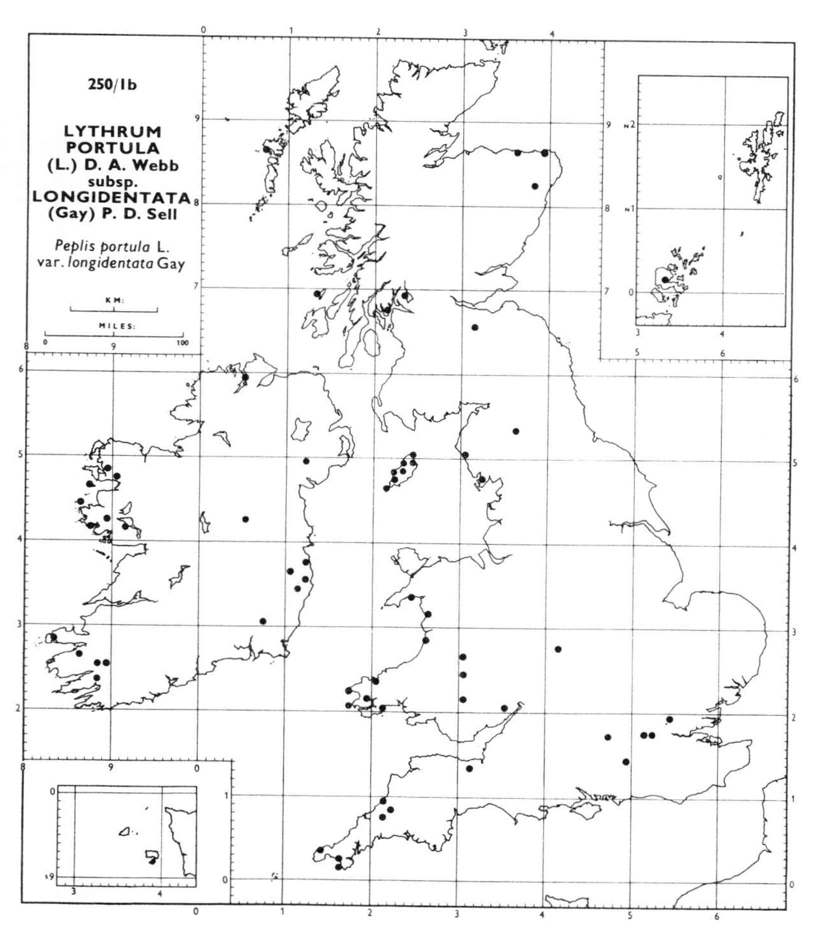

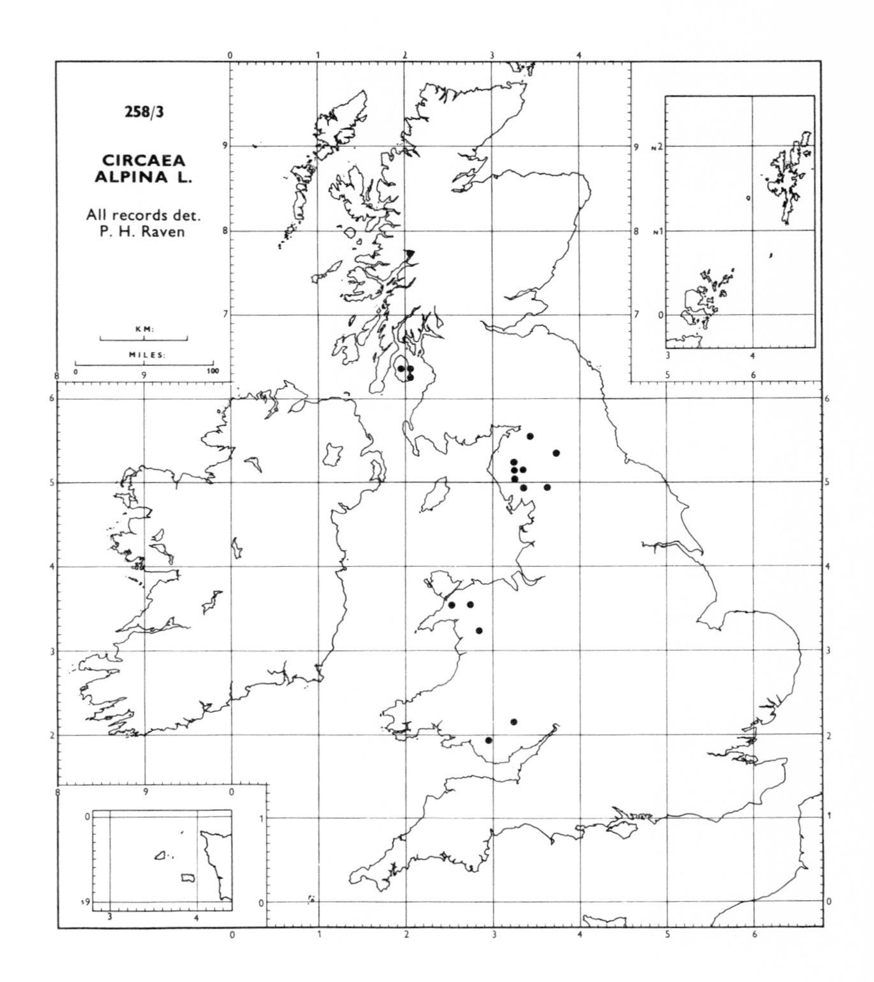

258/3 Circaea alpina L.

ABD, BIRM, BM, CGE, DBN, E, GL, K, MANCH, NMW, OXF, STA, Lousley, Roberts, Sandwith, Wallace
This map is based on data collected by P. H. Raven and previously published (Raven, 1963). All the records have been verified by him, including the recent addition from v.-c. 66.

The map of *C. alpina* agg. in the *Atlas*, p. 149, may be accepted as a map of the distribution of the hybrid **C. × intermedia** Ehrh. Raven believes that the very restricted distribution of *C. alpina* compared with the hybrid (which is sterile) indicates that it was more widespread nearer the last glacial maximum, but has been eliminated, by subsequent changes in the climate, from Ireland and most of Scotland. It occurs throughout northern Europe and in mountains in the south.

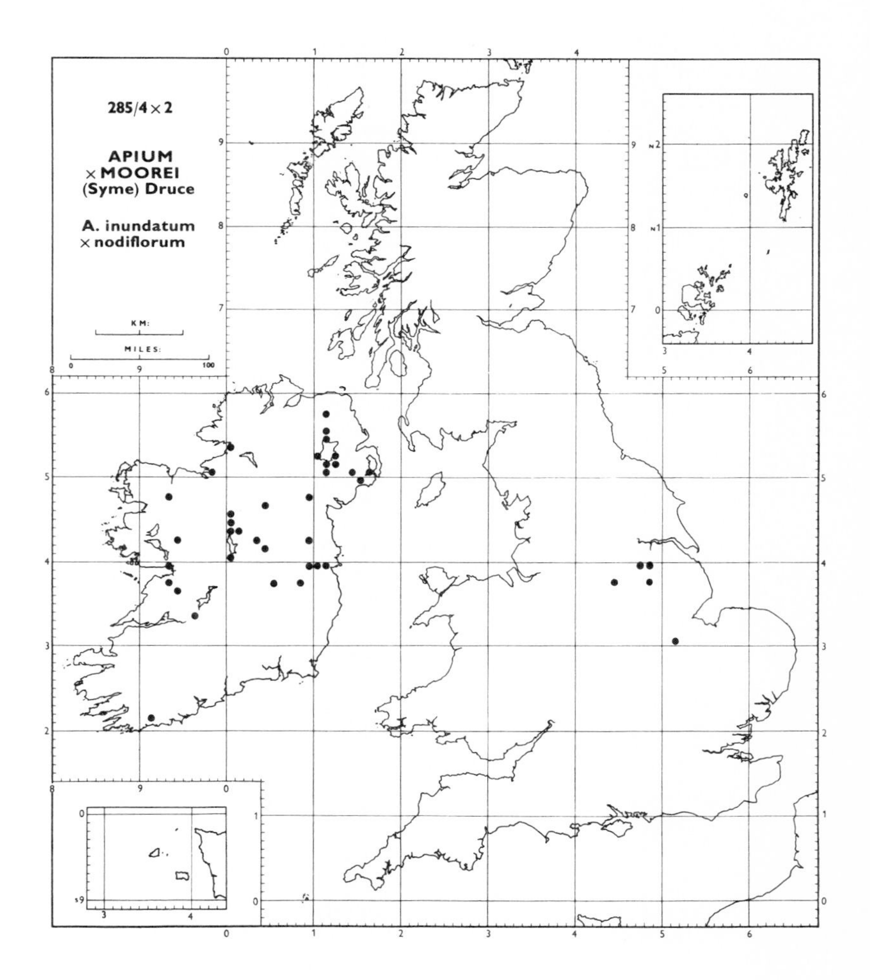

285/4 × 2 Apium × moorei (Syme) Druce
= A. inundatum × nodiflorum

BM, CGE, OXF, TCD, Lousley, Wallace: Floras
The much greater frequency of this hybrid in Ireland is surprising as both parents occur throughout England and Ireland (see *Atlas*, pp. 158–9). In this respect its distribution resembles *Equisetum* × *litorale* (q.v. p. 1).

314b/1b Daucus carota L. subsp. **gummifer** Hook. f.

BM, CGE, DBN, NMW, OXF, TCD

A complete range of intermediates exists between subsp. *gummifer* and subsp. *carota*. The extreme form of subsp. *gummifer* has the rays of the umbel densely hispid: they are also more or less equal in length, giving to the umbel a characteristic convex shape. This form is more or less restricted to the coasts of Devon and Cornwall (1–4). Further north and east less hairy forms occur, and coastal specimens with a few scattered hairs on the umbel rays have been noted from West Scotland and the east coast of England. These plants, however, approach subsp. *carota* in their unequal umbel ray branches and taller more slender habit, and have been excluded from the map. Outside Devon and Cornwall the records have all been verified from herbarium material seen by the editor.

It occurs elsewhere on the coasts of southern and western Europe.

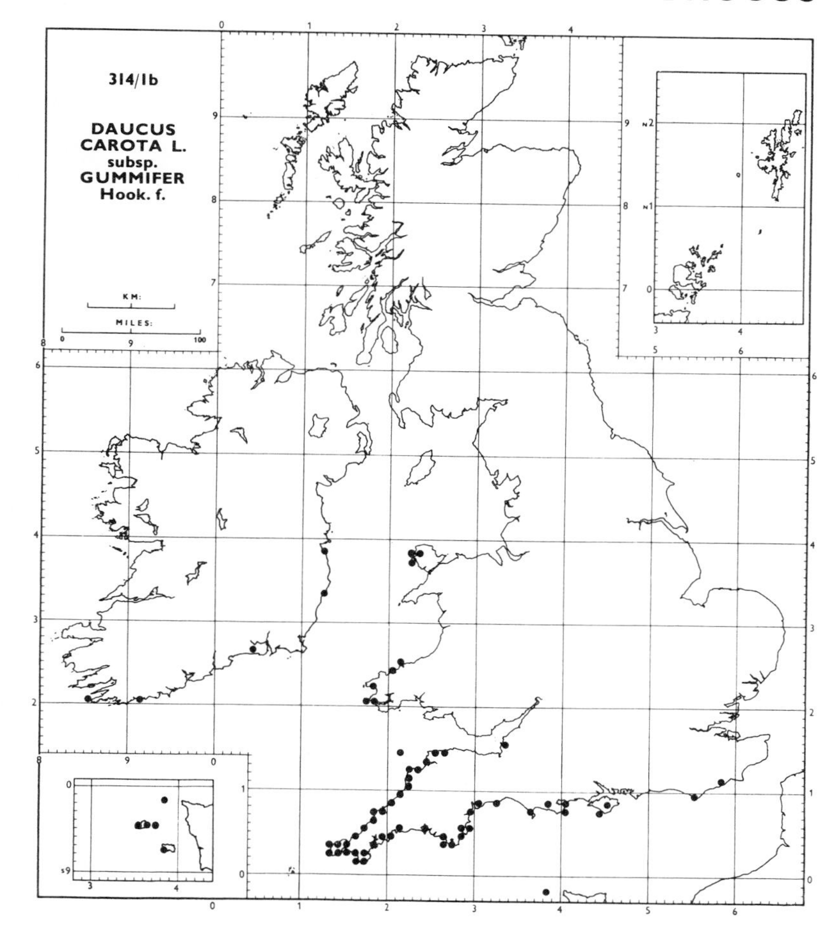

320/1 Polygonum aviculare L. *sensu lato*

BFT, BM, CGE, DBN, E, NMW, NWH, OXF, SHY, TCD

The taxonomy of the species of this group has recently been revised by Styles (1962), who has given an outline of their geographical distribution. Unfortunately it is not possible to include maps of the two very widespread segregates *P. aviculare* L. *sensu stricto* and *P. arenastrum* Bor. There is no evidence to suggest that both species do not occur throughout the range indicated for the aggregate in the map in the *Atlas*, p. 173, with the exception of the far north. In Orkney (111) *P. aviculare* is a local plant, being largely replaced by *P. boreale*, whereas *P. arenastrum* is frequent. In Shetland (112) *P. aviculare* is almost unknown.

The maps of **P. boreale** (Lange) Small and **P. rurivagum** Jord. ex Bor. are based on material seen by B. T. Styles and the editor. It is probable that *P. boreale* occurs more frequently than is indicated around the coast of north Scotland.

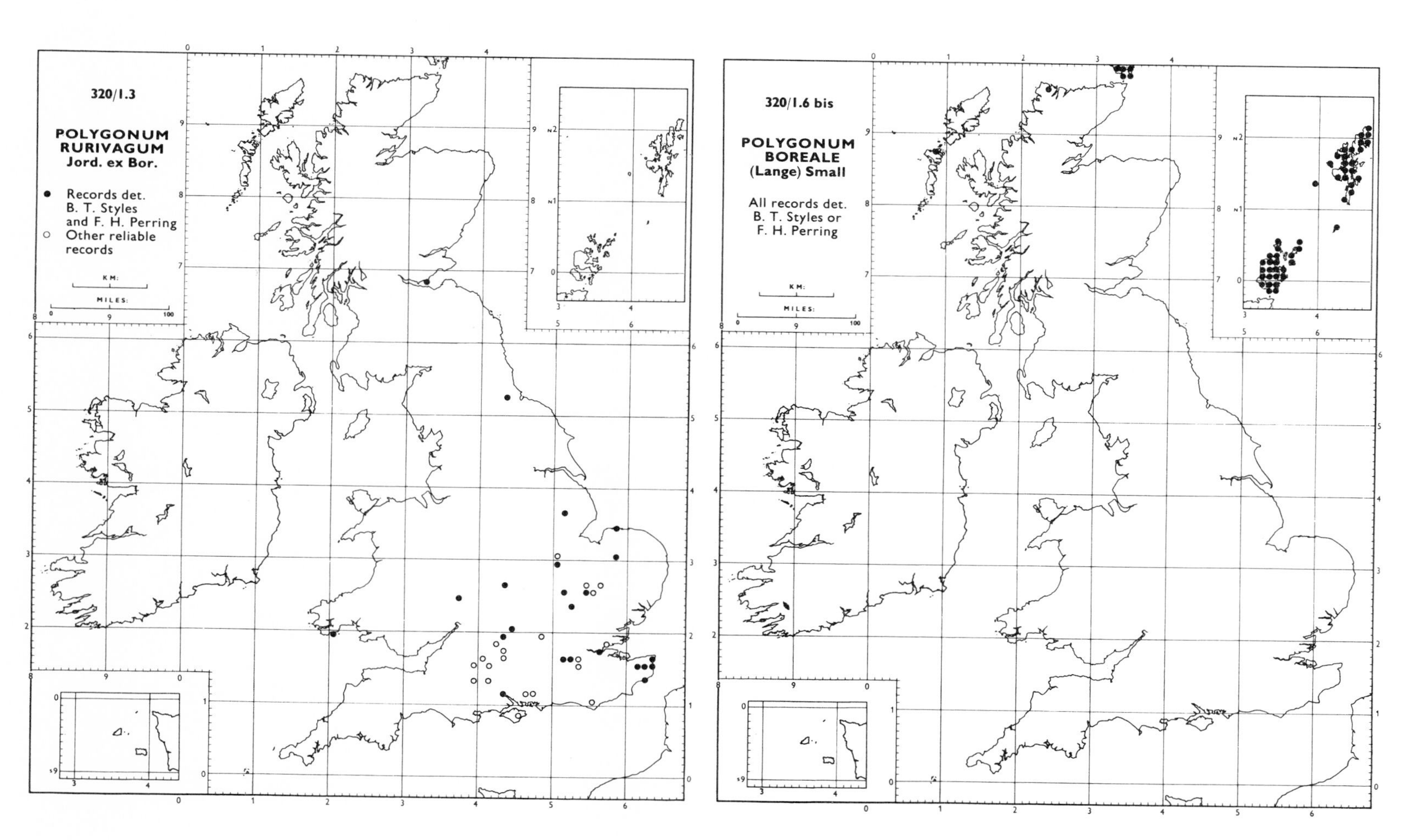

POLYGONACEAE

325/1.3 **Rumex tenuifolius** (Wallr.) Löve

BFT, BM, CGE, DBN, E, LCN, NMW, OXF, Lousley

On the advice of J. E. Lousley a very conservative view of this species has been taken and only specimens in which the *basal* leaves are at least seven times as long as broad and in which the stem is decumbent at the base, have been accepted. It is more or less confined to the poorest and most acid sands.

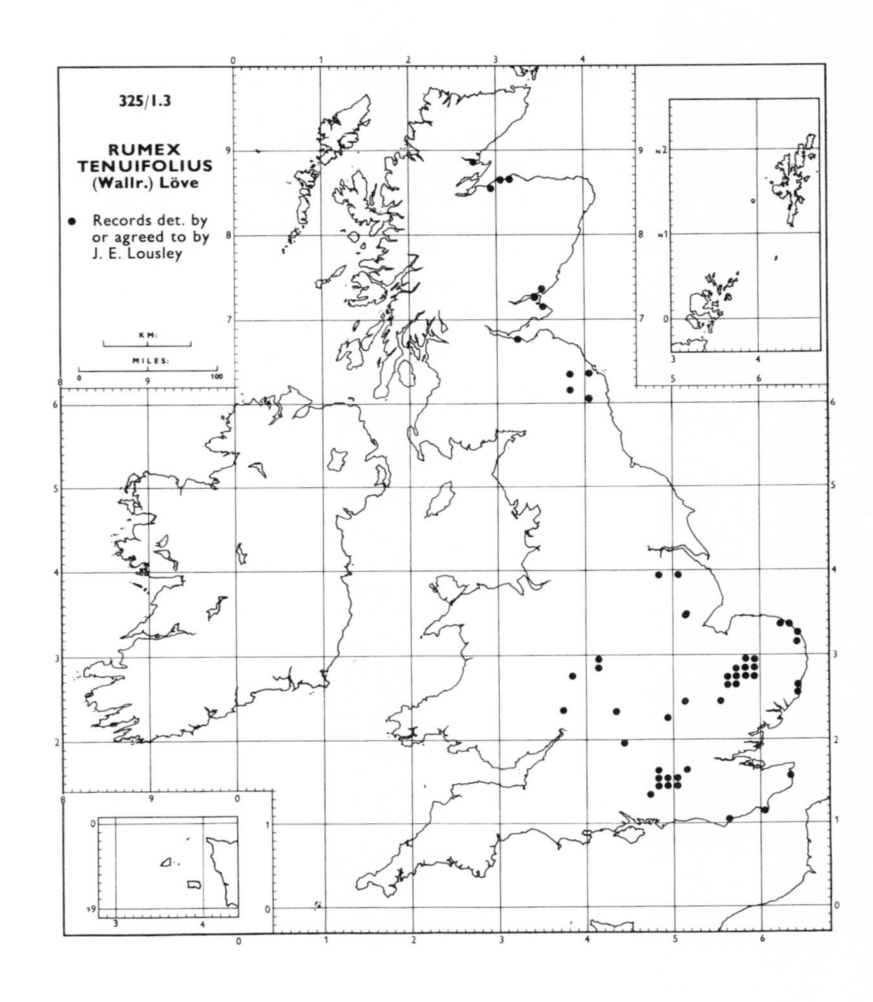

357/2×1 **Erica** ×**praegeri** Ostenfeld
=E. mackaiana × tetralix
E. stuartii auct.

CGE, TCD

This is a sterile hybrid which D. A. Webb (1954) suggests is formed by pollen of *E. mackaiana* reaching the stigma of *E. tetralix*, since the former species appears to set no seed in Ireland and since hybrid plants may be found at least 1·5 km. from the nearest known station for *E. mackaiana*. A plant found in north-west Mayo (H. 27) was tentatively identified by Webb as *E.*×*praegeri* (Lamb, 1964), but further examination suggests that it is a variant of *E. tetralix*. (Nomenclature, cf. Sell, 1967.)

357/3×1 **Erica** ×**watsonii** Benth.
=E. ciliaris × tetralix

BM, CGE, K, OXF, Lousley, Wallace: Floras

The map is probably complete. This hybrid is only known to occur in areas where the two parent species occur together (see *Atlas*, p. 194).

358/2×1 **Vaccinium** ×**intermedium** Ruthe
=V. myrtillus × vitis-idaea

BM, CGE: Floras

Although the two parents overlap over a wide area of north and west Britain, this hybrid is only frequent in Stafford (39) and Derby (57).

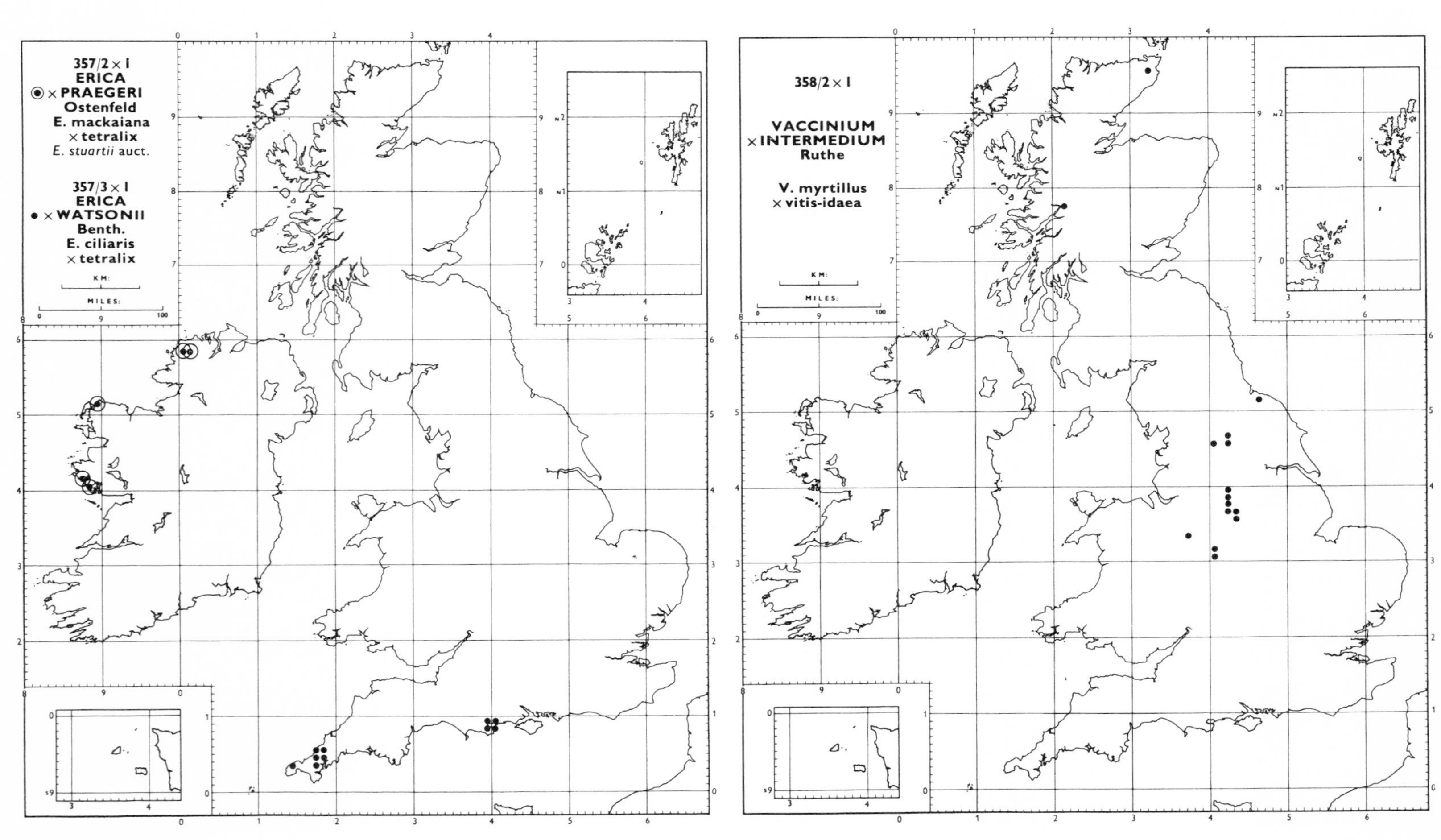

359/3b Pyrola rotundifolia L.
subsp. **maritima** (Kenyon) E. F. Warb.

This map is based on the observations of A. J. Farmer. He has either determined herbarium material or made field records for all squares, with the exception of that in Devon (4).

If these records were omitted from the map of the species (*Atlas*, p. 197) it would show the distribution of subsp. **rotundifolia**.

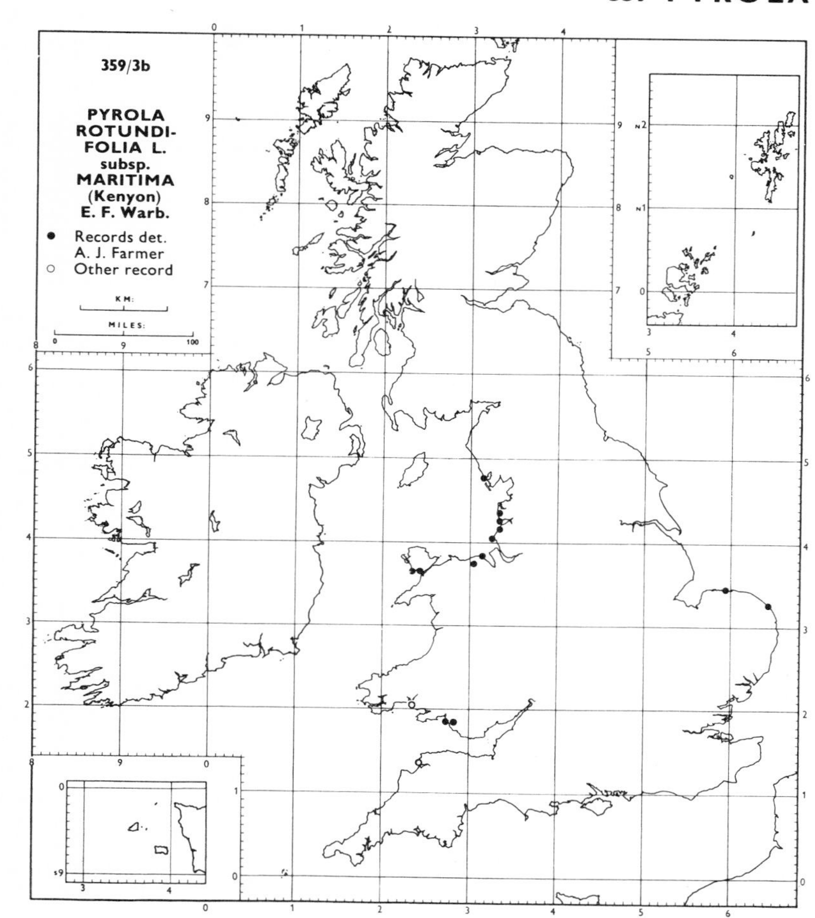

362/1.1 Monotropa hypopitys L.
subsp. **hypopitys**
362/1.2 subsp. **hypophegea** (Wallr.) Soó

CGE, DBN, SHY, TCD: Floras

Recent work by D. J. Wicker on *Monotropa* in Britain suggests that there is much introgression between these two taxa. The majority of the specimens are glabrous and have 1–6 flowers. In the past such material has been referred to *M. hypophegea*. There are very few specimens which could be referred to *M. hypopitys*, as it is recognized in continental Europe, where it has the inside of the corolla, the filaments and style covered in stiff hairs, and has between 6 and 25 flowers. It has a more continental distribution than *M. hypophegea*. Because of this difference in geographical distribution and the widespread occurrence of intermediates, it seems easiest to regard these taxa as subspecies.

All records of subsp. *hypophegea* have been accepted, but only specimens determined by Miss Wicker have been accepted for subsp. *hypopitys*, though some of the records marked as intermediate may be this taxon. Though the maps are obviously incomplete, lacking information from Ireland and northern England where the species occurs (see *Atlas*, p. 198), it is clear that the only form found in dunes on the west coast is subsp. *hypophegea*. Both subspecies may be found growing within a few metres of each other in beech woods in southern England.

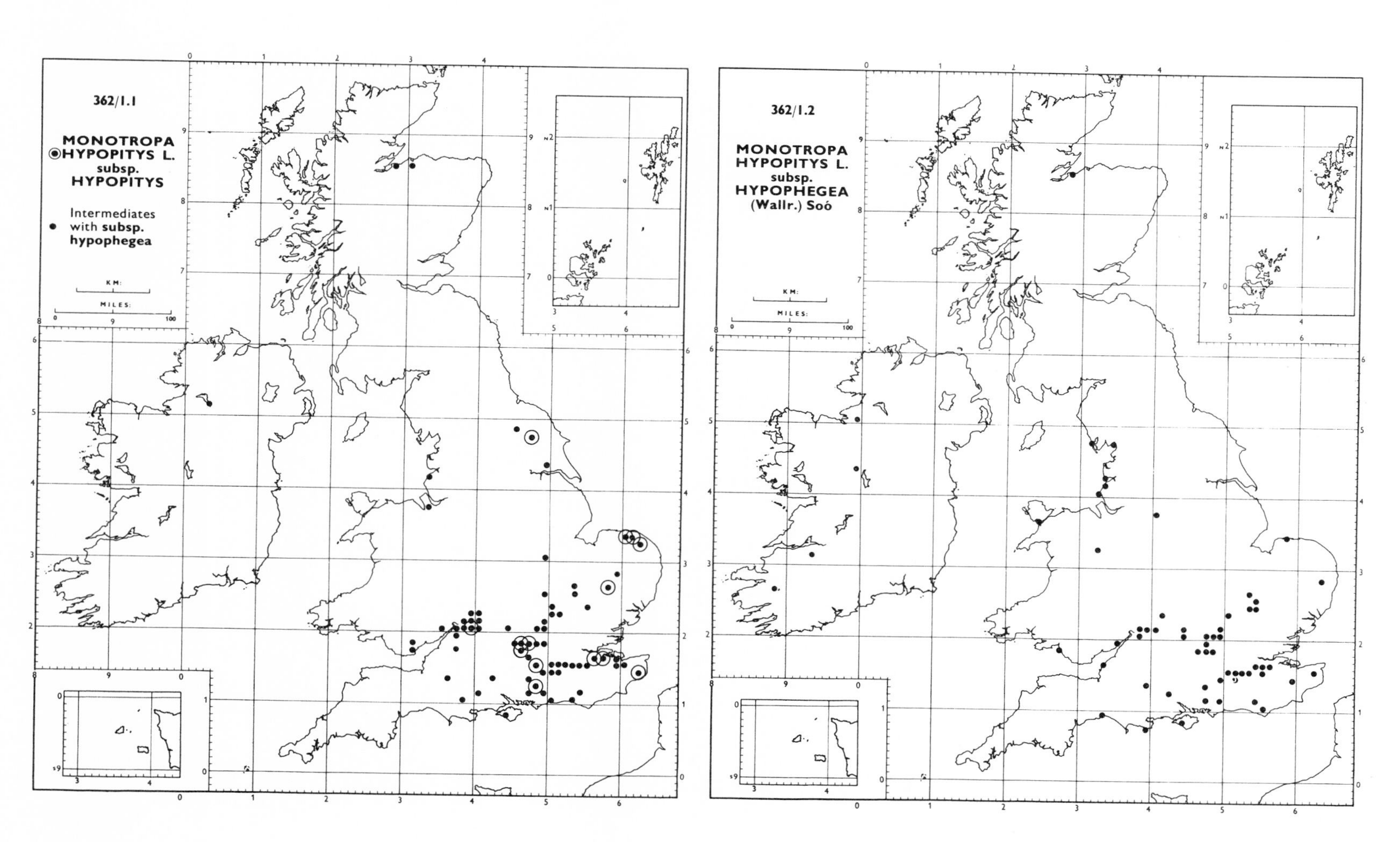

PLUMBAGINACEAE

365/5 **Limonium binervosum** (G. E. Sm.) C. E. Salmon *sensu lato*

BM, CGE, NMW, OXF, Lousley, Wallace: Floras

L. binervosum *sensu stricto* (*Atlas*, p. 199) is a very variable species with many named varieties: it has probably given rise to the three, probably apomictic, segregates included on the map, which are all apparently endemic.

L. recurvum C. E. Salmon occurs only at Portland, Dorset (9). **L. transwallianum** (Pugsl.) Pugsl. occurs only on maritime cliffs in Pembroke and Clare (45, H. 9). The Pembrokeshire plants have a chromosome number of 2n=35. The populations from Clare consist of similar but more robust plants with a different chromosome number, 2n=27. **L. paradoxum** Pugsl. occurs in Pembroke and with less certainty in east Donegal (H. 34).

367/3 × 5 **Primula veris** L. × **vulgaris** Huds.

BM, DBN, E, NMW, SHY, TCD: Floras

This very widespread fertile hybrid occurs throughout the range in which the two parents overlap: this is in effect throughout the range of *P. veris* (see *Atlas*, p. 201). There is some evidence that the hybrid is more frequent in the west of the range of *P. veris* where both species occur in the open. In the east *P. vulgaris* is confined to woodlands.

367/5 **Primula vulgaris** Huds.

The map of this species is reproduced here as the map in the *Atlas*, p. 210, erroneously omitted a large number of Irish records.

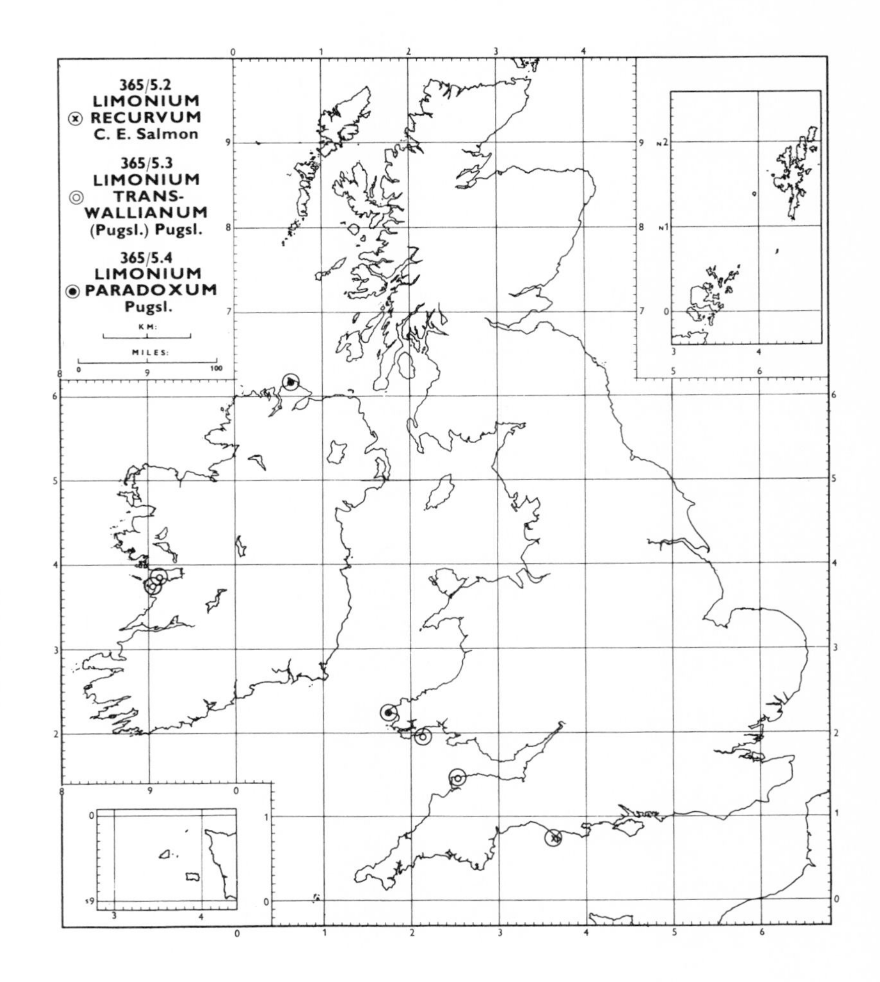

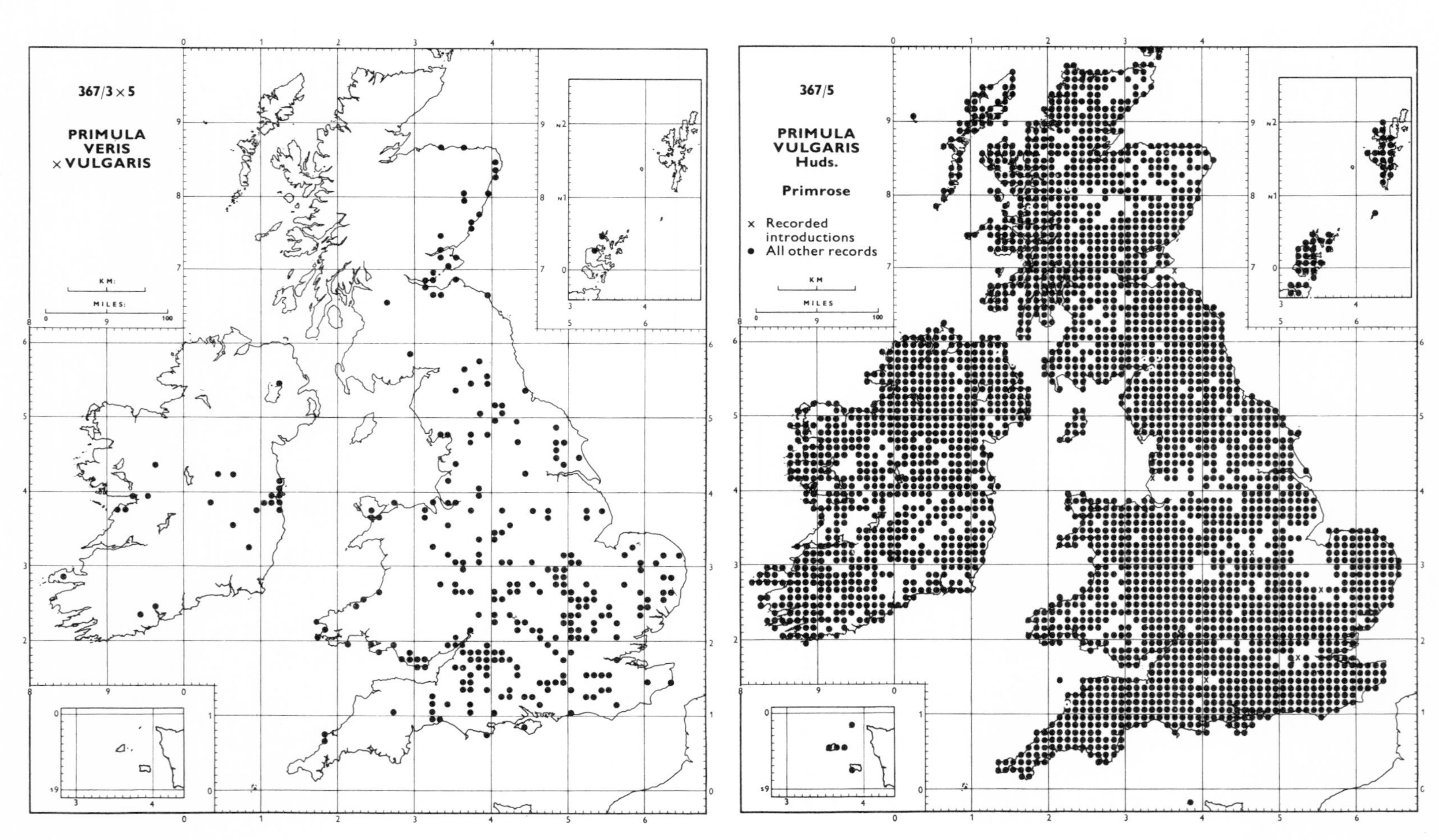

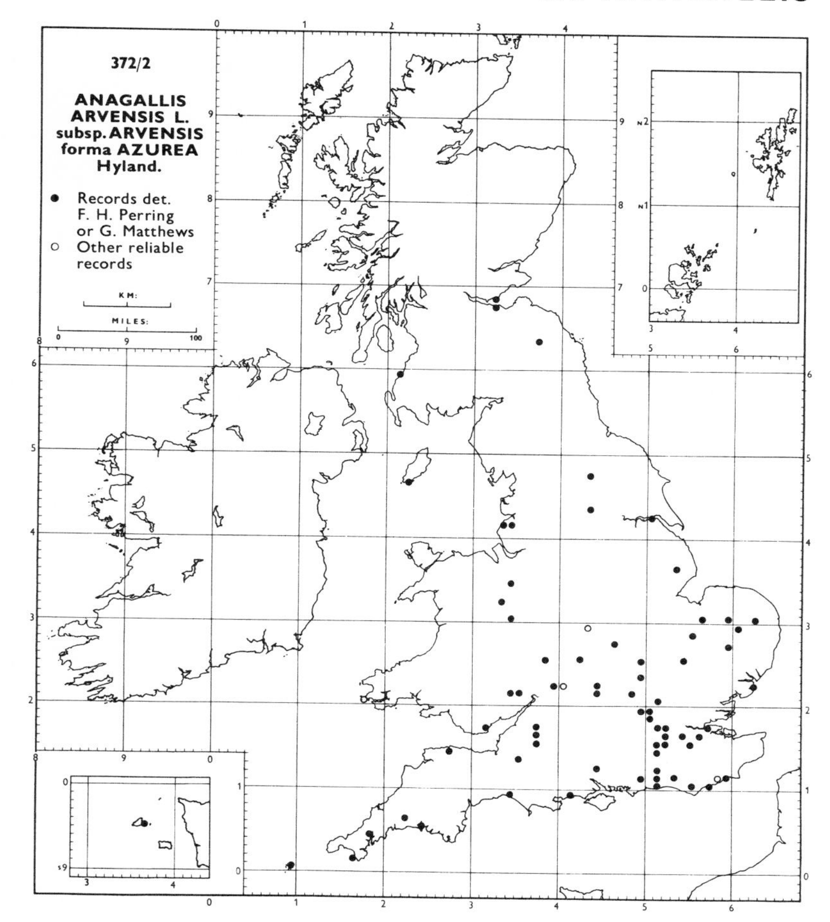

372/2 Anagallis arvensis L.
subsp. **arvensis** forma **azurea** Hyland.
372/3 subsp. **foemina** (Mill.) Schinz & Thell.

BFT, BM, CGE, DBN, E, NMW, SHY, TCD
There has long been confusion between these two blue-flowered *Anagallis* taxa, but whereas forma *azurea* is merely a colour variant of *A. arvensis*, subsp. *foemina* differs in having shorter pedicels, and narrower petals (less than 3·5 mm. broad) fringed with few, scattered glands.

These two maps are based on material seen by the editor in the above-mentioned herbaria, and by G. Matthews in the British Museum. To these records have been added those made recently by reliable recorders since the two taxa became more clearly understood. (Nomenclature, cf. Sell, 1967.)

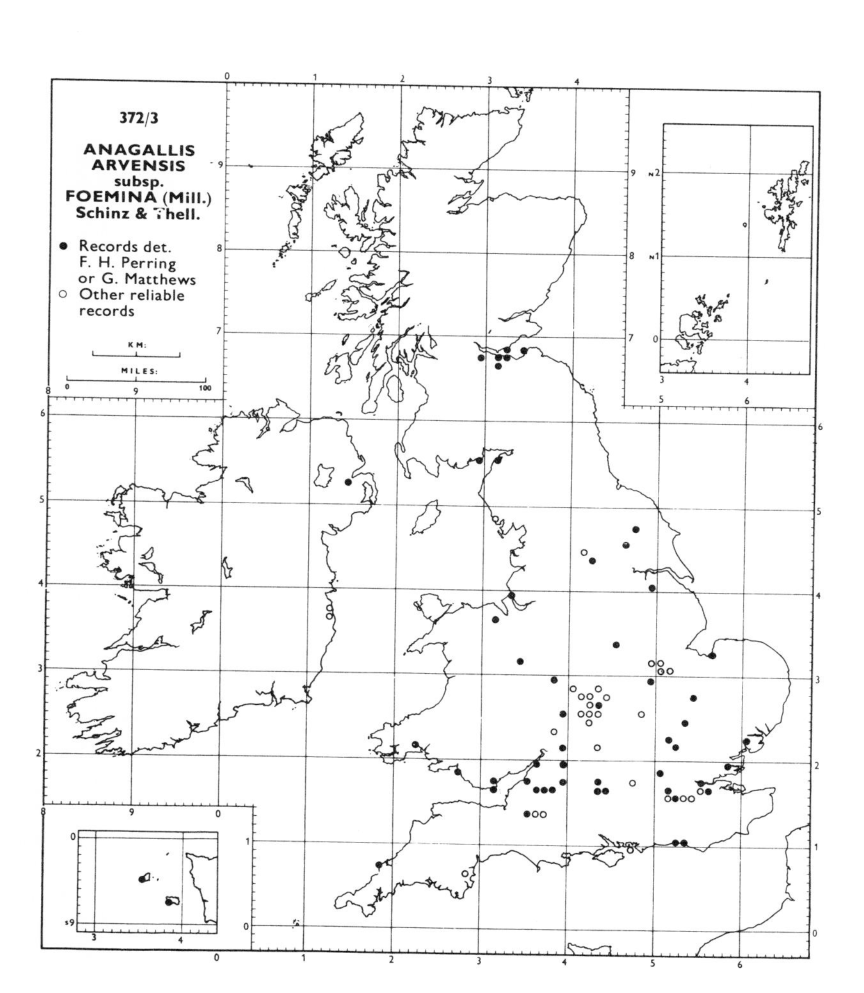

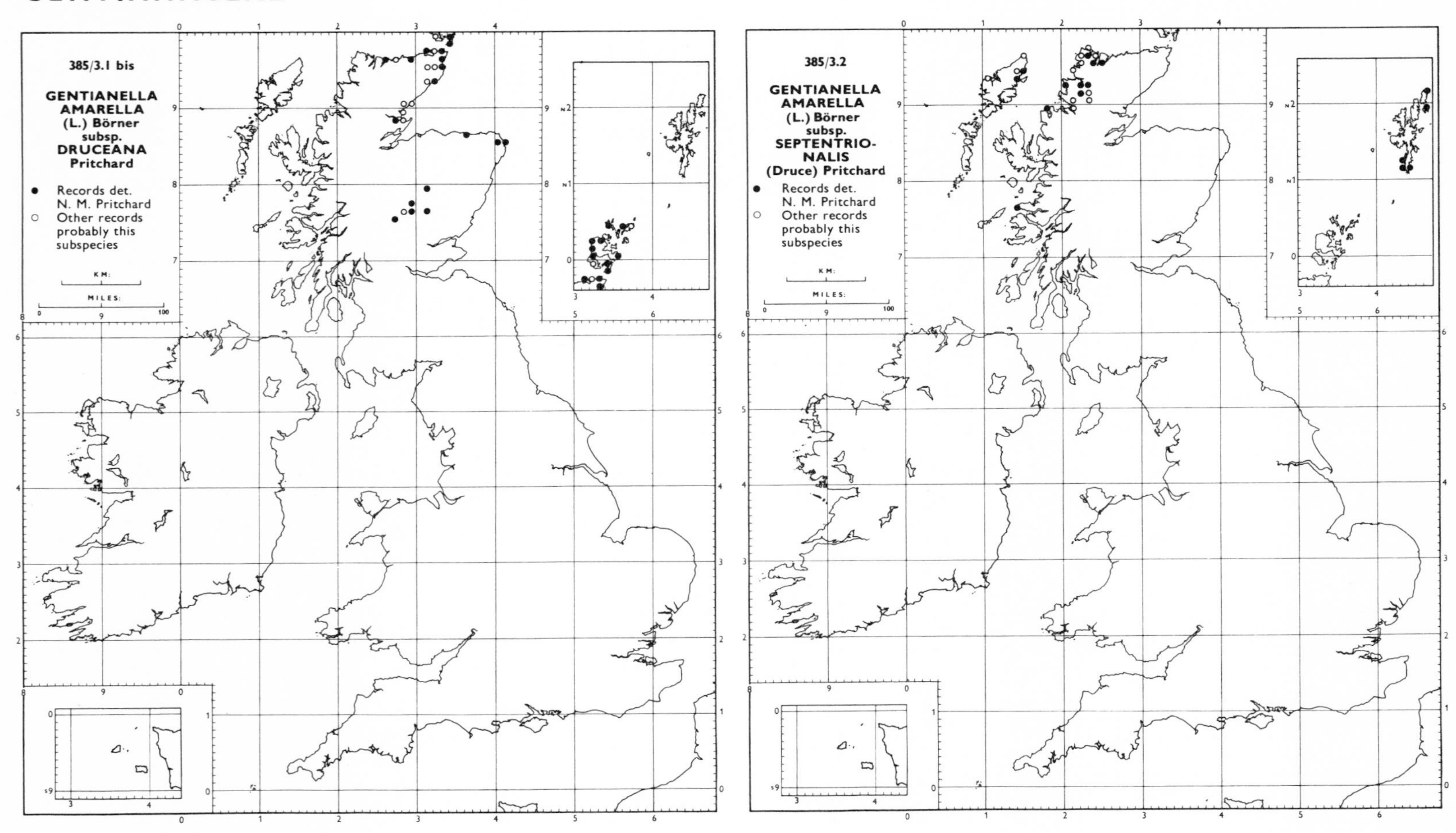

385/3 Gentianella amarella (L.) Börner
subsp. **druceana** Pritchard
subsp. **septentrionalis** (Druce) Pritchard

As far as is known the endemic subsp. *septentrionalis* and subsp. *druceana* are the only infraspecific taxa of this species occurring in north and central Scotland (Pritchard, 1960). N. M. Pritchard has supplied the information from which the maps of these two subspecies have been prepared. All other records given in the *Atlas* (p. 209) for Great Britain are referable to subsp. **amarella**. As far as is known, all the records for Ireland are referable to the endemic subsp. **hibernica** Pritchard (Pritchard, 1959).

387/1 Nymphoides peltata (S. G. Gmel.) Kuntze

CGE, OXF
This species is heterostylous, and it may be of significance that only the "pin" form has been found in the Fens (v.-cs 26, 28, 29, 31 and 53). It raises the question whether or not this species was originally introduced into that area and now largely reproduces vegetatively. That it is a widespread introduction can be seen from the map in the *Atlas*, p. 210. The map now published is based on material determined by Miss B. Mann.

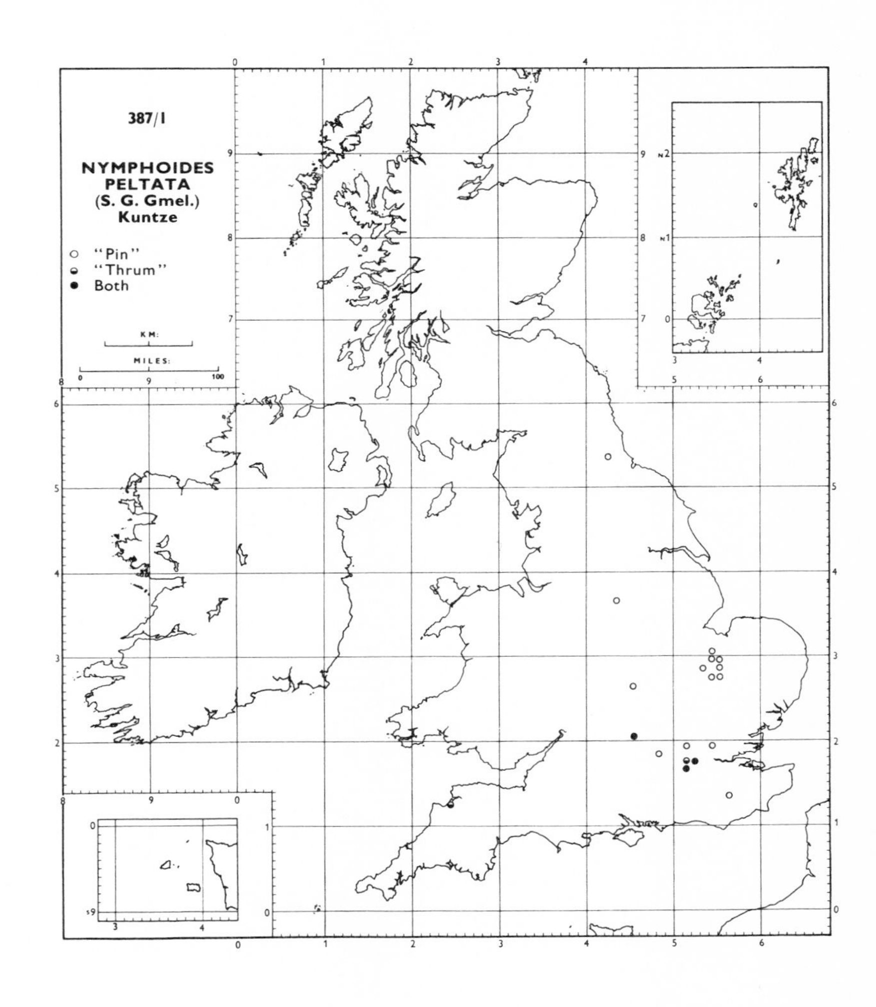

392/2×1 Symphytum ×uplandicum Nyman =S. asperum × officinale

This taxon is very variable in flower colour and leaf decurrence and probably arose as a result of hybridization between *S. officinale* L. and *S. asperum* Lepech. In the *Atlas* (p. 211) many of the records of *S. officinale* undoubtedly refer to *S.* × *uplandicum*. Pure *S. officinale* has cream-coloured flowers and is more or less confined to fens and river banks. *S.* × *uplandicum* occurs in a wide range of habitats, particularly roadsides and hedgebanks, long stretches of which may be occupied by a single clone. This is particularly true of a form with clear blue flowers and non-decurrent leaves, which may possibly be an F_1 hybrid. In view of the rarity of *S. asperum*, which is always an introduction in Britain, it seems likely that *S.* × *uplandicum* has been introduced from Europe, where it is grown as a crop plant. It occurs naturally in southern Russia where the ranges of the two parents overlap. The many variable forms of *S.* × *uplandicum* have possibly arisen as a result of back-crossing between the introduced *S.* × *uplandicum* and the native *S. officinale*.

This map is based on all records received, but must be regarded as provisional.

406/1b Calystegia sepium (L.) R. Br. subsp. roseata Brummitt

This map is based on data supplied by R. K. Brummitt. Subsp. *roseata* differs from subsp. **sepium** in having a pink corolla, a usually pubescent stem, and usually more acute leaf apex. Intermediates between subsp. *roseata* and subsp. *sepium* are not infrequent in lowland areas. As far as is known, subsp. *sepium* occurs throughout the range of the species in this country (see *Atlas*, p. 217). Subsp. *roseata* is found elsewhere on the Atlantic coasts of Norway, Sweden, the Netherlands and possibly also France, Spain and Portugal. It is also recorded from the coasts of Chile, New Zealand and Australia. (cf. Sell, 1967.)

406/2 Calystegia pulchra Brummitt & Heywood
C. dahurica auct.

The map published in the *Atlas*, p. 217, was very incomplete, and there was some confusion between this species and *C. sepium* subsp. *roseata*. For both these reasons a new map of *C. pulchra* is printed here. It is naturalized in this country, Holland, France, Germany, Poland, Austria, Czechoslovakia, Denmark and Sweden, and possibly elsewhere. Its origin is a mystery, but it is probably a garden hybrid which arose somewhere in Europe.

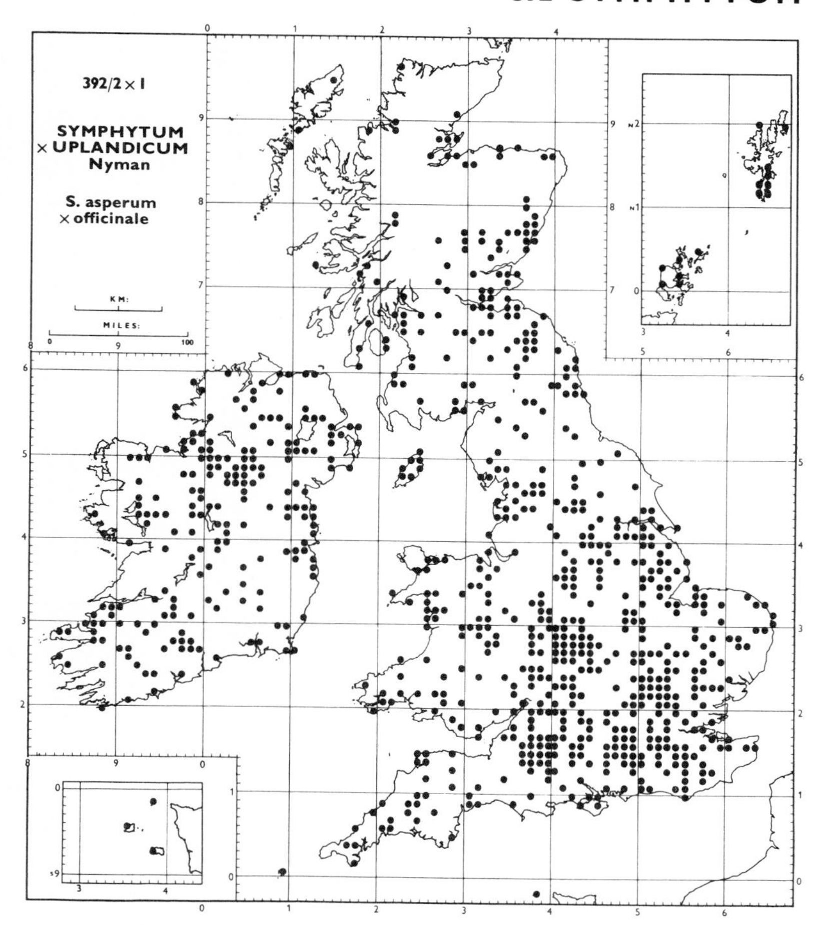

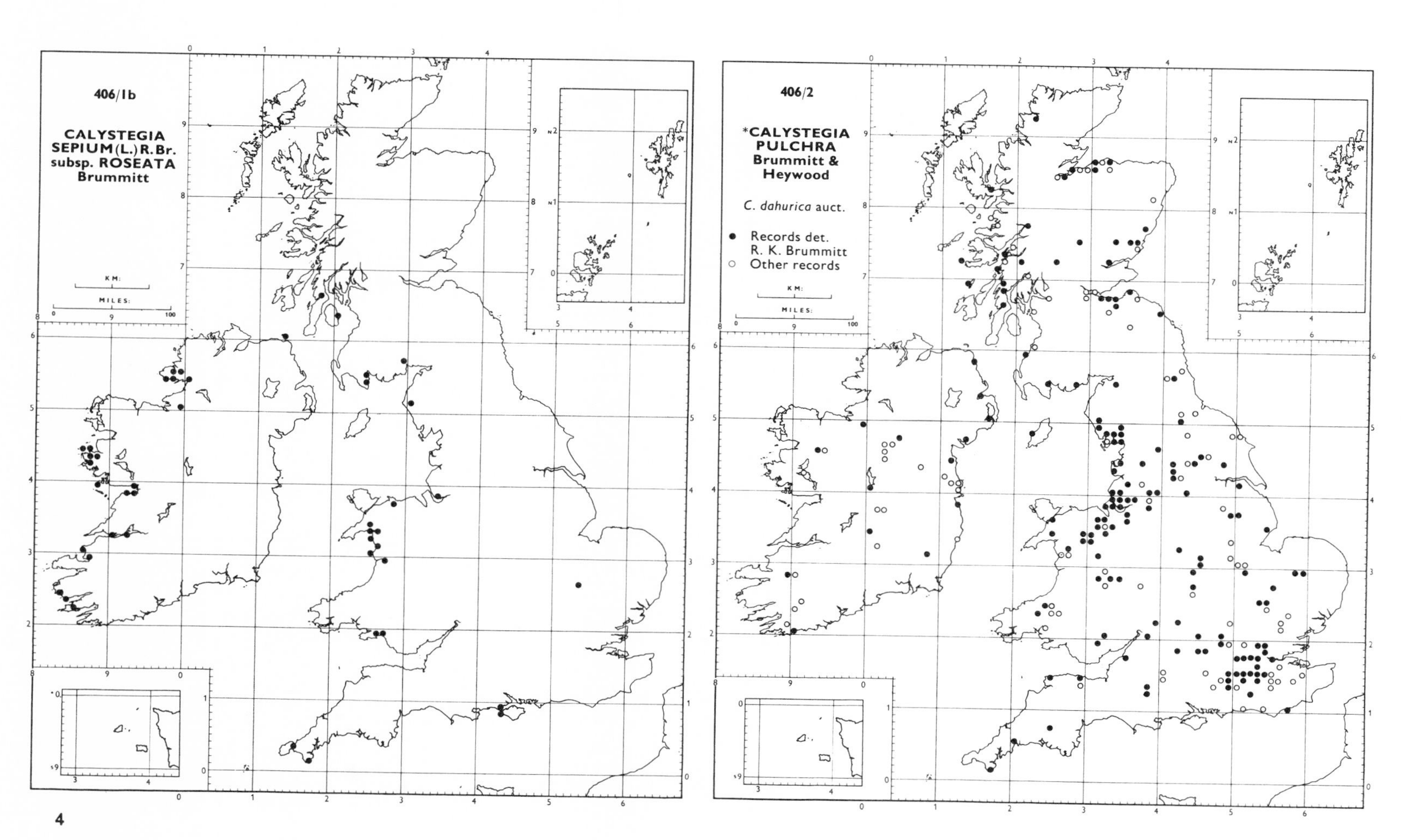

4

SOLANACEAE

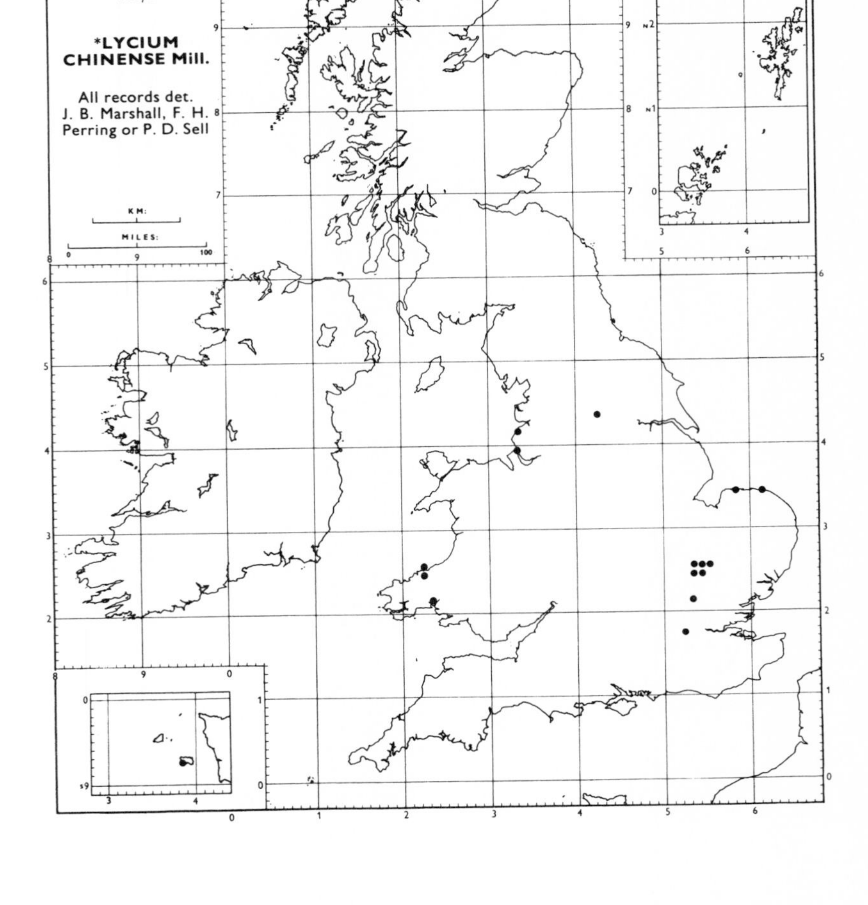

409/2 Lycium chinense Mill.

BFT, BM, CGE, DBN, LCN, NMW, OXF, SHY, TCD
This species and the closely related **L. barbarum** L. (*L. halimifolium* Mill.) have often been confused in the past. In preparing the map in the *Atlas*, p. 219, it was found to be impossible to separate the records of the two species satisfactorily. Inspection of herbarium material by the editor, by J. B. Marshall and by P. D. Sell has shown that *L. chinense* is much the rarer of these two naturalized species, and that much material referred to this taxon in the past was *L. barbarum*. *L. chinense* is a native of East Asia.

The map in the *Atlas* may therefore be taken as representing almost entirely the distribution of *L. barbarum* in the British Isles. (For nomenclature see Feinbrun and Stearn, 1964.)

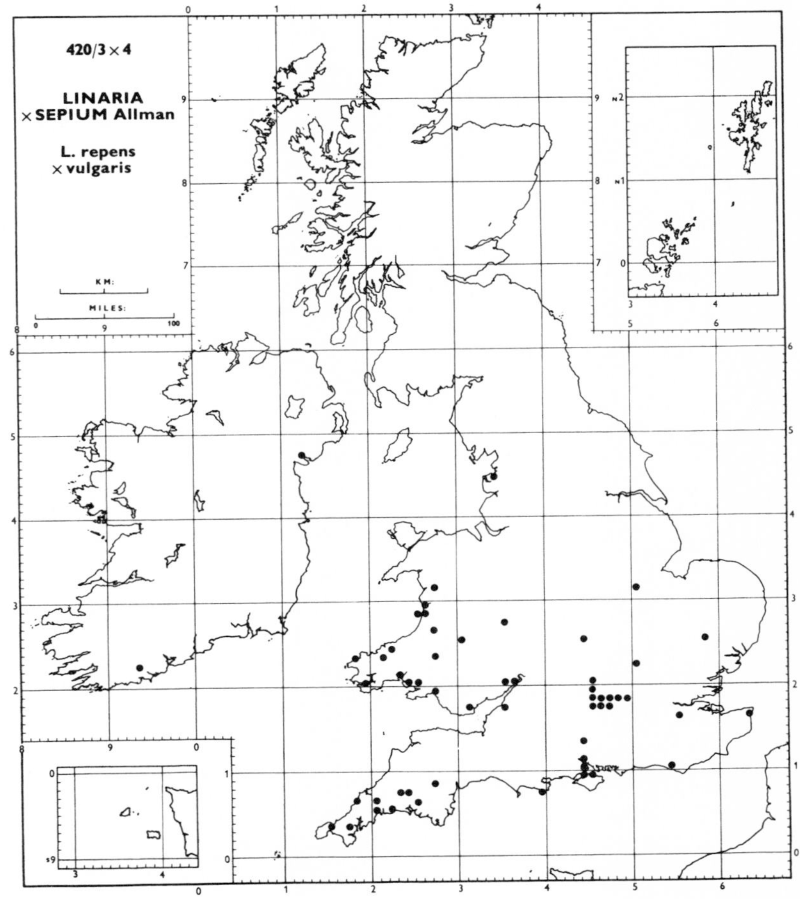

420/3×4 Linaria ×sepium Allman
=L. repens × vulgaris

BM, CGE, NMW, OXF: Floras
This hybrid does not occur throughout the range of overlap of the two parent species, both of which occur in northern England and in Scotland (see *Atlas*, p. 223). It usually occurs only where the two parents are growing together, when very variable hybrid swarms arise: this is particularly noticeable on railway lines (see Druce, 1897).

The record from v.-c. H. 4 has not been traced. *L.* × *sepium* also occurs in western France.

430/4 Veronica scutellata L.
var. **villosa** Schumach.

BFT, BM, CGE, DBN, E, LCN, NMW, OXF, SHY, TCD: Floras

This is a distinct variety, in which the stems are covered in dense long hairs. Var. *villosa* appears to be more continental than var. **scutellata**, but there is no evidence that it entirely replaces it in any part of its range. Both glabrous and hairy varieties may be found growing side by side in the same locality. (For map of species see *Atlas*, p. 228.)

430/8 Veronica spicata L.
subsp. **spicata**
subsp. **hybrida** (L.) E. F. Warb.

BM, CGE, E, NMW, OXF, Lousley, Wallace: Floras

As populations of these two subspecies do not overlap in the British Isles, hybridization cannot take place, and for this reason they were formerly considered to be distinct species. However, in view of the intermediates which occur on the European mainland, sub-specific rank is more appropriate.

430/13b Veronica serpyllifolia L.
subsp. **humifusa** (Dickson) Syme

BM, CGE, NMW, OXF: Floras

This locally distinct subspecies may be recognized by its blue flowers, and by the capsule, which is densely ciliate with gland-tipped hairs. It replaces subsp. *serpyllifolia* in some of our higher mountain areas. The record from v.-c. 50 has not been traced: that from v.-c. 37 was an error. It also occurs in north and central Europe.

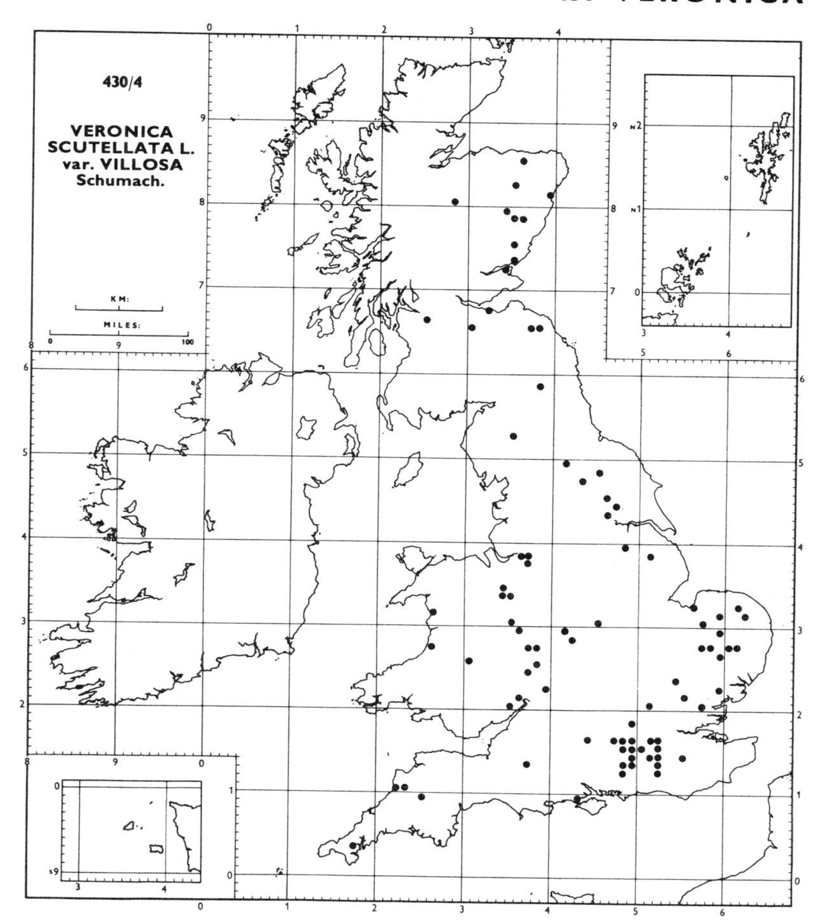

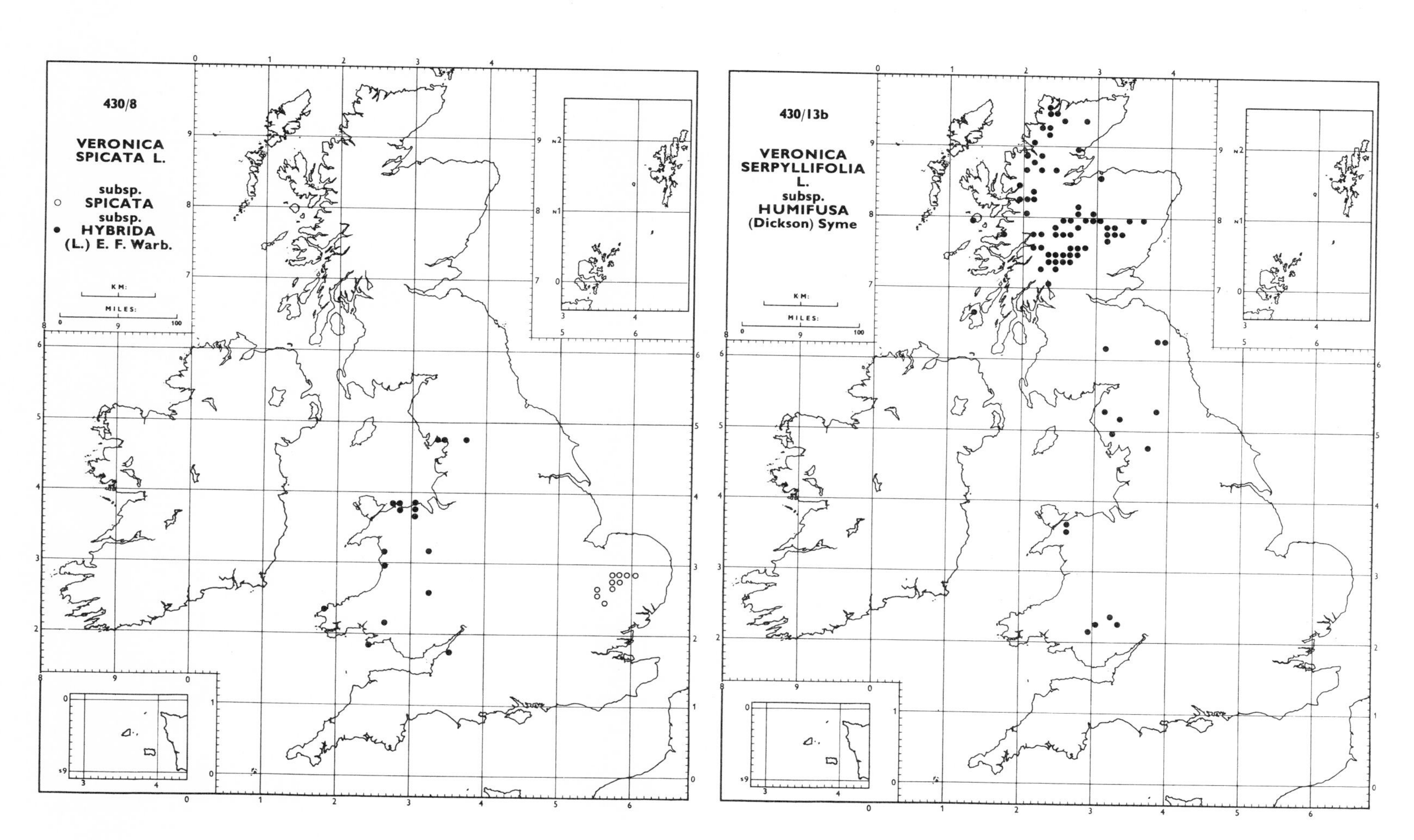

433/2 Rhinanthus minor L.
subsp. **minor**
subsp. **stenophyllus** (Schur) O. Schwarz
subsp. **monticola** (Sterneck) O. Schwarz
R. spadiceus Wilmott
subsp. **calcareus** (Wilmott) E. F. Warb.
subsp. **borealis** (Sterneck) P. D. Sell

BFT, BM, CGE, DBN, E, NMW, NWH, OXF, SHY, TCD
The treatment of this genus by Warburg in Clapham, Tutin and Warburg (1962) is followed here with the exception of subsp. *borealis*. This was included at specific rank by Warburg because, whereas all the other subspecies of *R. minor* have calyces which have hairs only on the margin, this taxon has a calyx which is hairy all over. However, in view of the fact that a series of hybrids exists between it and subsp. *monticola* and subsp. *stenophyllus* there seems no reason to give it a different rank from these and the other taxa included within the *R. minor* aggregate.

The maps are based upon material determined by the editor or E. F. Warburg, and other records from adjacent areas made by reliable recorders. The maps of subsp. *monticola*, subsp. *borealis* and subsp. *calcareus* are probably almost complete, but those for subsp. *minor* and subsp. *stenophyllus* are under-recorded, particularly in Ireland, as can be seen by comparison with the map for the species (*Atlas*, p. 233).

Records of subsp. *calcareus* for Clare (H. 9) have been omitted from the map as material seen by the editor and E. F. Warburg from there does not appear to belong to the same taxon as the original material from the south of England. Subsp. *monticola* is doubtfully recorded from v.-cs 81, 102, 103 and H. 33. Records of subsp. *borealis* from v.-cs 100 and 109 have not been traced: those from v.-cs 72 and 102 are probably errors.

(Nomenclature, cf. Sell, 1967.)

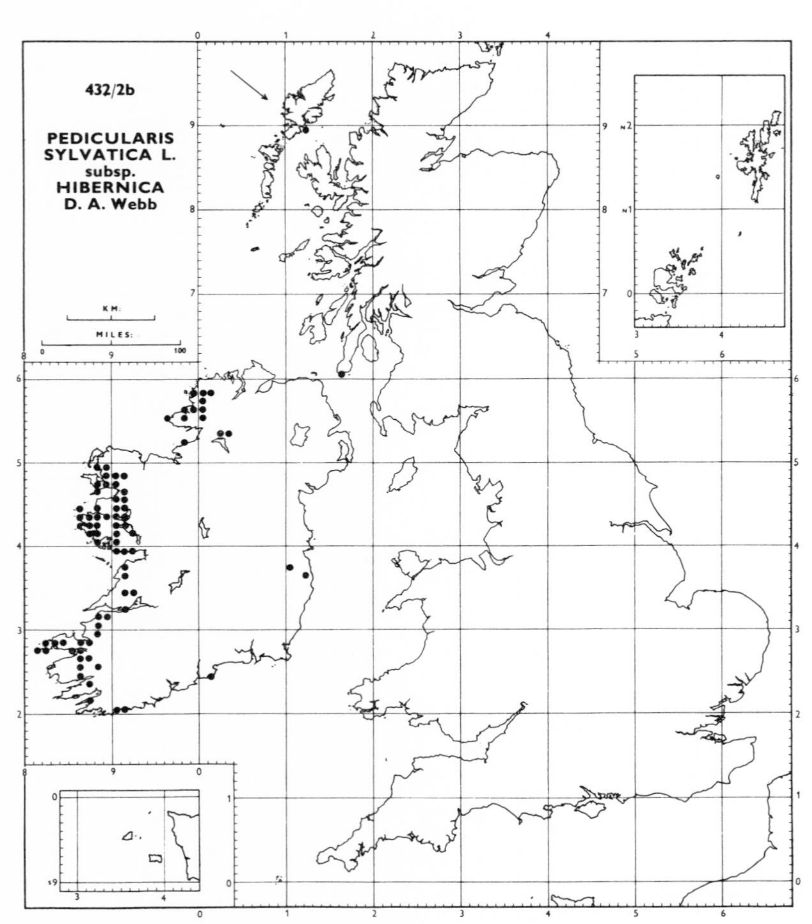

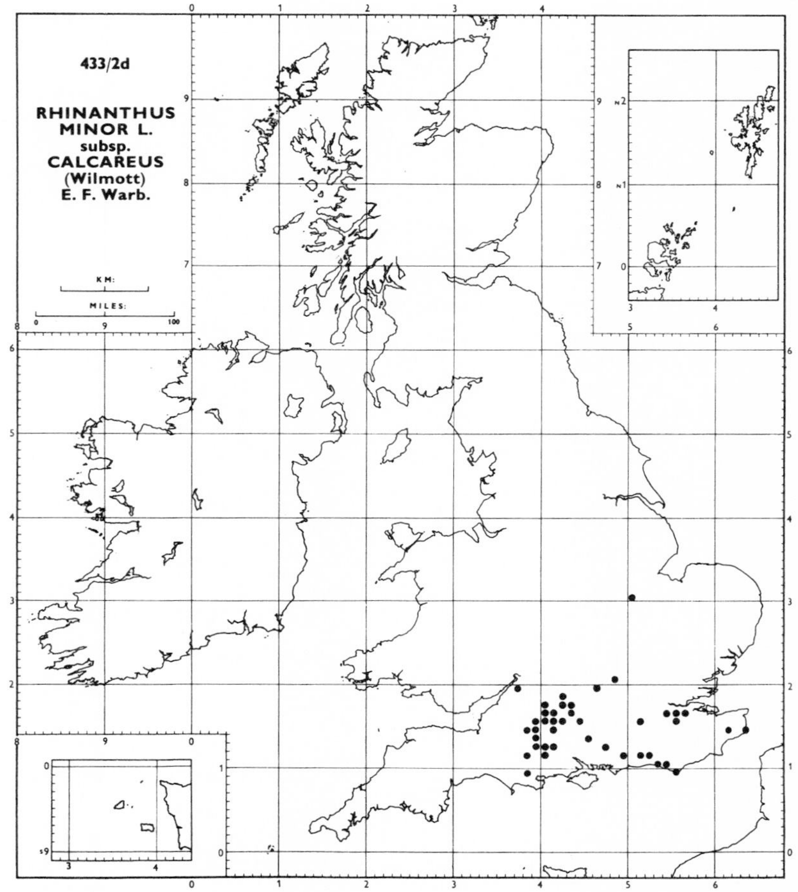

432/2b Pedicularis sylvatica L.
subsp. **hibernica** D. A. Webb

BM, CGE, DBN, K, OXF, TCD
This endemic subspecies differs from subsp. *sylvatica* in that the calyx and pedicels are uniformly clothed with long white curled hairs. The map is based upon that published by Webb (1956), with the addition of a few further records which have come to his notice since that date.

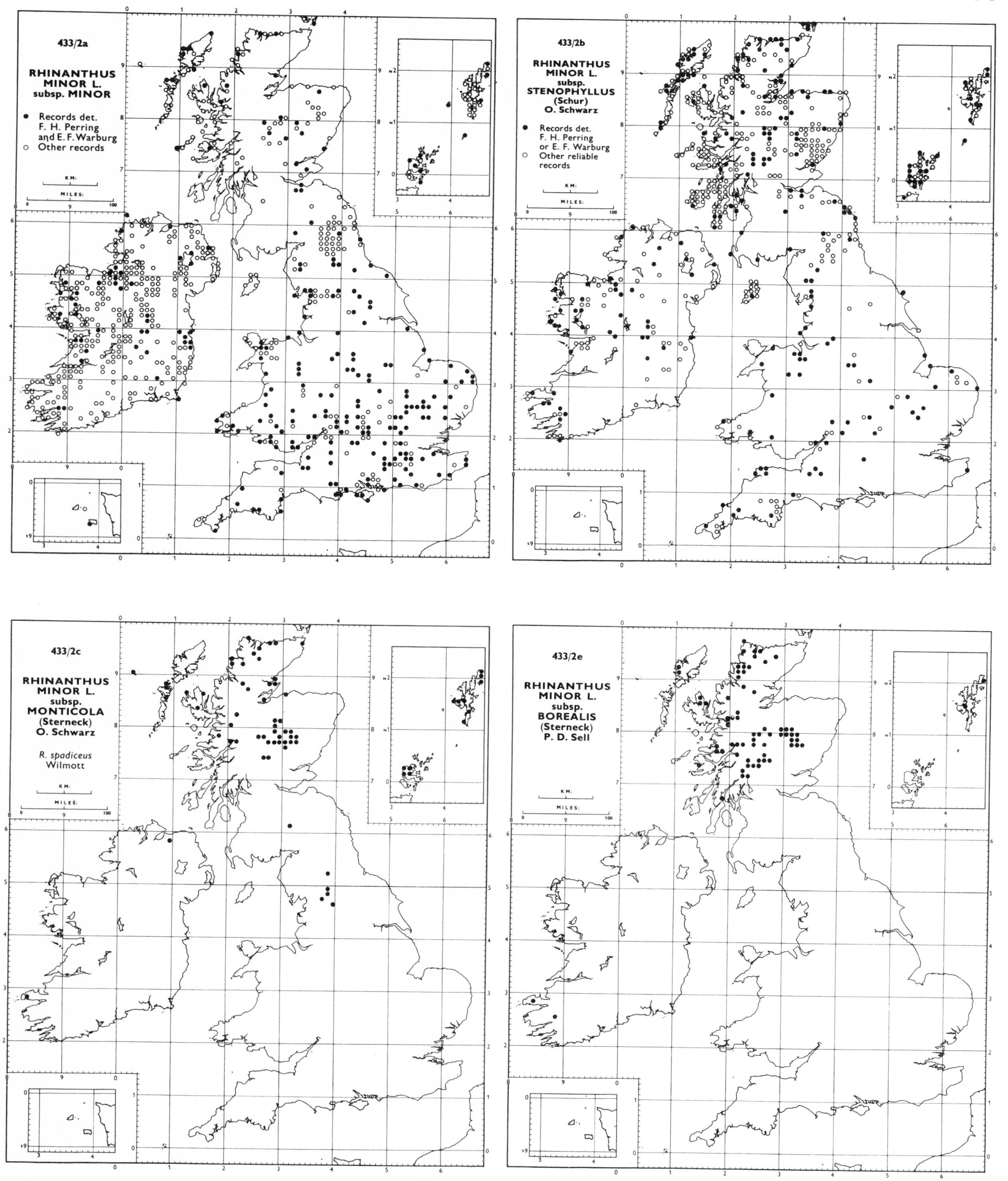

433/2a
RHINANTHUS MINOR L. subsp. MINOR
Records det. F. H. Perring and E. F. Warburg
Other records
KM:
MILES:
433/2b
RHINANTHUS MINOR L. subsp. STENOPHYLLUS (Schur) O. Schwarz
Records det. F. H. Perring or E. F. Warburg
Other reliable records
KM:
MILES:
433/2c
RHINANTHUS MINOR L. subsp. MONTICOLA (Sterneck) O. Schwarz
R. spadiceus Wilmott
KM:
MILES:
433/2e
RHINANTHUS MINOR L. subsp. BOREALIS (Sterneck) P. D. Sell
KM:
MILES:

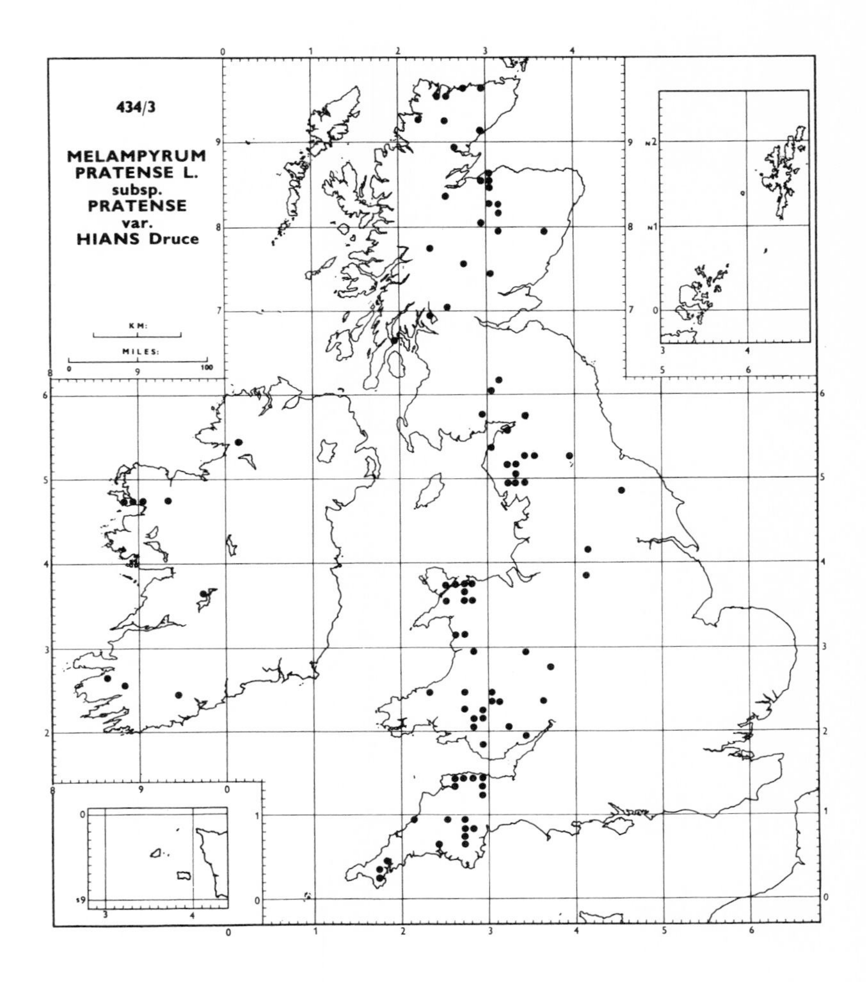

434/3 Melampyrum pratense L.
subsp. **pratense** var. **hians** Druce
subsp. **commutatum** (Kerner) C. E. Britton

An account of these taxa has been published by A. J. E. Smith (1963), who has supplied or approved the data on which these maps are based. Subsp. *commutatum* is a plant of calcareous habitats, whereas subsp. *pratense* is more or less confined to acid habitats. The map in the *Atlas*, p. 234, may be taken as illustrating the distribution of subsp. *pratense* with very minor exceptions. Subsp. *commutatum* occurs in similar habitats on the Continent.

Subsp. *pratense* var. *hians*, with golden-yellow flowers, appears to be confined to the palaeogenic regions of Britain. A record of it from Sussex is unconfirmed. It is possibly endemic.

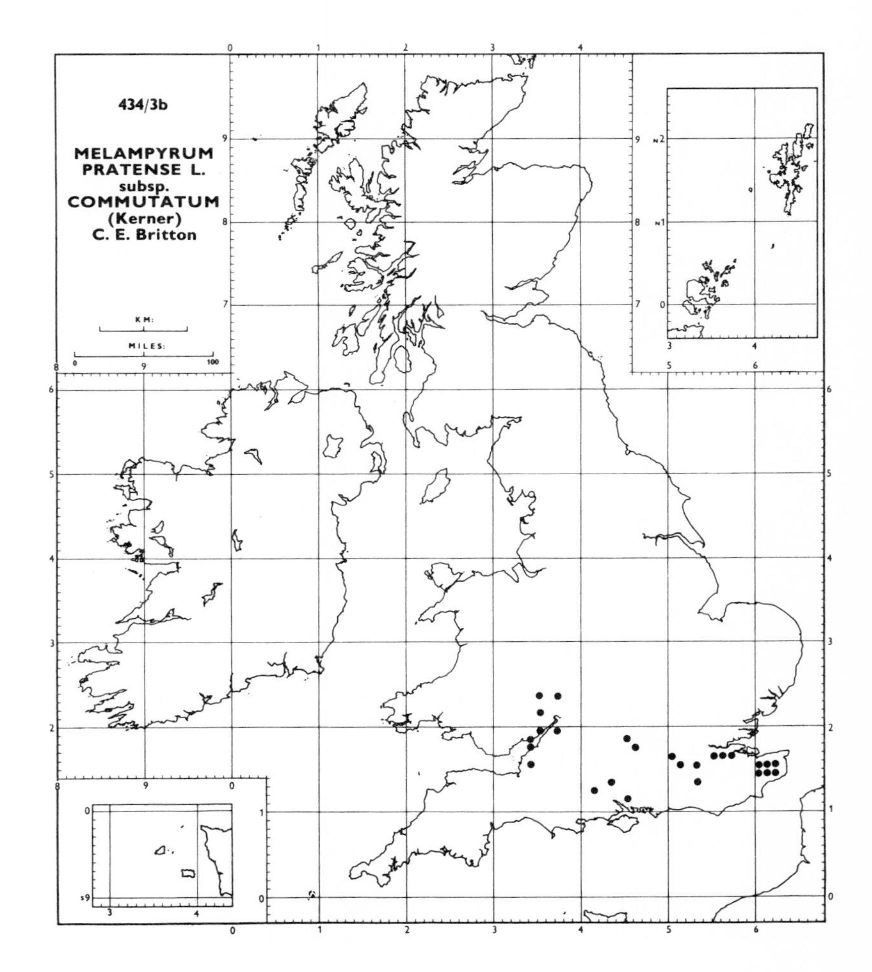

435 EUPHRASIA L.

The preparation of the maps of this difficult genus would not have been possible without the great contribution made by P. D. Sell, E. F. Warburg and P. F. Yeo in determining the many specimens which have been brought to them, or which they have themselves collected during the last decade. I am especially grateful to Dr Yeo, who had the misfortune to be the most accessible expert to me, but who bore the brunt of the identification, and in addition generously made available extensive notes and a card index of records which he has collected together from the literature and from herbarium material which he has seen.

The maps are based on material seen by these three experts, and by H. W. Pugsley in the past. Data from the monograph by Pugsley (1930) have been included as well as some records supplied by E. O. Callen. For a number of common and widespread species field records have been accepted when they are from a 10 km. square adjacent to a square from which a record passed by an authority has been obtained.

The notes on the species which follow were prepared in collaboration with P. F. Yeo.

435/1.1

EUPHRASIA MICRANTHA Reichb.

○ Records unconfirmed by specialists

435/1.1 Euphrasia micrantha Reichb.

This is an easily recognizable species of acid heaths which is normally found in association with heather, *Calluna vulgaris*. For these reasons it has been possible to accept field records beyond the limits laid down in the introduction to the genus. *E. micrantha* is readily distinguishable from the closely related *E. scottica* except in certain coastal areas of north and west Scotland. It is remarkable that the species does not appear to occur in East Anglia, where seemingly suitable habitats exist. It may now be extinct in south-east England.

The apparent rarity of this species in central Ireland may be due to under-recording. It is widespread in west and central Europe.

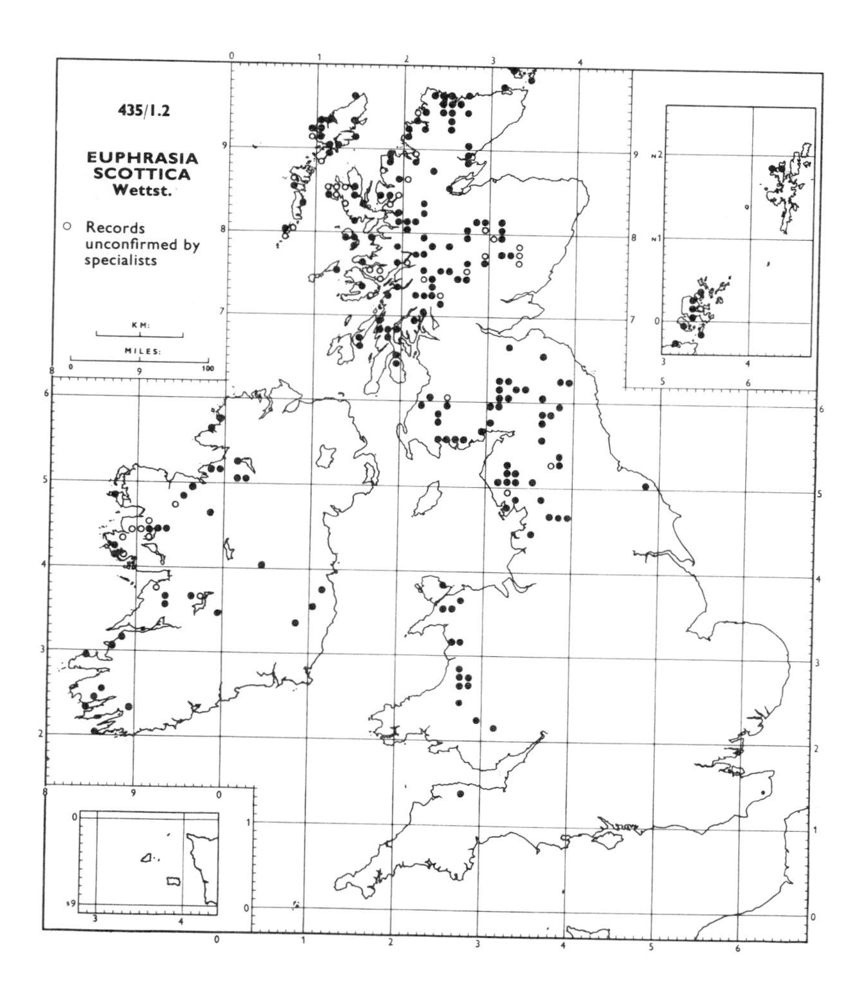

435/1.2 Euphrasia scottica Wettst.

A characteristic species of flush communities in hilly areas of the north and west, not normally associated with *Calluna vulgaris*. Elsewhere it occurs in Scandinavia and the Faeroes.

SCROPHULARIACEAE

435/1.3 **Euphrasia rhumica** Pugsl.

This supposed endemic species confined to Rhum (104) may only be a locally formed hybrid segregate. See also *E. eurycarpa* (435/1.6).

435/1.4 **Euphrasia frigida** Pugsl.

This Arctic species is more or less confined to altitudes above 1,000 ft (320 m.), where it occurs on damp rock ledges and in short open turf.

435/1.5 **Euphrasia foulaensis** Townsend ex Wettst.

A characteristic species of cliff-top and salt-marsh communities found outside Britain only in the Faeroes.

435/1.8 **Euphrasia rotundifolia** Pugsl.

This species is very closely related to *E. marshallii* with which it has a sympatric distribution and of which it may be only a variant. Recorded outside Britain only in Iceland.

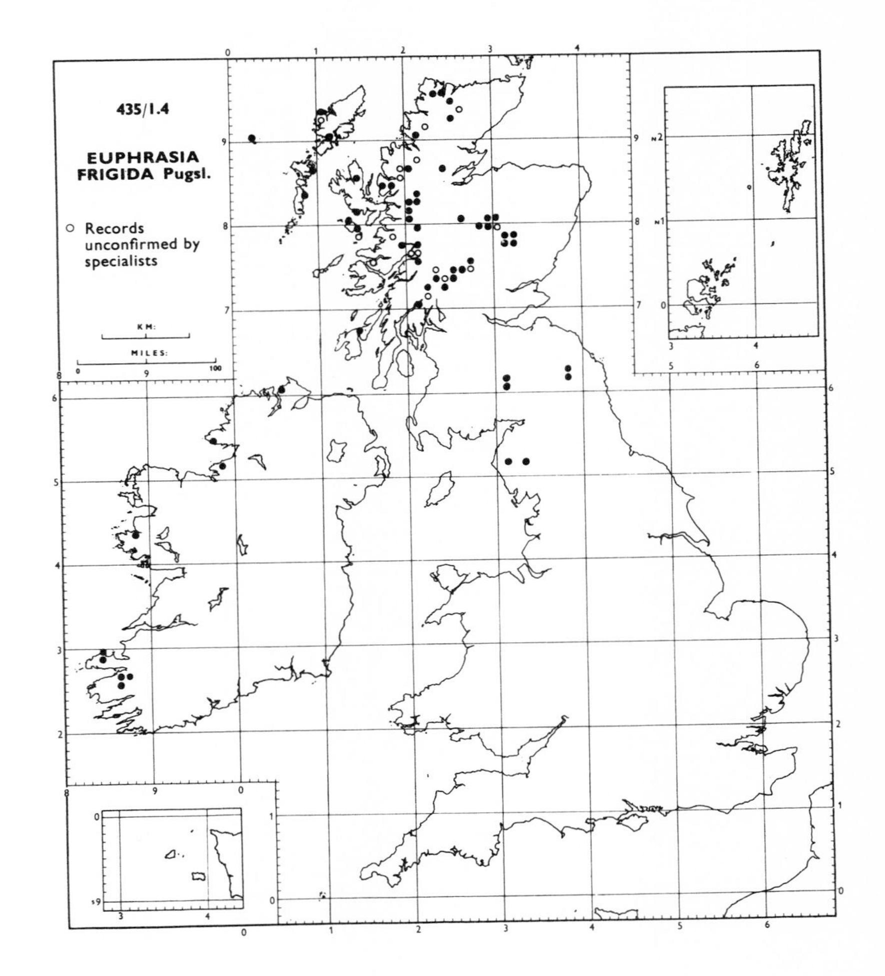

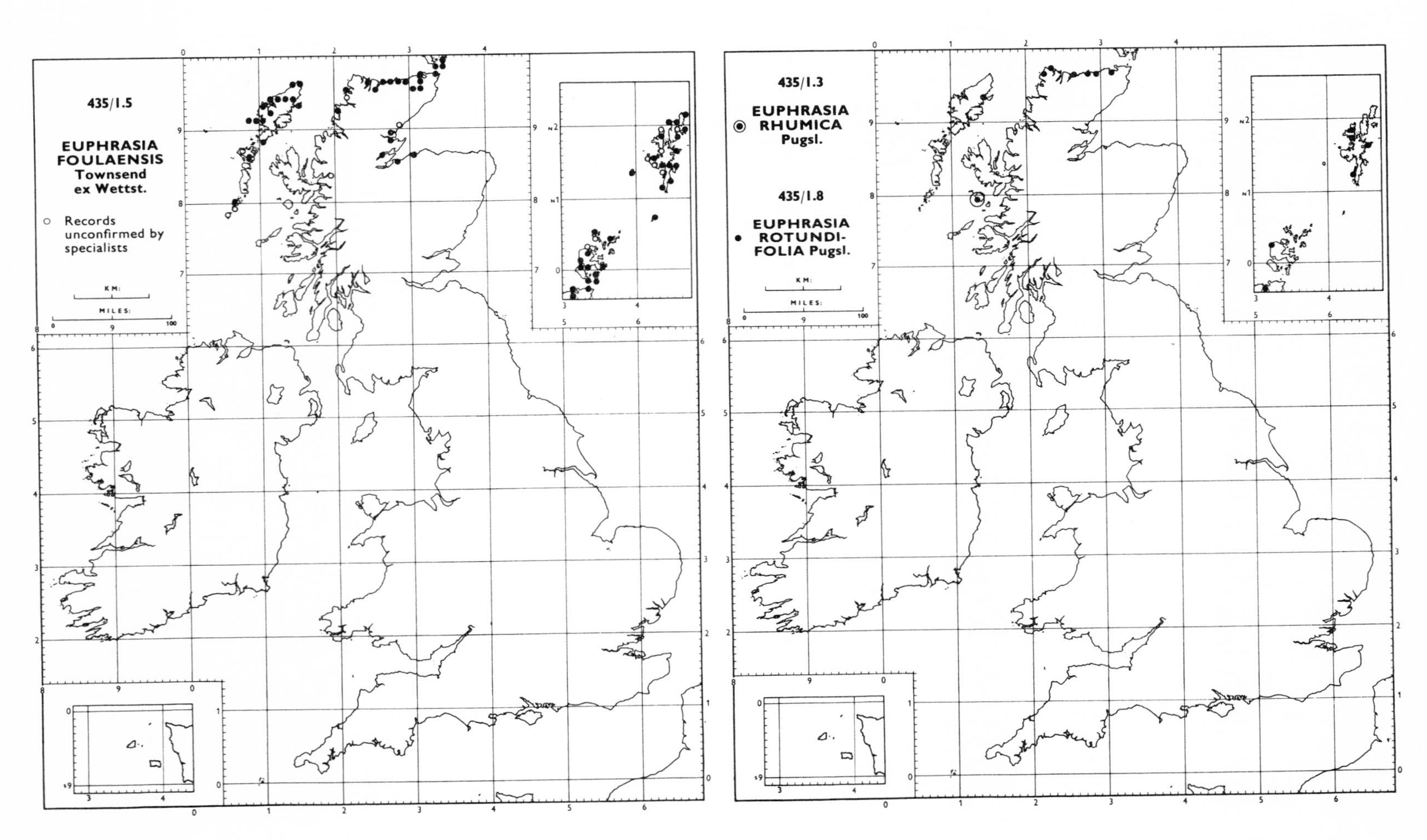

435/1.6 **Euphrasia eurycarpa** Pugsl.

A supposed endemic species confined to Rhum (104), probably of similar origin to *E. rhumica* (435/1.3) q.v.

435/1.7 **Euphrasia campbelliae** Pugsl.

A local and uniform taxon, confined to the coast of Lewis in the Outer Hebrides (110), though it may be of recent hybrid origin. The morphology suggests that the parents might have been *E. micrantha* and *E. marshallii*.

435/1.9 **Euphrasia marshallii** Pugsl.

This species may be found on cliff-tops and in salt-marshes in similar habitats to *E. foulaensis*, with which it often occurs and hybridizes. It is endemic.

435/1.10 **Euphrasia curta** (Fries) Wettst.

This is a rare and local species, the populations of which vary considerably in different parts of its range. It usually occurs in rocky, or sandy, open habitats. Plants from the mountains of north Wales, which are dwarf and slender, are var. **rupestris** Pugsl. Plants in the west of Scotland tend to have smaller flowers and broader capsules, whilst those from the north coast, although falling within the range of variation of *E. curta*, may have originated as the result of hybridization between *E. marshallii* and *E. nemorosa*. The isolated record from Oxfordshire (23) is based on plants which somewhat resemble *E. nemorosa*. The published map excludes some of the records included in Pugsley (1930), where these referred to glabrous or only slightly hairy specimens, which should probably be placed in other species, or are the results of hybridization. Outside the British Isles it occurs in north and east-central Europe, the Faeroes, Iceland and north-east North America.

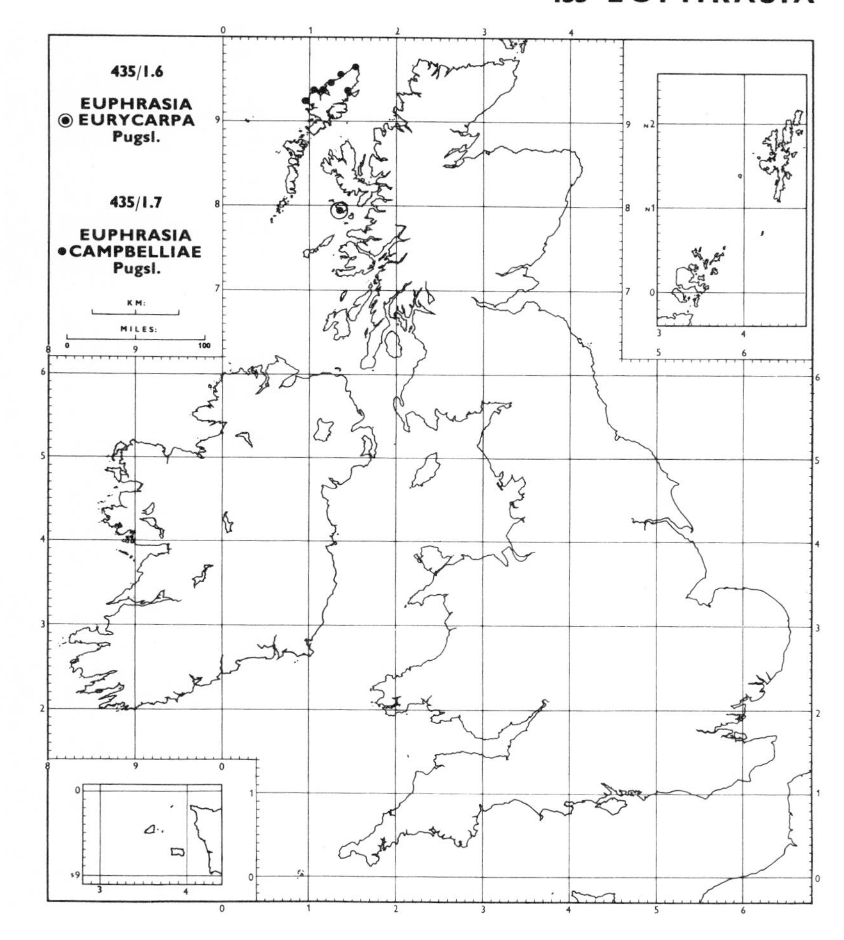

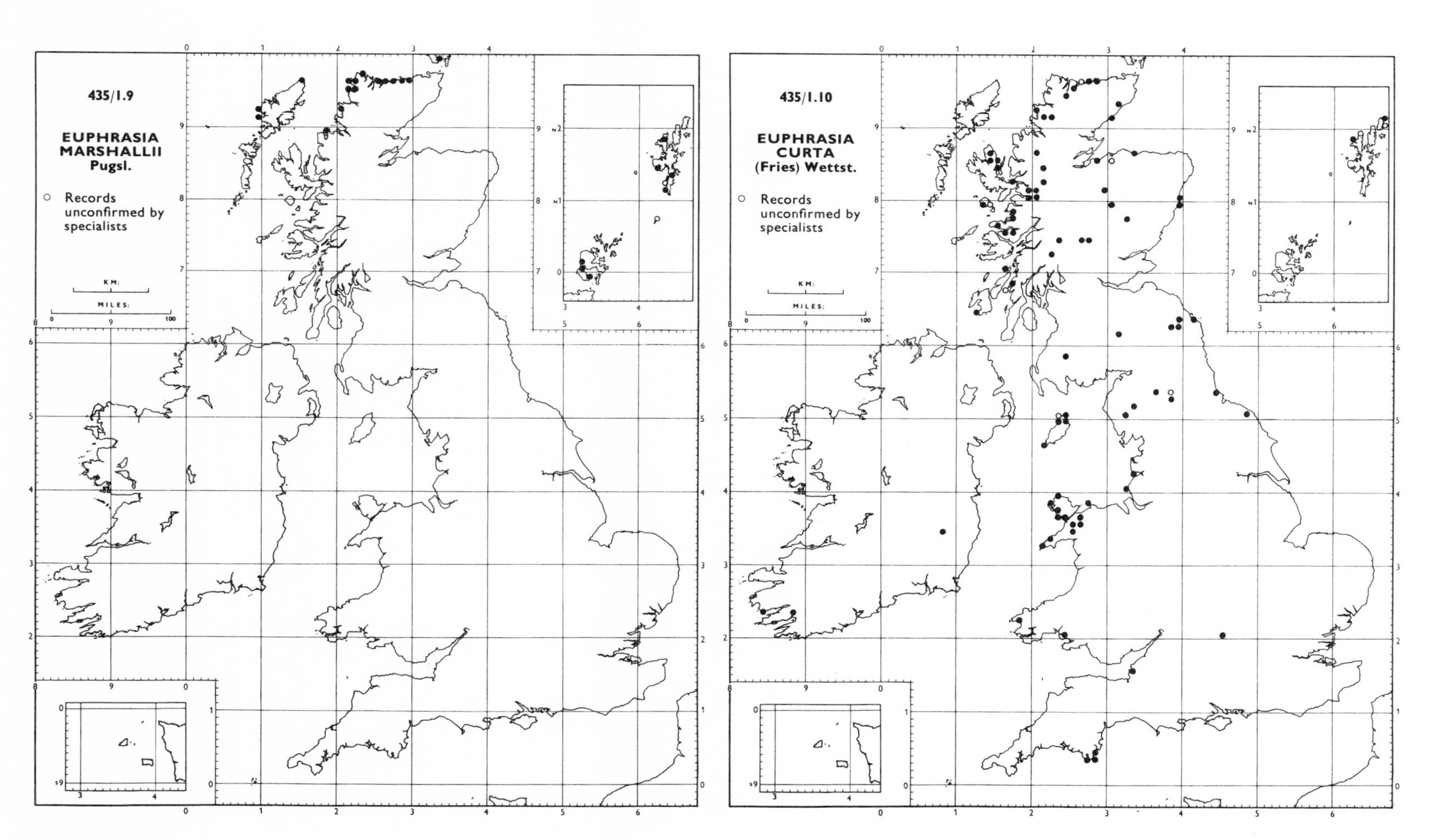

435/I.11 **Euphrasia cambrica** Pugsl.

This endemic species takes the place of *E. frigida* in the mountains of Caernarvon (49), where, however, it is local. Records for Brecon (42) and Westmorland (69) have not been confirmed.

435/I.12 **Euphrasia tetraquetra** (Brébisson) Arrondeau
E. occidentalis Wettst.

There is some doubt about the northern limits of this species, as in north Scotland it becomes difficult to distinguish from *E. foulaensis* and dwarf forms of *E. nemorosa*. Records from this area have therefore been omitted. Although mainly a species of grassy coastal cliffs and dunes, it occurs occasionally in limestone grassland inland. Outside the British Isles known only from the coasts of north and north-west France, and Quebec and Maine, North America.

435/I.13 **Euphrasia nemorosa** (Pers.) Wallr.

This variable species is the commonest in lowland grassland. It extends to the north of Scotland, where it occurs mainly on dunes. These northern plants differ from the southern plants by having broader leaves, with larger teeth, and branches which are more spreading. This form can be difficult to distinguish from *E. tetraquetra* (435/I.12) q.v. Similar forms occur in the west of Ireland. Elsewhere it occurs in western and central Europe, and as a probable introduction in North America.

435/I.14 **Euphrasia heslop-harrisonii** Pugsl.

A little-known endemic species which needs study in the field.

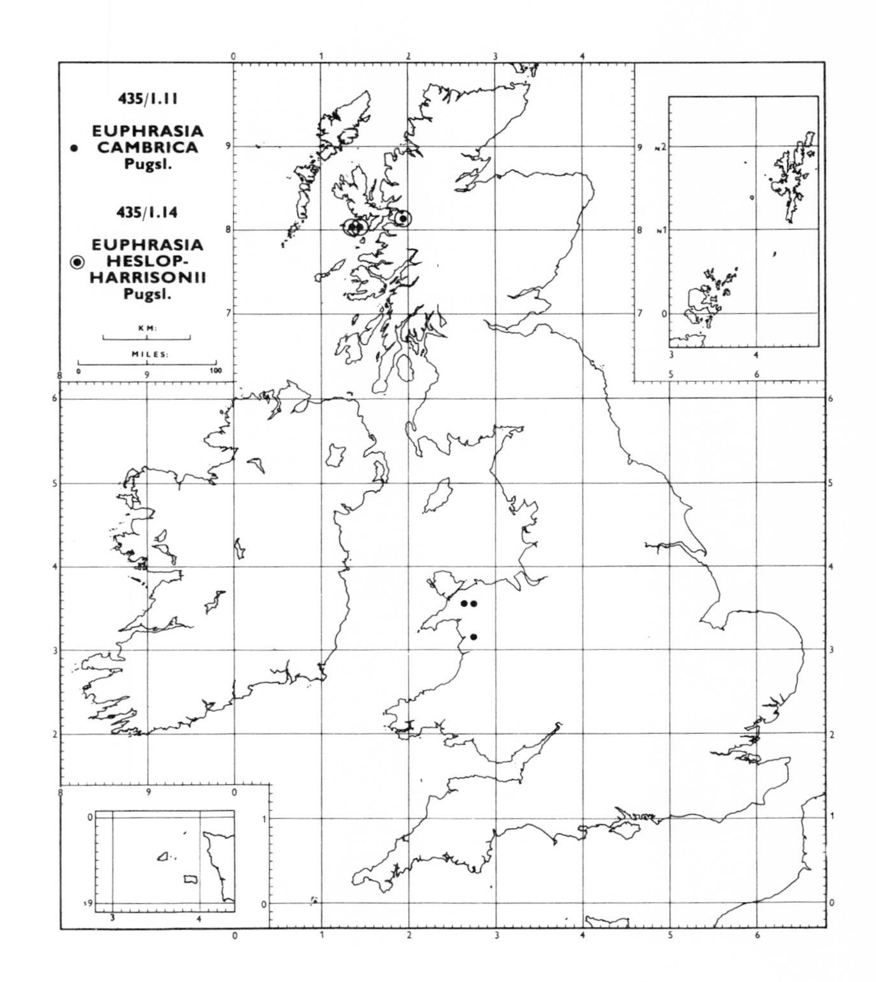

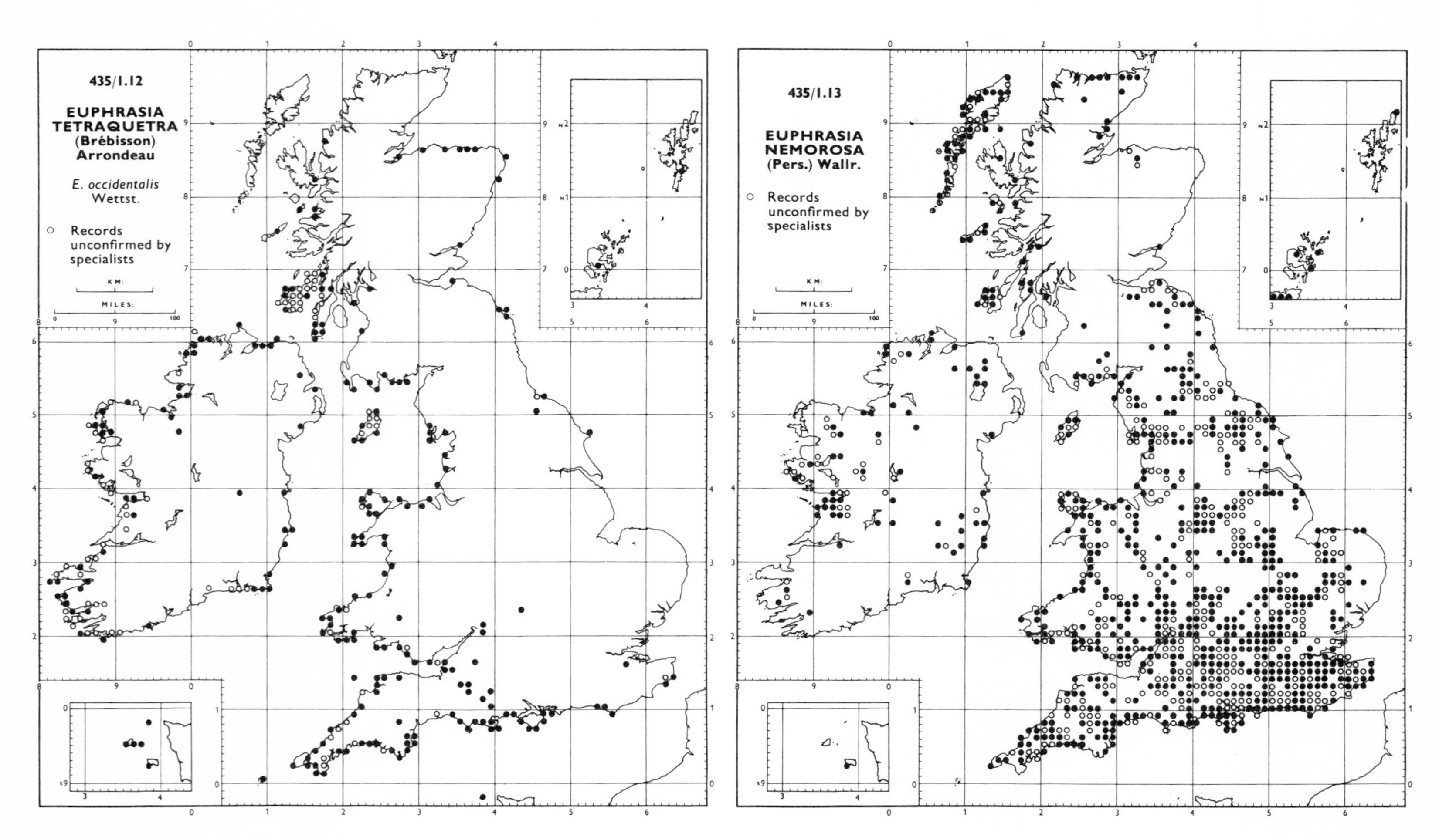

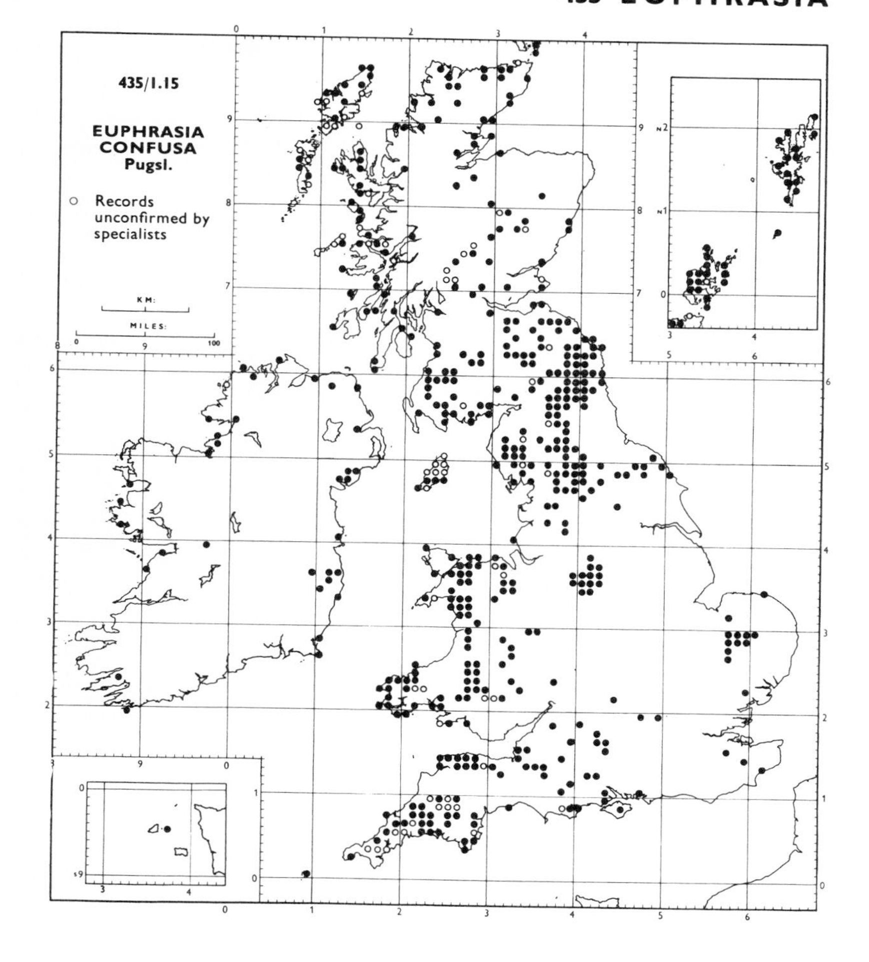

435/1.15 Euphrasia confusa Pugsl.

This is the characteristic eyebright of short turf on a wide variety of soils in the upland zone, including the carboniferous limestone. The species is particularly well adapted by its decumbent habit to withstand intense grazing pressure. Its distribution pattern may be related to the concentration of sheep-walks in Britain.

The populations found on sandy soils in east and south-east England can be distinguished from populations elsewhere by their darker flowers and longer internodes. However, intermediates with typical plants are sometimes found growing with them.

The species is only recorded from the Faeroes outside the British Isles.

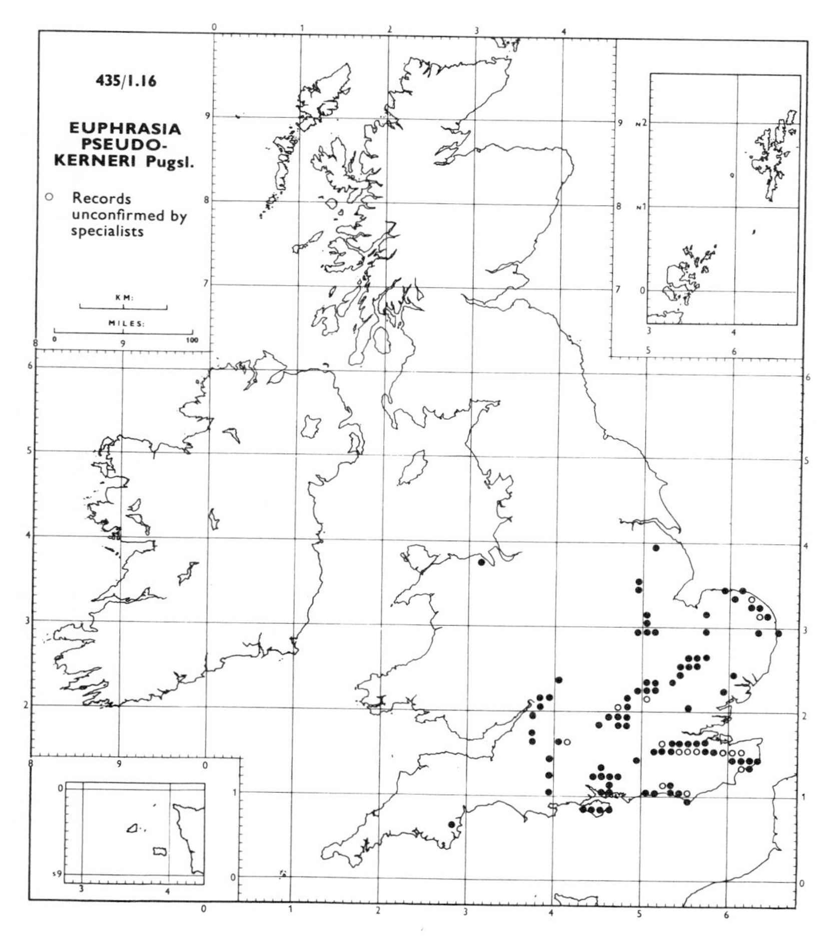

435/1.16 Euphrasia pseudokerneri Pugsl.

This endemic species is characteristic of chalk and limestone grassland in south-east England. It can be distinguished from *E. nemorosa*, which occurs in the same habitat, by its larger flowers and later flowering period. In Norfolk (27, 28) it occurs in fens.

A taxon which occurs on limestone in the west of Ireland has many of the characteristics of this species and further investigations are needed to determine whether the resemblance is retained in cultivation.

E. pseudokerneri is recorded from isolated localities in south Devon (3) and Flint (51).

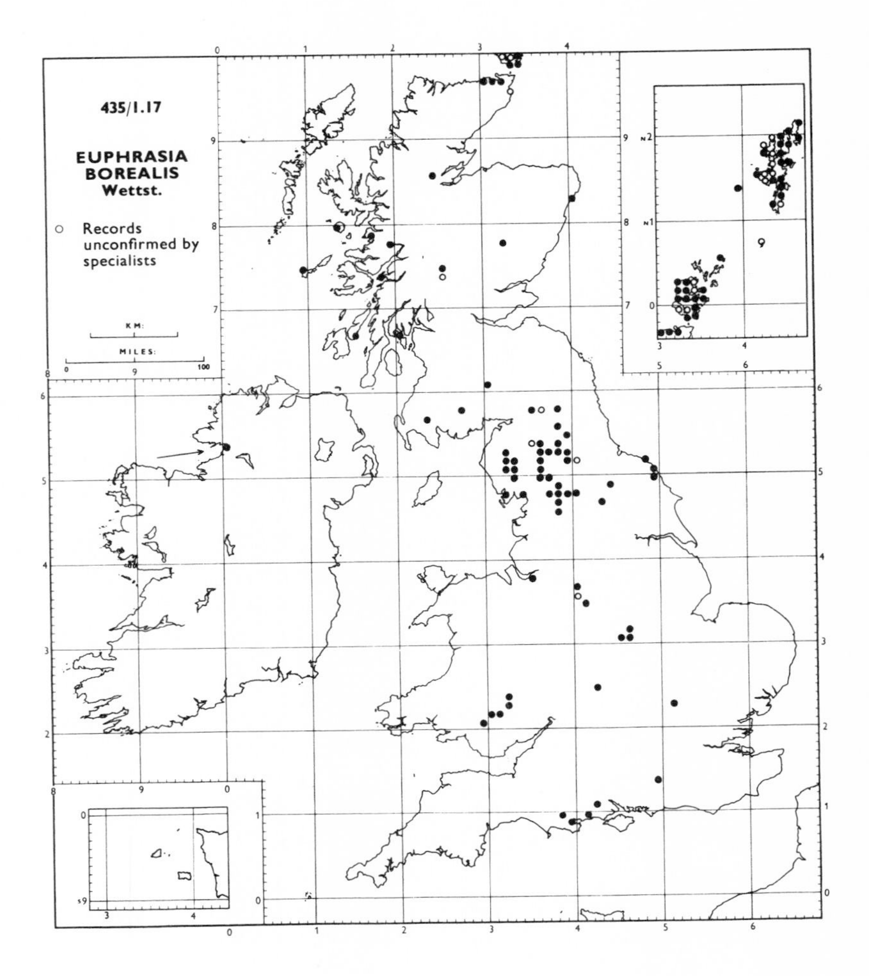

435/I.17 **Euphrasia borealis** Wettst.

The map shows the majority of the records of plants called *E. borealis* Wettst. A few that clearly belong to *E. brevipila* have been omitted.

The plants of Orkney and Shetland may represent a distinct species which also occurs in the Faeroes. If so, the correct name for this plant is **E. arctica** Lange ex Rostrup.

The mainland taxon is most distinct in north England and south Scotland, but it grades into *E. brevipila* in other areas, and is perhaps best regarded as a variety of that species.

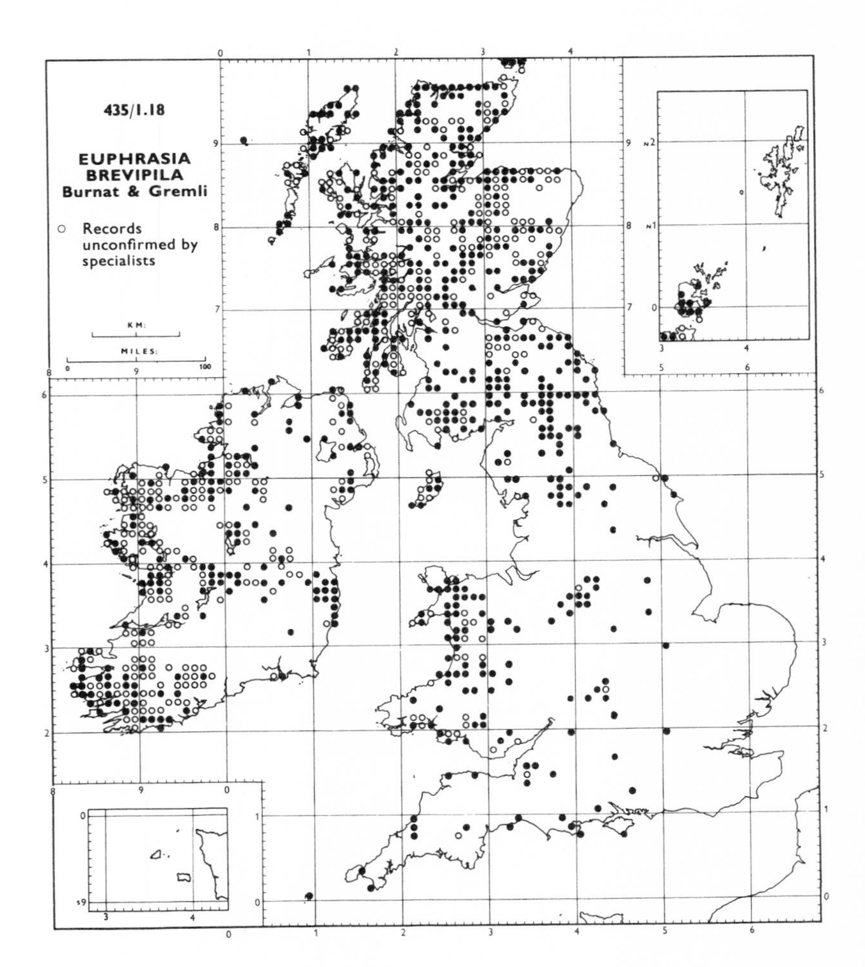

435/I.18 **Euphrasia brevipila** Burnat & Gremli

This variable species is common in meadows and pastures, and along roadsides, in the north and west. Two distinct varieties have been described from Scotland, var. **notata** (Towns.) Pugsl. in the central Highlands, and var. **reayensis** Pugsl. from the north coast. Both varieties grade into var. **brevipila**.

435/1.19 & 24 **Euphrasia rostkoviana** Hayne
E. hirtella auct. brit.

This species occurs with *E. brevipila* in hay meadows and rough pastures, particularly in Ireland and Wales.

Plants referred to *E. hirtella* from north Wales have not been refound and specimens seen from Scotland are not clearly distinguishable from *E. rostkoviana*. The majority of British plants have flowers which are smaller than those of specimens from central Europe.

435/1.20 **Euphrasia montana** Jord.

This species largely replaces *E. rostkoviana* in the meadows of the Lake District and the northern Pennines (64–7, 69, 70). There are isolated records from Brecon (42), and Caernarvon (49). It is widespread in central Europe.

435/1.21 **Euphrasia rivularis** Pugsl.

A characteristic endemic species by mountain streams in the Lake District (69, 70) and north Wales (48, 49).

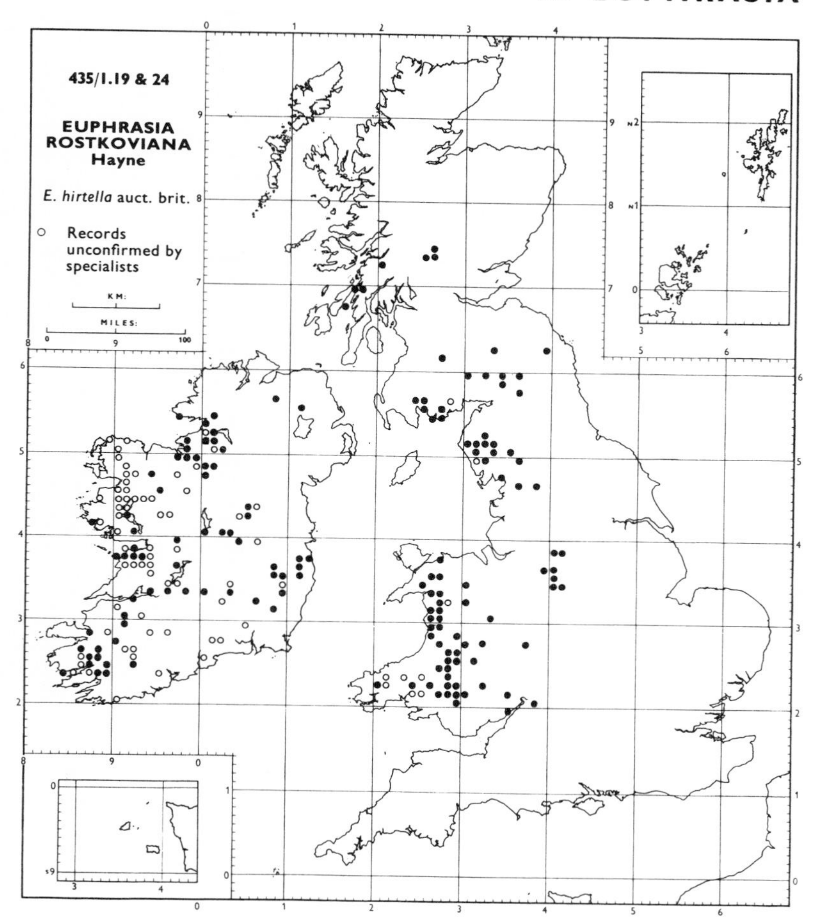

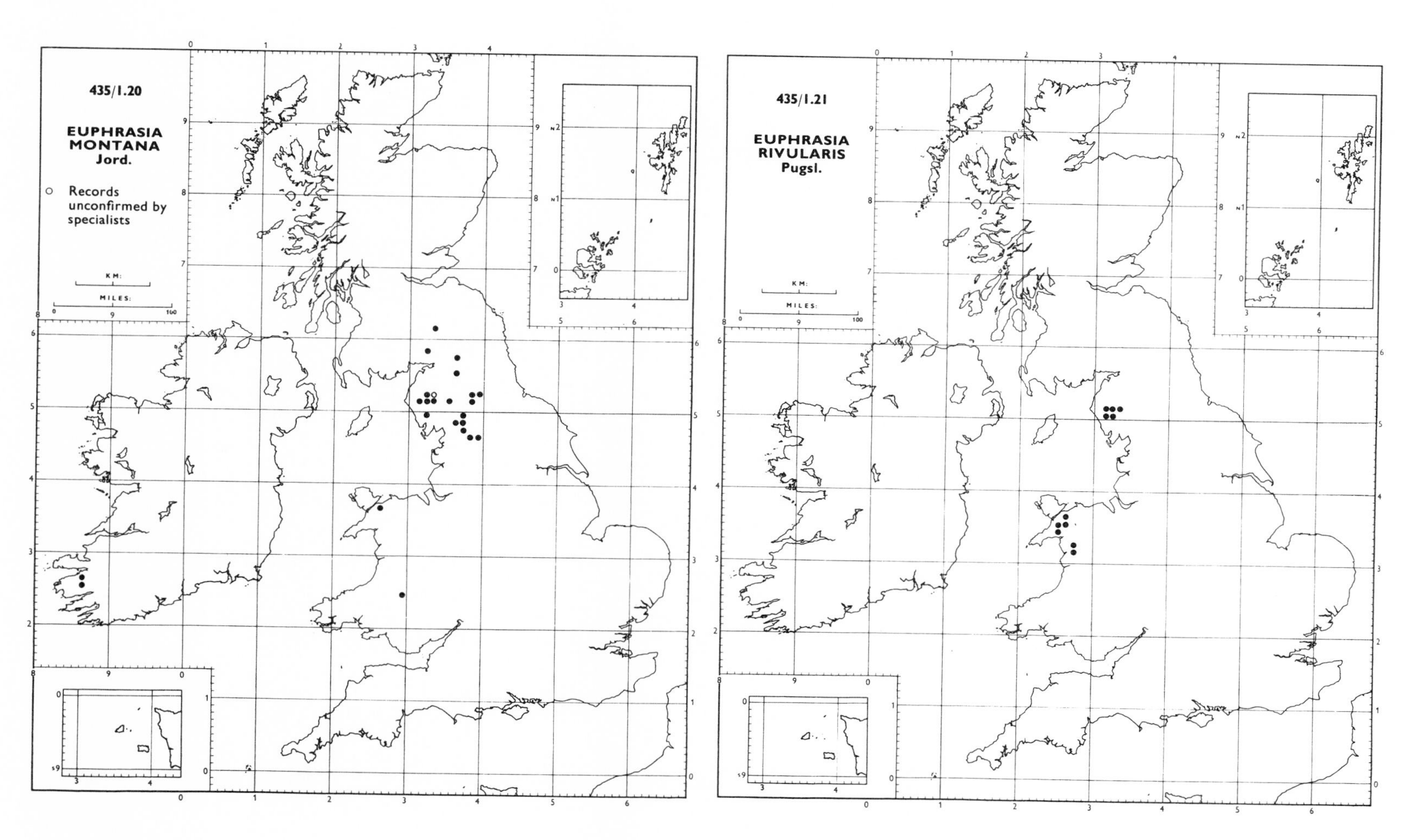

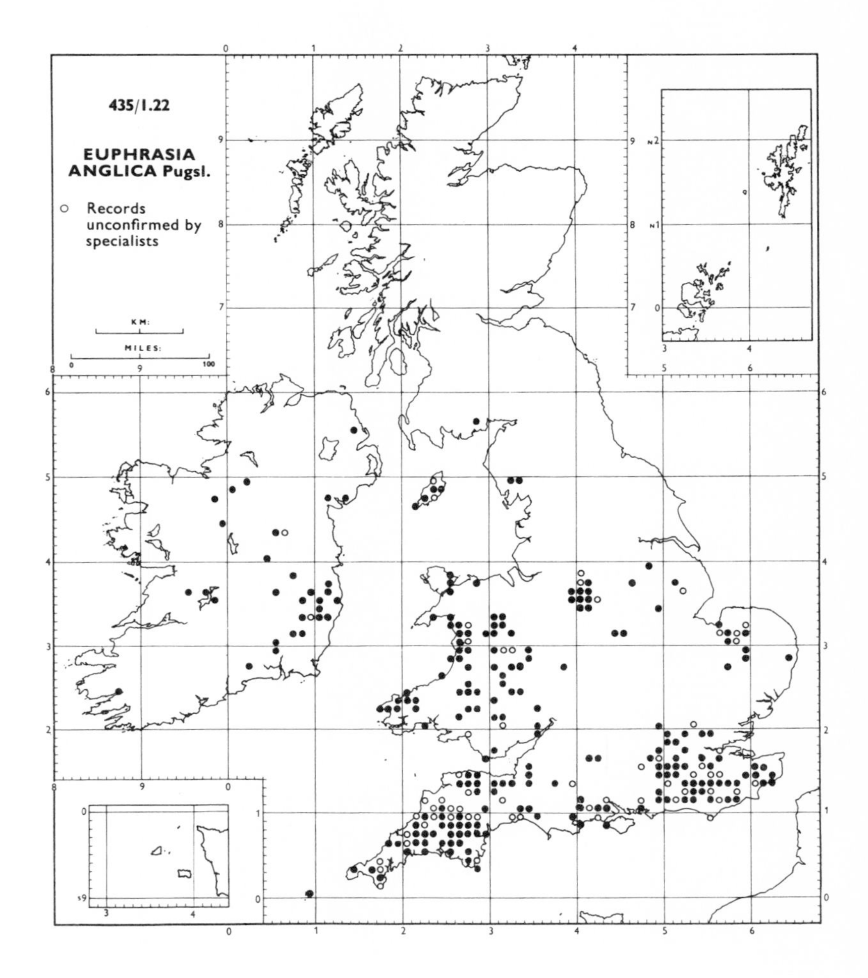

435/1.22 **Euphrasia anglica** Pugsl.

This endemic eyebright is widespread in heather and rough pastures in southern Britain; like *E. confusa* it is particularly a species of short, grazed turf, and is well adapted by its habit to withstand intense grazing pressure.

In areas where *E. anglica* and *E. rostkoviana* both occur intermediates may be found.

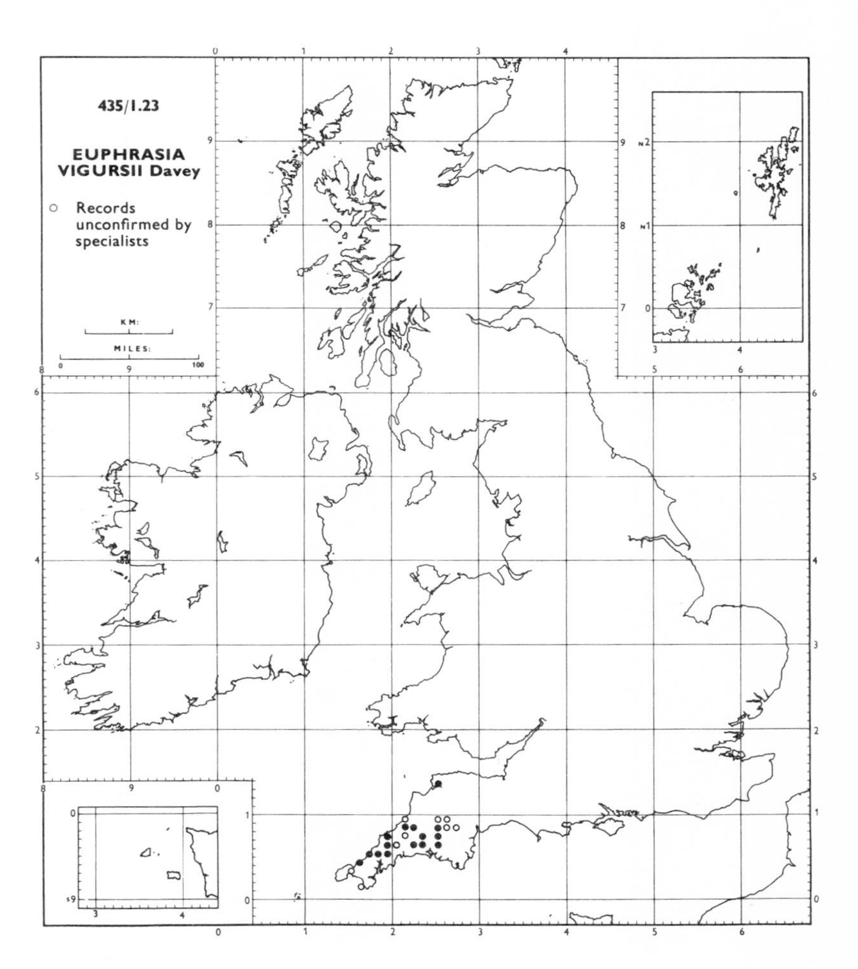

435/1.23 **Euphrasia vigursii** Davey

Despite the fact that this species probably originated by hybridization between *E. anglica* and *E. micrantha* (see Yeo, 1956), it is restricted to a very limited part of the area of overlap of the distribution of these two species. This restriction may be due to limited occurrence of the habitat to which it is best adapted—a mixed heath of *Ulex gallii* and *Agrostis setacea*.

436/1 Odontites verna (Bellardi) Dumort.
subsp. **pumila** (Nordst.) A. Pederson
Bartsia odontites var. *litoralis* auct. brit.
subsp. **verna**
subsp. **serotina** Corb.

BFT, BM, CGE, DBN, NWH, OXF, SHY, TCD

There is much overlap in the morphology of the last two subspecies and many intermediates occur. Reliance on the angle of branching alone has in the past meant that specimens which are best referred to subsp. *serotina* have been placed in subsp. *verna*. For these maps material of subsp. *verna* accepted by the editor, P. D. Sell and E. F. Warburg had branches spreading at an angle of less than 45° to the stem, and bracts longer than the flowers.

The maps are incomplete and unsatisfactory, but they do reflect the main trends: subsp. *serotina* is common in the south, but rare or absent in most of Scotland, whereas subsp. *verna* is rare in the south and more frequent in north Scotland.

The third taxon, subsp. *pumila*, is a dwarf often unbranched plant with internodes shorter than the leaves, which occurs in a few grassy places near the sea in the north and west of Scotland. It occurs in similar habits in south Sweden, Denmark and Holland (cf. Sell, 1967).

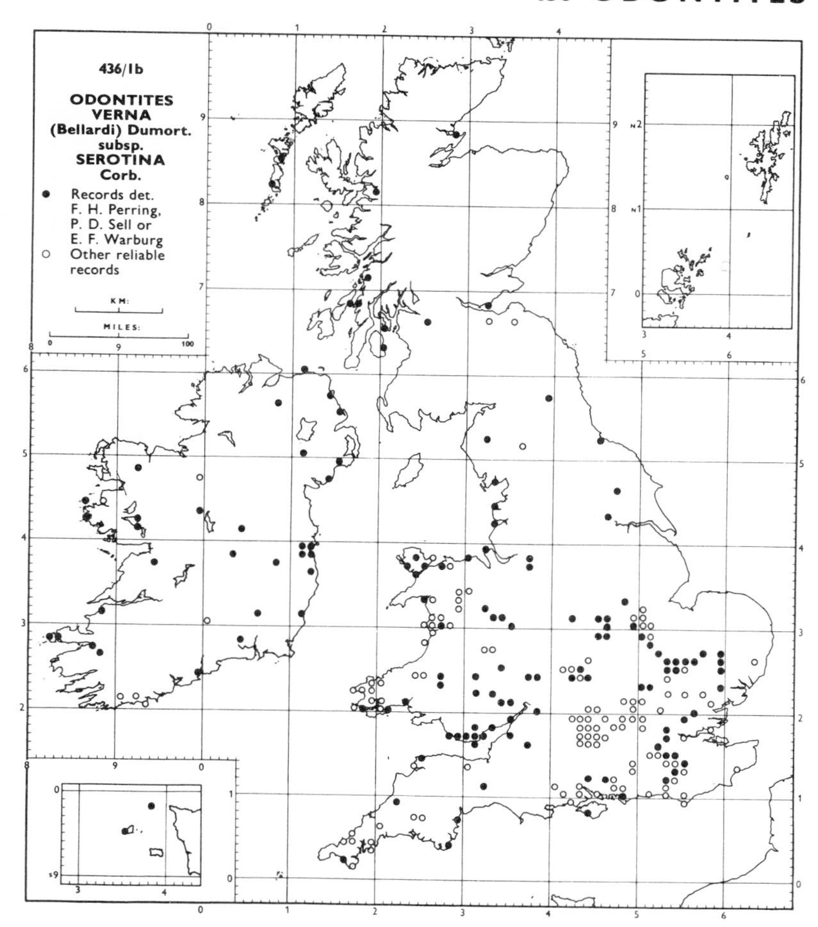

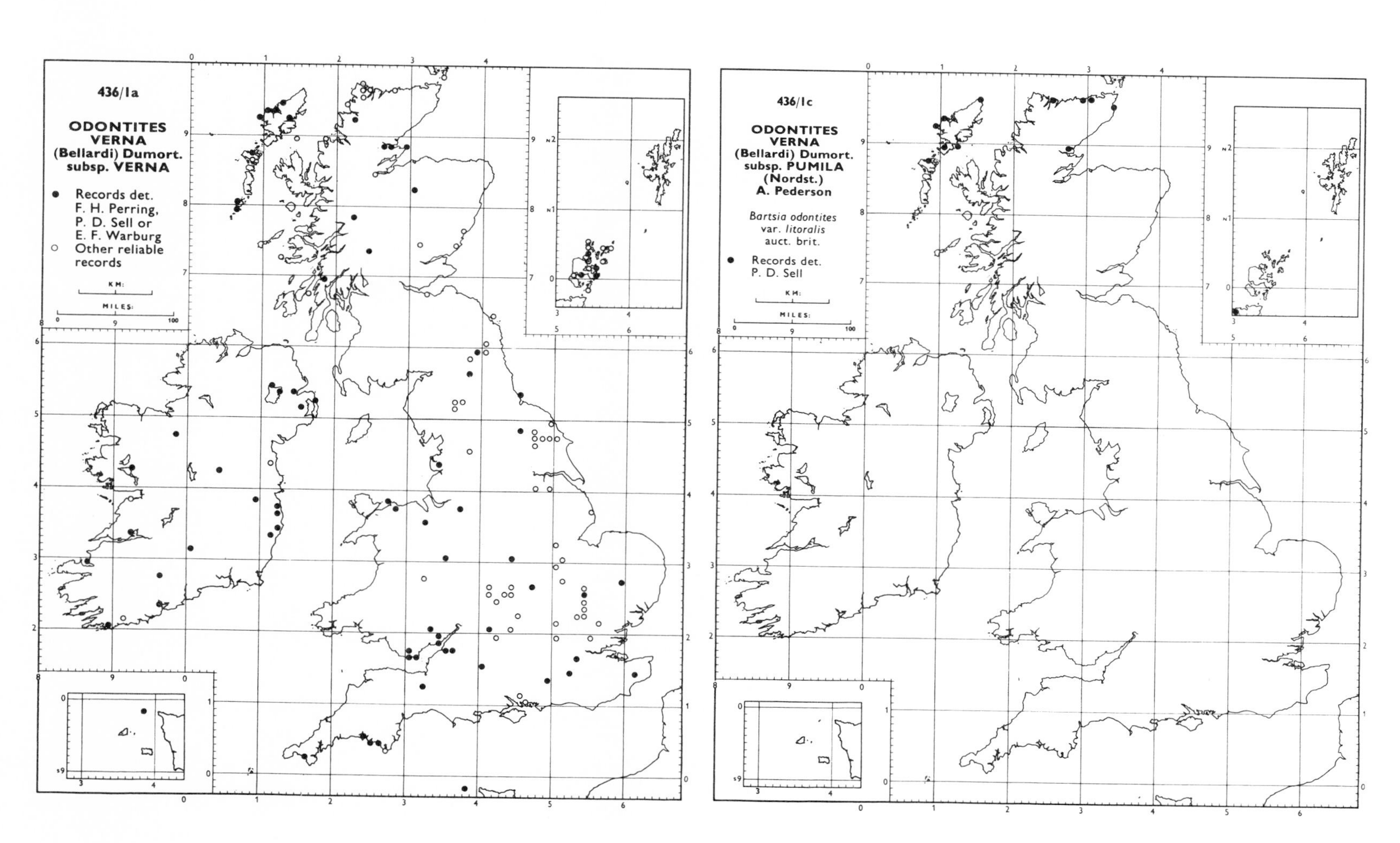

LENTIBULARIACEAE

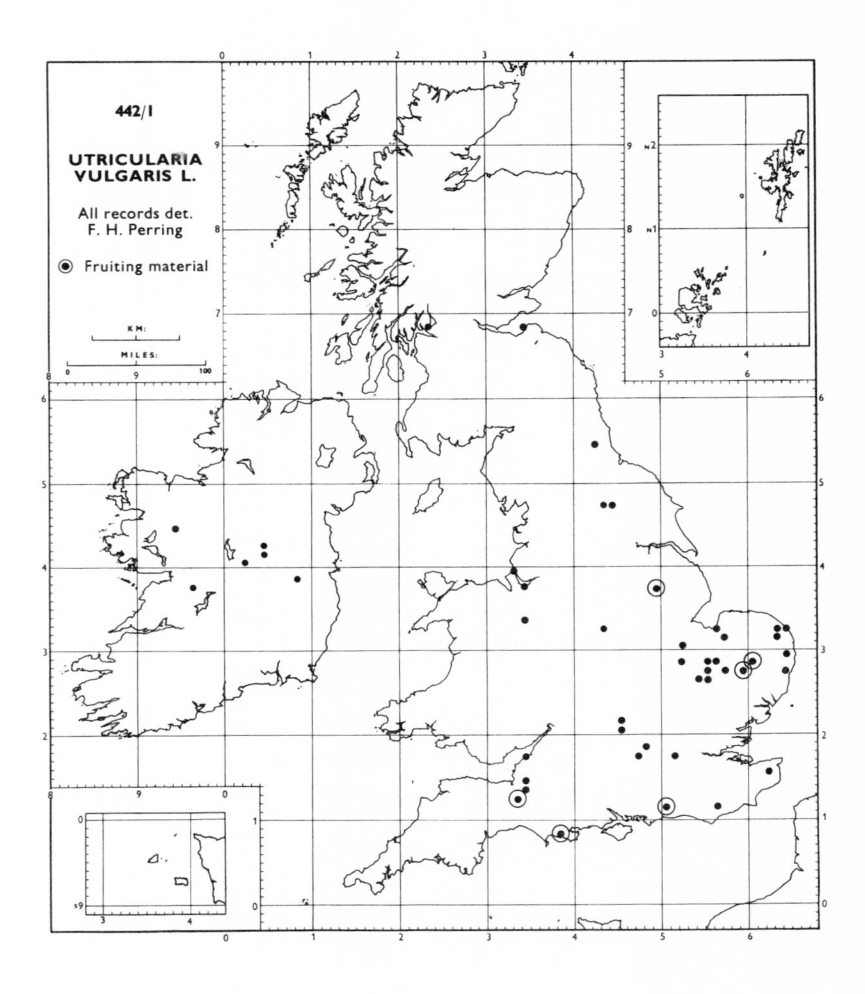

442/1 Utricularia vulgaris L.

442/2 Utricularia neglecta Lehm.

These maps are based entirely on herbarium material determined by the editor, who named flowering specimens only, as the two taxa cannot be distinguished with certainty on the basis of vegetative characters alone. As neither species flowers in the northern part of its range the maps are incomplete; nevertheless they seem to reflect a difference in the ecology of the two taxa. *U. vulgaris* is the species of base-rich waters, and is particularly common in the East Anglian fens, whereas *U. neglecta* is more frequent in acid conditions and is perhaps the only one of the two which occurs in Wales and south-west England. As far as it is possible to judge from vegetative material most of the records from Scotland (see *Atlas*, p. 239) are *U. neglecta.*

No fruiting material of *U. neglecta* was seen. *U. vulgaris* appears to fruit only in the southern half of its range.

Both species are widespread in Europe.

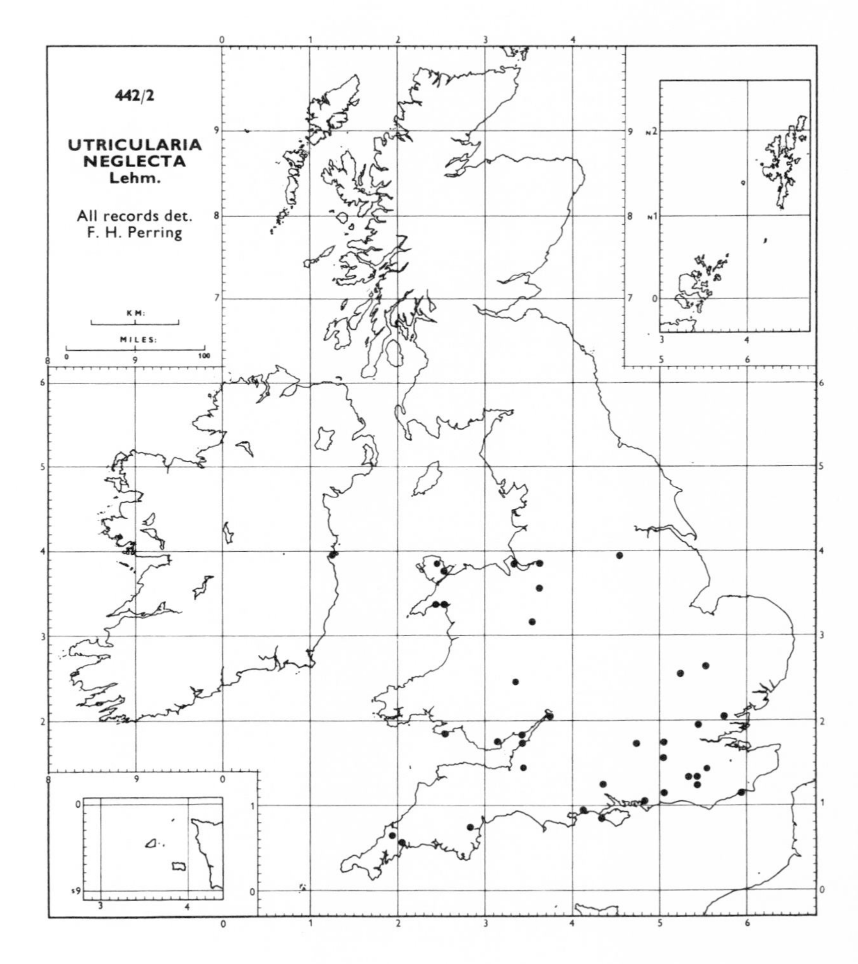

445 MENTHA L.

The preparation of the maps of this difficult genus would not have been possible without the work of the late R. A. Graham. He identified most of the material in the main British herbaria, from which the editor has collected the records. Graham also prepared record cards from his own extensive collection, which is now in the possession of R. M. Harley. To the latter I am greatly indebted for taxonomic advice, and for the examination of a great deal of material, particularly in the Herbarium of the School of Botany in Oxford.

445/1 **Mentha requienii** Benth.

BM, CGE, OXF: Floras
An introduction which is occasionally naturalized as in east Sussex (14) and Armagh (H. 37). It is a native of Corsica and Sardinia. A record for v.-c. 70 has not been traced.

445/1
*MENTHA REQUIENII Benth.

445/3×5 **Mentha ×gentilis** L.
=M. arvensis × spicata
M. gracilis Sole; *M. cardiaca* (S. F. Gray) Baker

CGE, NMW, OXF, Graham: Floras
This is a very variable hybrid which often occurs in the absence of either parent. Published records for v.-cs 32, 66, 85, H. 7 and 33 have not been localized. Those for H. 1 and 2 were errors.

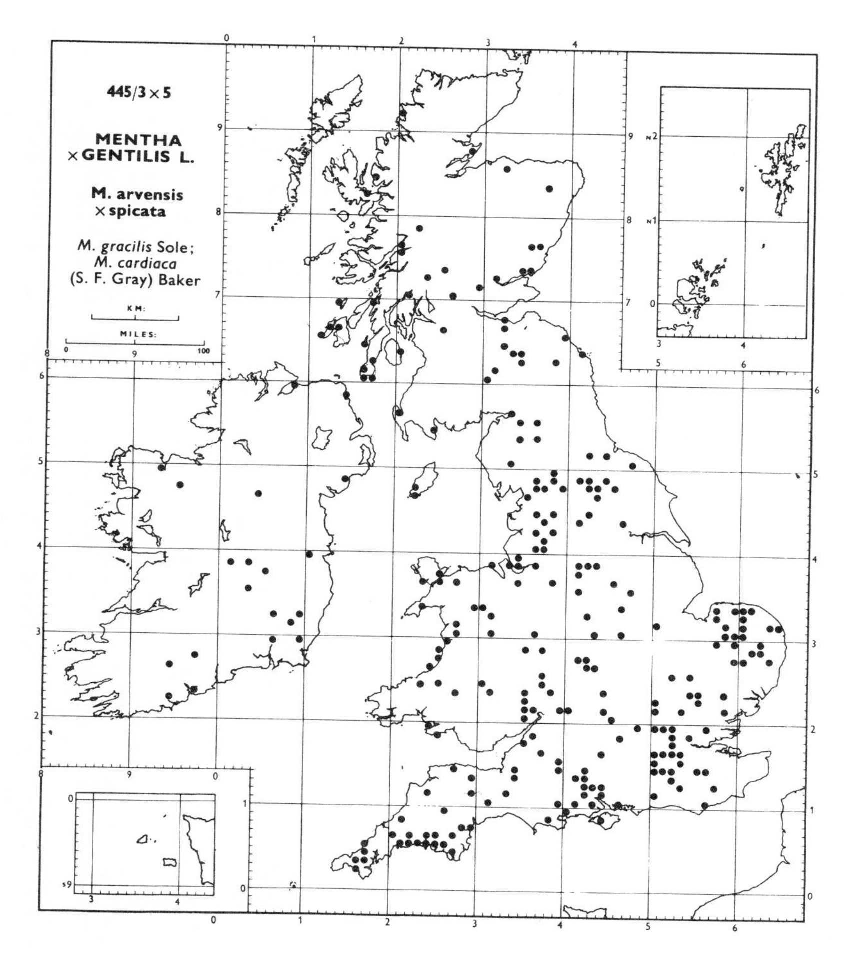

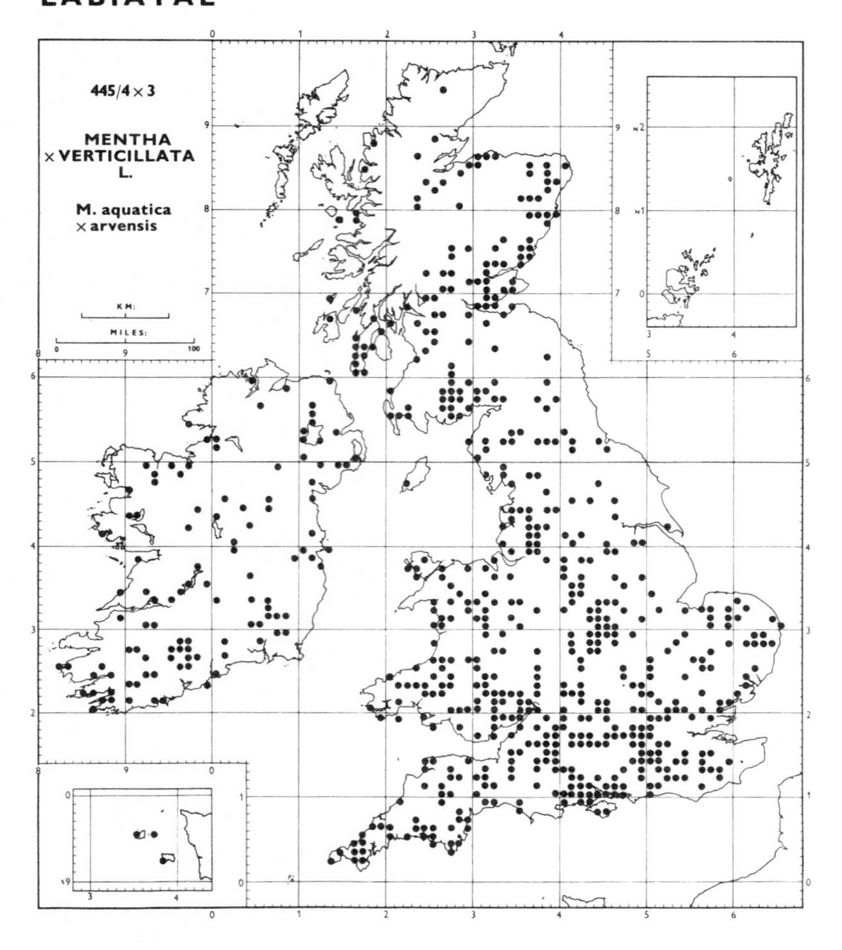

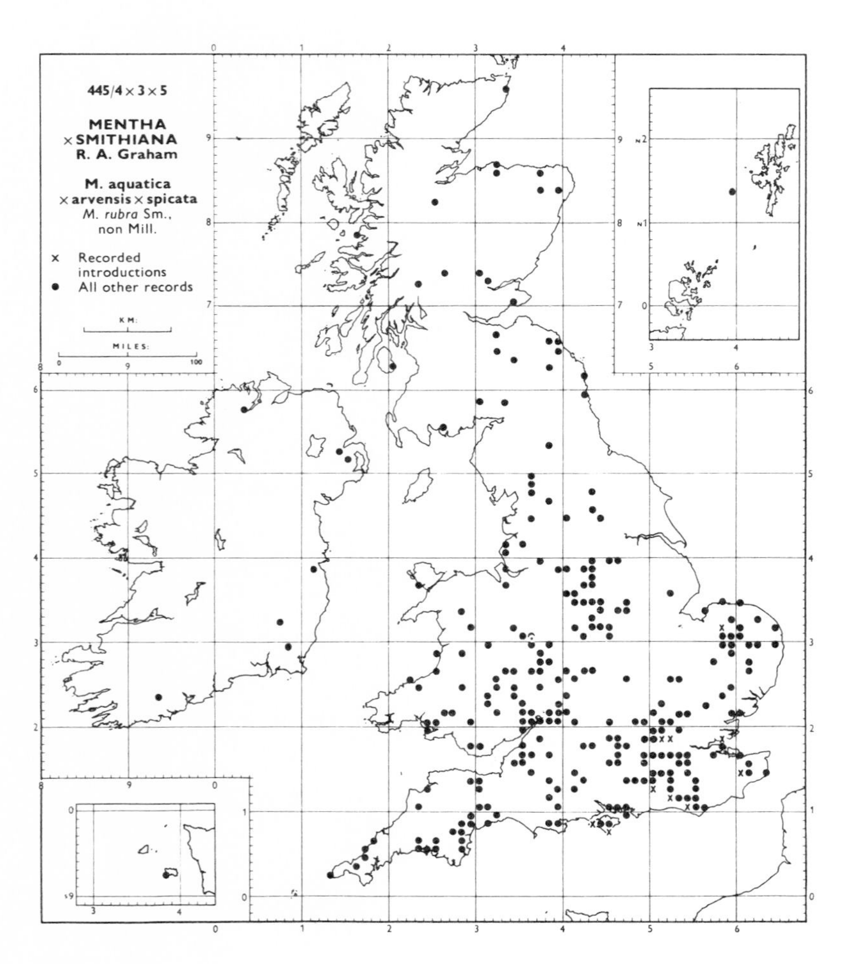

445/4×3 Mentha ×verticillata L.
=M. aquatica×arvensis

DBN, E, NMW, OXF, Graham
This is the commonest hybrid mint in the British Isles and is undoubtedly native. It occurs throughout the range of overlap of the two parents, and occurs locally in the absence of one or both of them. As the hybrid is readily recognizable, the map is based on all records received. Those for v.-cs 68, 79–81, 98, 103 and 111 have not been traced.

445/7×5 Mentha ×niliaca Juss. ex Jacq.
=M. rotundifolia×spicata

There are a number of hybrids which have this parentage. Some are local in their distribution and are presumably represented by a single clone.

var. **alopecuroides** (Hull) Briq.

CGE, MANCH, OXF, Graham
This robust hybrid with oval or suborbicular leaves is widely cultivated. It often escapes and is recorded erroneously as *M. rotundifolia*. The map is based only on herbarium material and the taxon is probably more widespread than this suggests.

var. **nicholsoniana** (Strail) J. Fraser

CGE, Graham
Well established by streams in the west country. For a description of this taxon and details of its distribution, see Fraser (1927).

var. **webberi** J. Fraser

CGE, OXF, Graham
A very local hybrid of stream banks, roadsides and waste places. For a description of this taxon, see Fraser (1934).

445/7 bis Mentha scotica R. A. Graham
M.×niliaca var. *sapida* auct.

CGE, OXF, Graham
Established by streams in eastern Scotland. For a description of this species and details of its distribution, see Graham (1958).

445/4×3×5 Mentha ×smithiana R. A. Graham
=M. aquatica×arvensis×spicata
M. rubra Sm., non Mill.

CGE, E, NMW, OXF, Graham: Floras
Despite its hybrid origin, this taxon has a very distinct distribution, and often occurs in the absence of one parent, *M. spicata*. It probably arose originally as a result of hybridization between *M.×verticillata* and *M. spicata*. Records from v.-cs 31, 49, 53, 70, 74, 90, 91, 98 and 111 have not been localized. The record from H. 1 is an error.

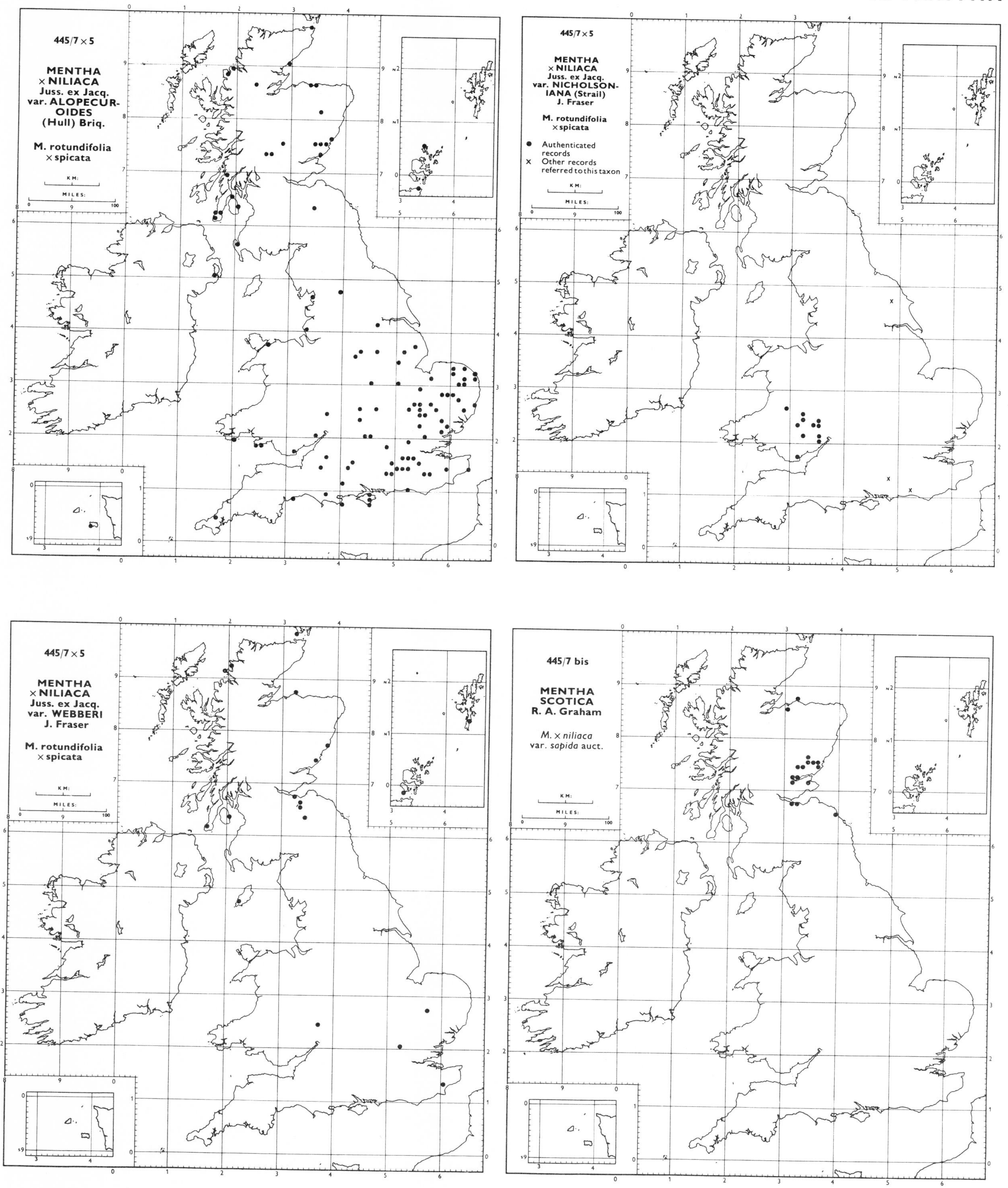
445/7 × 5
MENTHA
× NILIACA
Juss. ex Jacq.
var. ALOPECUROIDES
(Hull) Briq.
M. rotundifolia
× spicata
445/7 × 5
MENTHA
× NILIACA
Juss. ex Jacq.
var. NICHOLSONIANA (Strail)
J. Fraser
M. rotundifolia
× spicata
Authenticated records
Other records referred to this taxon
445/7 × 5
MENTHA
× NILIACA
Juss. ex Jacq.
var. WEBBERI
J. Fraser
M. rotundifolia
× spicata
445/7 bis
MENTHA
SCOTICA
R. A. Graham
M. × niliaca
var. sapida auct.
KM:
MILES:

445/3×7 Mentha ×muellerana F. W. Schultz
=M. arvensis×rotundifolia

CGE, OXF, Graham
A very rare hybrid found only in south Devon (3).

445/4×7 Mentha ×maximilianea F. W. Schultz
=M. aquatica×rotundifolia

CGE, OXF, Graham
This very rare hybrid is known only from Cornwall, Devon and Jersey (1, 3, 4, S). Plants from different localities differ considerably.

445/4×5 Mentha ×piperita L.
=M. aquatica×spicata

var. **piperita**
including *M.*×*dumetorum* Schult.

BM, CGE, E, NMW, OXF, SHY, Graham: Floras
Hairy forms of this hybrid, which have in the past been referred to *M.*×*dumetorum*, have probably arisen spontaneously, whereas the glabrous plants are mainly cultivated forms. The plant frequently escapes from cultivation and occurs in the absence of one or both of the parents. Records from v.-cs 74, 79, 86 and H. 32 have not been localized.

var. **citrata** (Ehrh.) Briq.

CGE, E, O, Graham
This rather distinct variety is always an escape from cultivation. Records from v.-c. 57 and north Wales have not been localized.

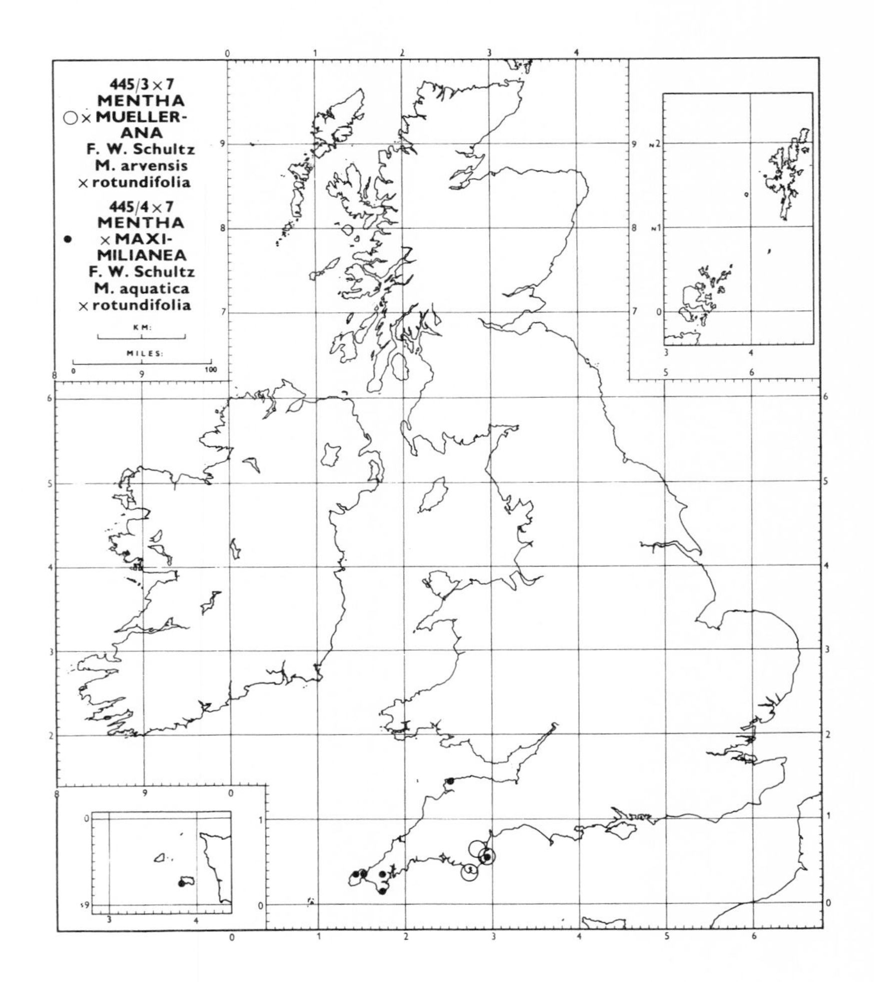

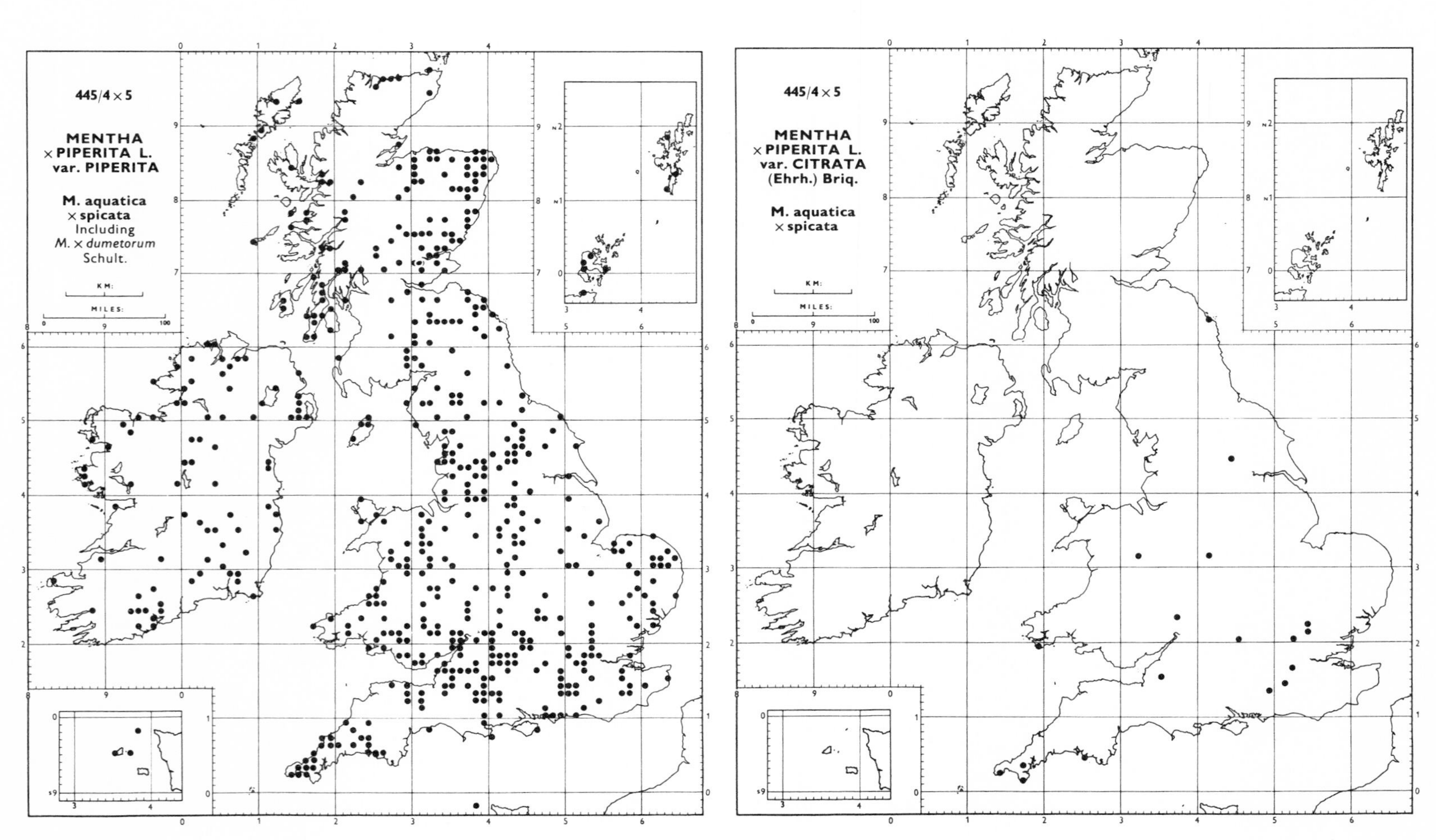

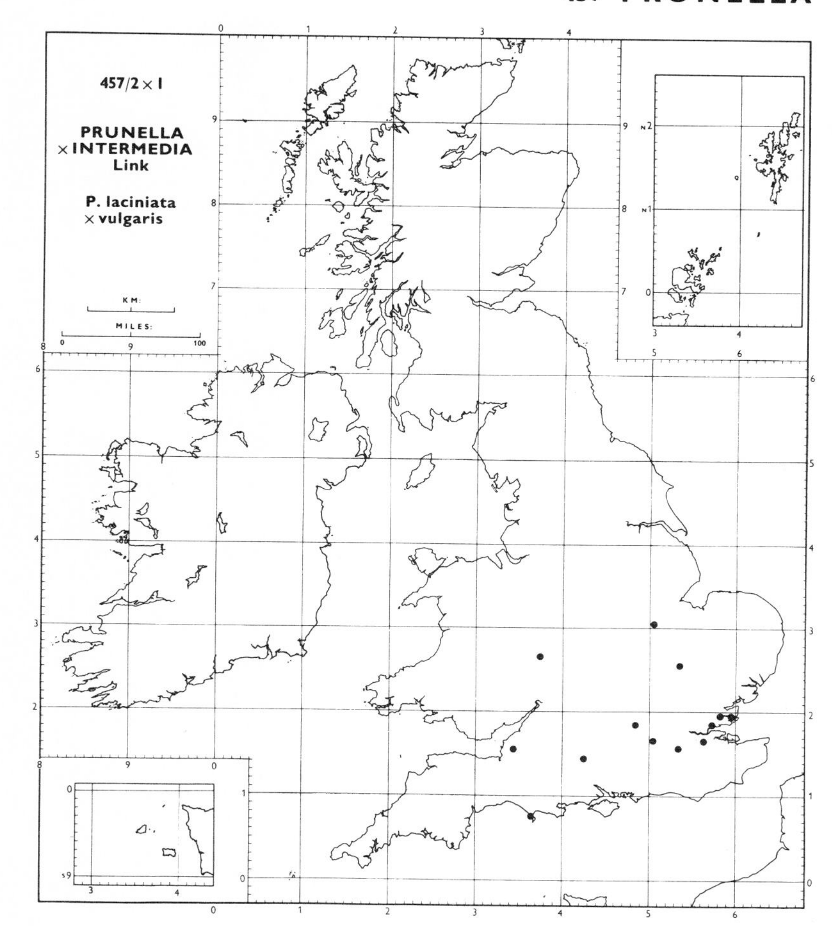

457/2 × 1 Prunella × intermedia Link
= P. laciniata × vulgaris

CGE, Lousley: Floras
As far as is known, this hybrid is only found where the two parents occur together. It is widespread on the Continent of Europe.

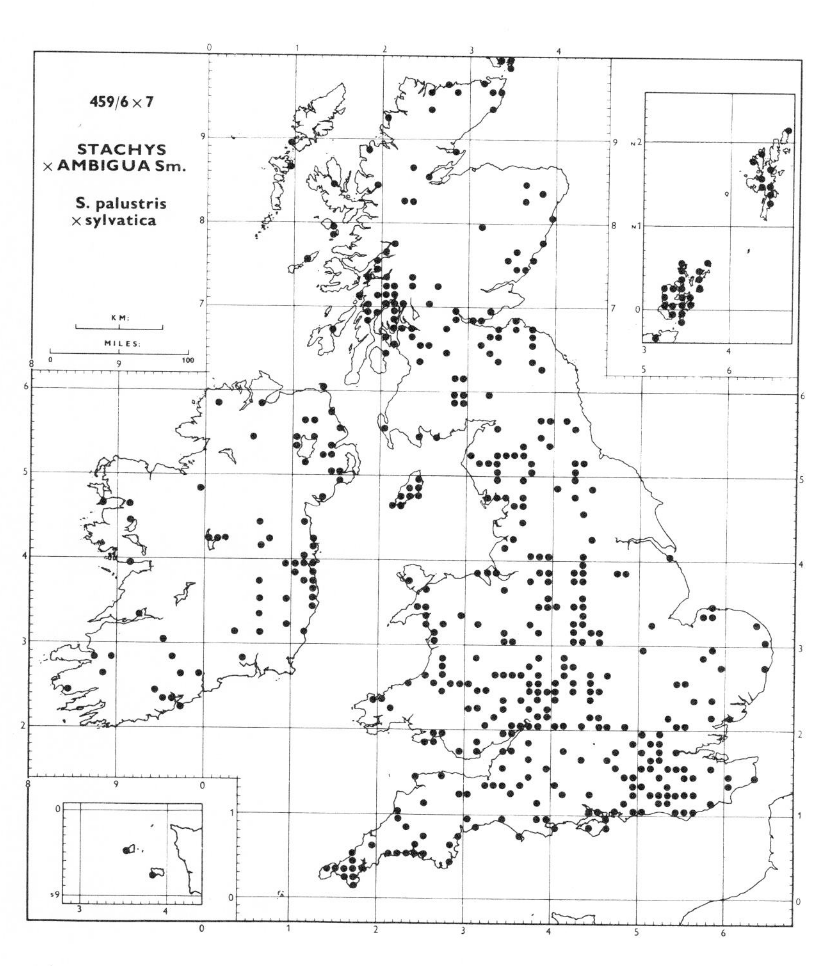

459/6 × 7 Stachys × ambigua Sm.
= S. palustris × sylvatica

BM, CGE, DBN, E, NMW, NWH, OXF, SHY, TCD
This hybrid can be distinguished from its parents by the length/breadth ratio of its leaves. In *S. sylvatica* the ratio is about 3:2 whereas in *S. palustris* it is about 4:1. In the majority of intermediate populations the ratio varies between 2:1 and 3:1. However, there appears to be considerable overlap between *S. palustris* and *S. × ambigua* which suggests that, though the hybrid is said to be sterile, occasional back-crossing to *S. palustris* does occur. There is scarcely any overlap between *S. × ambigua* and *S. sylvatica*, but very occasionally intermediates have been found with length/breadth ratio between 1·5:1 and 2:1: they are confined almost entirely to Scotland.

In parts of western England, the north and west of Scotland, and particularly in the Outer Hebrides, Orkney and Shetland (110–12) *S. × ambigua* occurs in the absence of *S. sylvatica* and often appears to be a relic of cultivation. The apparent abundance of *S. sylvatica* in Orkney in the *Atlas*, p. 247, is misleading; most of these records undoubtedly refer to a broad-leaved form of *S. × ambigua*.

469/1 × 2 Scutellaria × hybrida Strail
= S. galericulata × minor
S. nicholsoni Taub.

BM, CGE, DBN, E, NMW, NWH, OXF, SHY, TCD: Floras
It is noteworthy that this hybrid is confined to southern England and Ireland, as the parent species have a much wider range of overlap in their distribution (see *Atlas*, pp. 251–2). Known also from France.

485/3 × 4 Galium × pomeranicum Retz.
= G. mollugo × verum
G. ochroleucum Wolf ex Schweigg.

CGE, NMW: Floras
This hybrid only occurs within the area of overlap of the two parent species (see *Atlas*, p. 260); but it does not occur *throughout* this area, being rare or absent from most of Ireland and northern Scotland where *G. mollugo* is probably only an introduction. It is most frequent along the south coast of England and on chalk or limestone away from the sea. It is widespread in continental Europe.

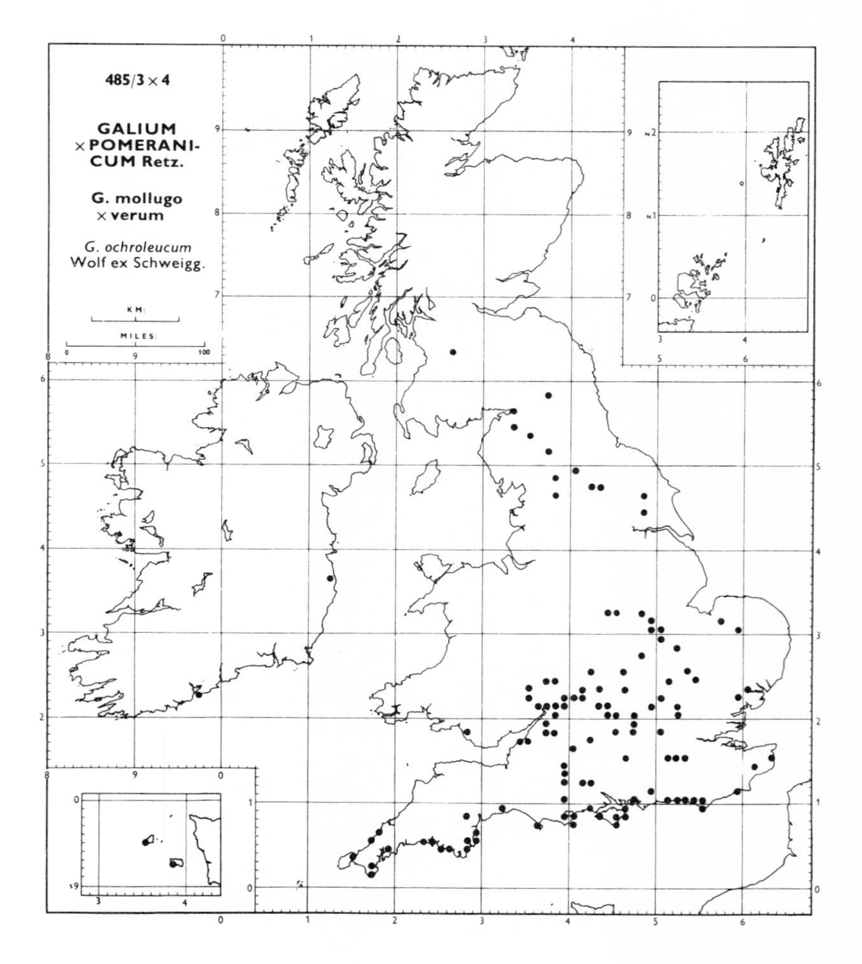

494/1b **Valerianella locusta** (L.) Betcke subsp. **dunensis** (D. E. Allen) P. D. Sell

BM, CGE, DBN, K, NMW.
Attention was first drawn to this taxon by Allen (1961). Subsequent investigations of further material have shown that it has a widespread distribution in the therophyte community of grey dunes. This map is probably very incomplete and many further records can be expected. M. Le Gall mentions a dwarf form, which may be this taxon, as occurring on dunes in Brittany. (Nomenclature, cf. Sell, 1967.)

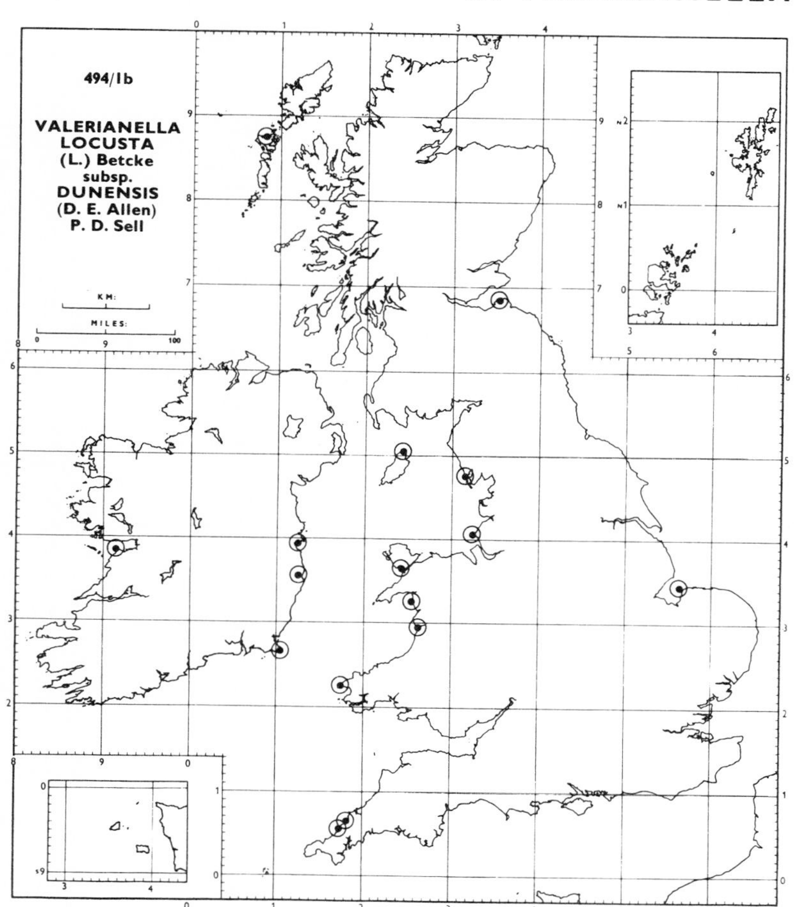

506/8a **Senecio vulgaris** L. (Plants with ray-florets)

BM, CGE, DBN, NMW, NWH, TCD: Floras
This map includes records of all specimens of *Senecio vulgaris* in which ray-florets have developed, including those which appear to be a back-cross with the type.

Two distinct taxa with ray-florets occur in the British Isles. Plants that are simple or branched only above, arachnoid hairy, and with spathulate, sinuate or laciniate, amplexicaul leaves and usually 1–8 capitula with small and often strongly revolute ray-florets, are referable to subsp. **denticulatus** (O. F. Muell.) P. D. Sell. It occurs in natural maritime habitats, such as dunes, cliffs and sandy fields near the sea, on the coasts of west and south-west Britain, the Channel Islands, western France, the east and north Friesian Islands, western and northern Jutland, Lolland, Bornholm, the south and east coasts of Sweden and perhaps Iceland. It retains its characters when grown from seed. Plants from inland localities are only slightly hairy, much more branched, have much more deeply cut leaves and usually larger, spreading ray-florets. They in fact only differ from subsp. *vulgaris* in having ray-florets and should be called subsp. **vulgaris** forma **ligulatus** D. E. Allen.

Both of these taxa have been called *S. vulgaris* var. *radiatus* Koch, which is a superfluous *nomen illegitimum* for *S. vulgaris* var. *denticulatus* (O. F. Muell.) Hyland. For a detailed account of these taxa, see Allen (1967).

On the map definite records of subsp. *denticulatus* have been distinguished. Almost all the other records are certainly forma *ligulatus*. (Nomenclature, cf. Sell, 1967.)

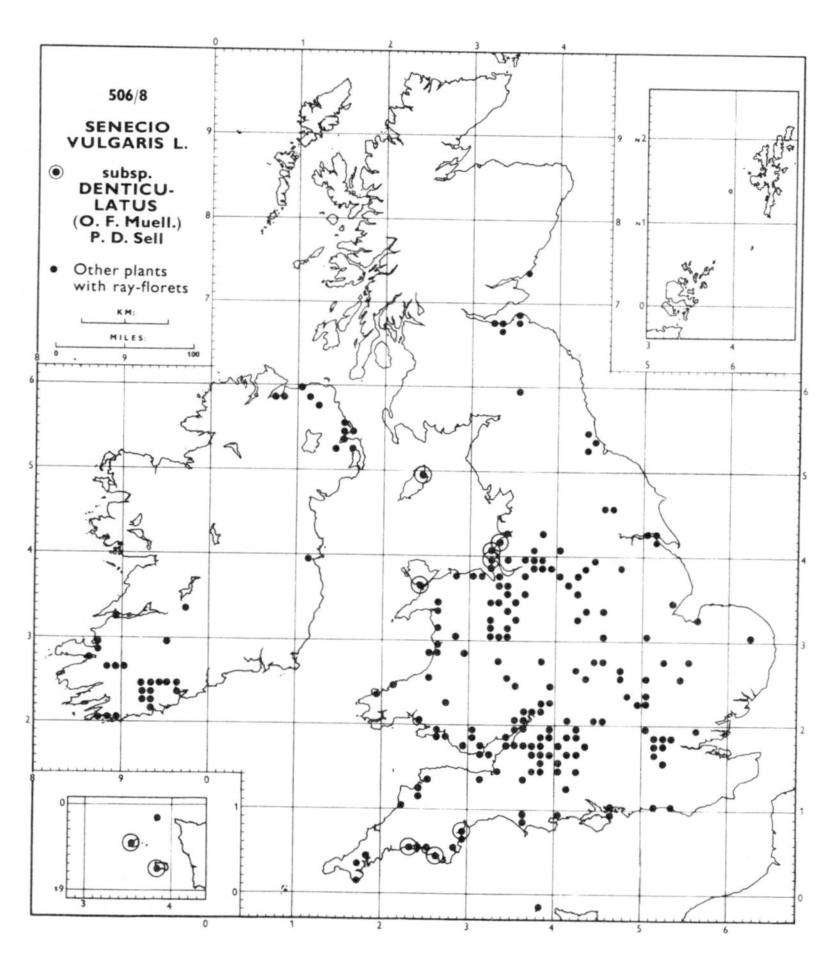

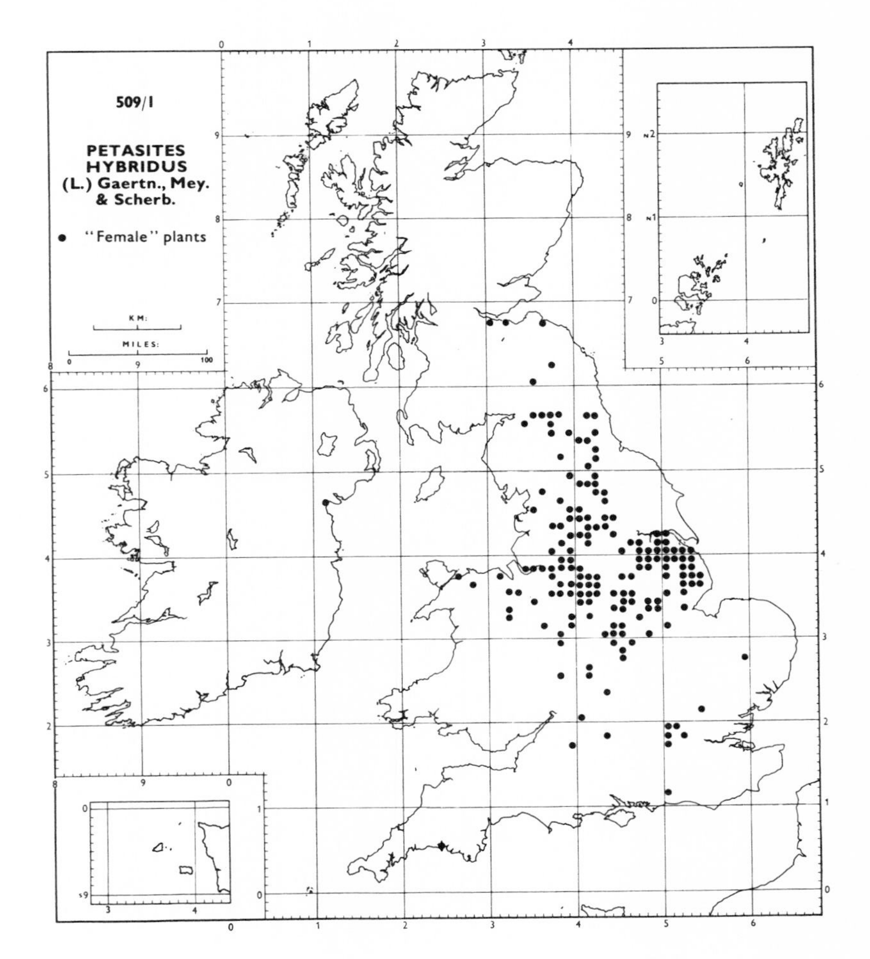

509/1 Petasites hybridus (L.) Gaertn., Mey. & Scherb.
"Female"

BM, BRIST, CGE, DHM, K, MANCH, NMW, OXF: Floras
Information from most of the sources listed above was collected together and published in two papers by D. H. Valentine (1946, 1947). Professor Valentine has also kindly made available MS. notes which have accumulated since the publication of these papers. The "female", in which the inflorescence lengthens after flowering, is so conspicuous that it is unlikely that it has been overlooked, and the map is therefore almost complete.

It is clear that in the midland counties (v.-cs 54–60 and 63–4) it is generally much more abundant than the "male" and frequently occurs without it.

A record from v.-c. 72 has not been traced: those from v.-cs 77 and 92 were probably errors.

Apparently in central Europe the "female" occupies wetter portions of the habitat when growing together with the "male".

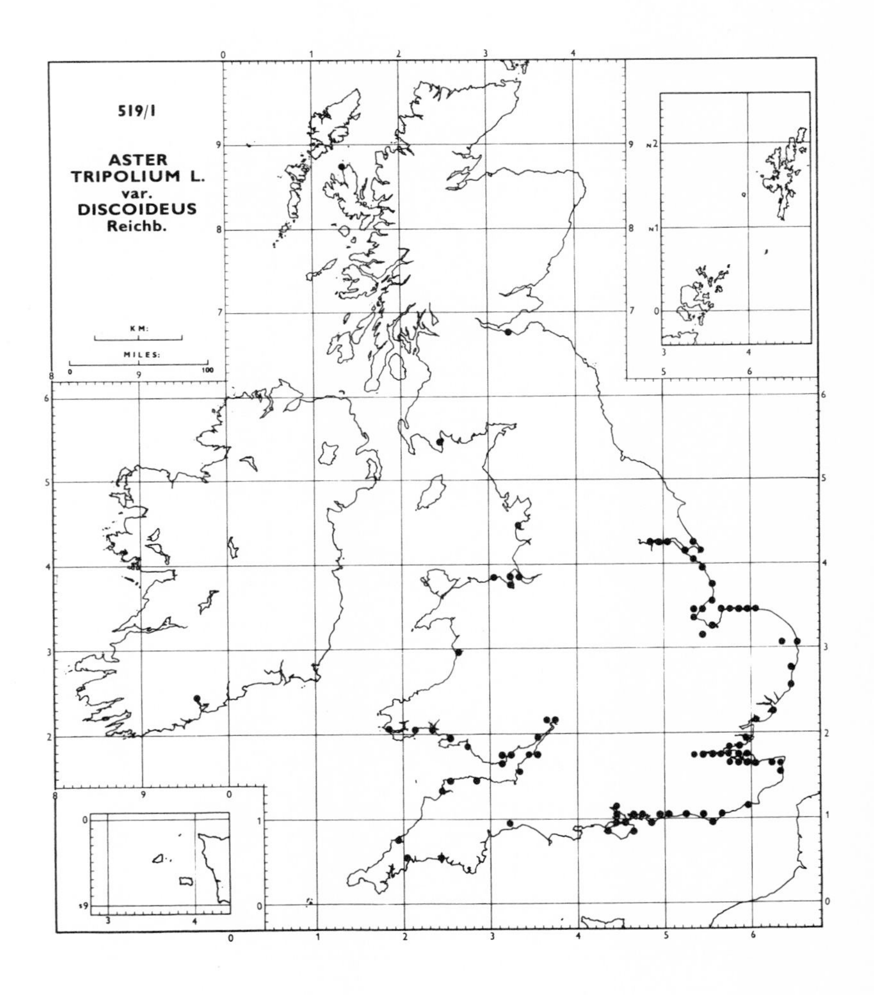

519/1 Aster tripolium L.
var. **discoideus** Reichb.

BM, CGE, NMW: Floras
In addition to the sources listed above many records were collected by A. J. Gray, nevertheless more localities are probable along the east coast of England. As far as is known this variety does not occur in the absence of the type, which is found all round the coasts of the British Isles and is occasional inland.

538/2 & 4 **Arctium minus** Bernh.
subsp. **minus**
subsp. **nemorosum** (Lejeune) Syme
A. nemorosum Lejeune; *A. vulgare* auct.

BFT, CGE, DBN, E, NMW, SHY, TCD
These two maps are a simplification of a very complex situation. In the north of Ireland and in Scotland most plants of *Arctium* are similar in appearance. They have single upright stems with long arching branches; the fruiting capitula when mature exceed 3·5 cm. in overall width, and are sessile and clustered towards the end of the branches; and the petals of the open flowers do not protrude beyond the involucral bracts. These are referred to subsp. *nemorosum*.

In south-west England most plants of *Arctium* are bushy in appearance with many stems and branches; the mature fruiting capitula are less than 2·5 cm. in overall width and are borne on short stalks well distributed along the length of the stem; and the petals of the open flowers protrude beyond the involucral bracts. These are referred to subsp. *minus*.

Elsewhere in the British Isles there occurs a variable taxon which in the midlands is the commonest *Arctium*. It includes the type specimen of *A. pubens* Bab. (*A. minus* subsp. *pubens* (Bab.) J. Arènes). Twenty populations of *Arctium* were grown from fruits collected wild and studied by the editor and P. D. Sell at Cambridge. Their study showed that each population was uniform, but differed from every other population. It was discovered that the genus was autogamous, fertilization probably nearly always occurring before the flowers opened. It seems likely that outbreeding occurs occasionally to produce fertile variants which by constant inbreeding produce a great variety of almost pure lines. It has not proved possible to map *A. pubens*.

The situation is further complicated by the possibility that *A. lappa* L. also hybridizes with the two subspecies of *A. minus*.

These maps are based on material seen by the editor, or upon data on capitulum width measurements of mature specimens, which were sent to him during the autumns of 1959 and 1960. Although incomplete, the maps do give, in outline, the main differences between the distribution patterns of these two taxa in the British Isles. They are widespread on the Continent, but their detailed distribution is not well known.

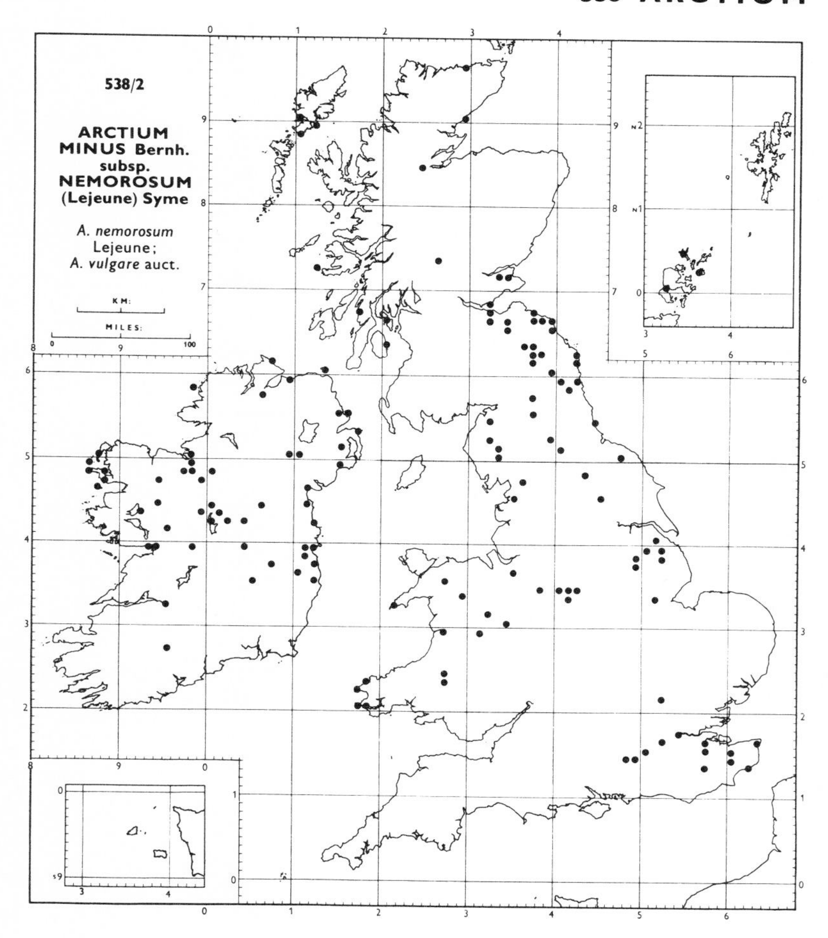

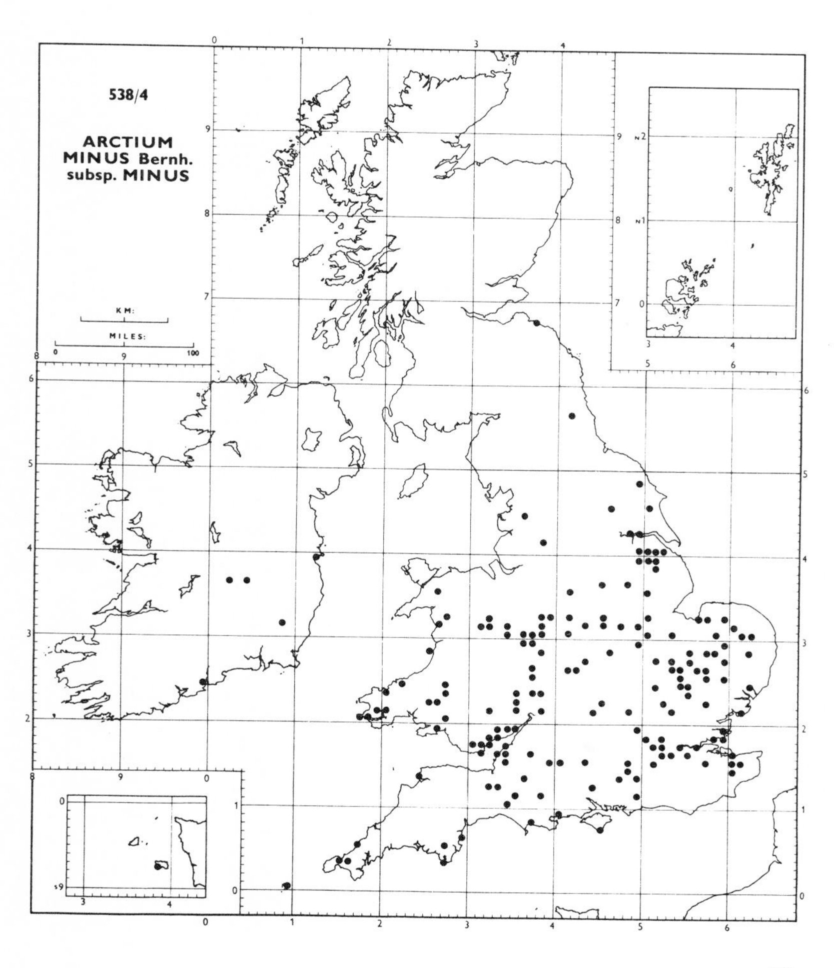

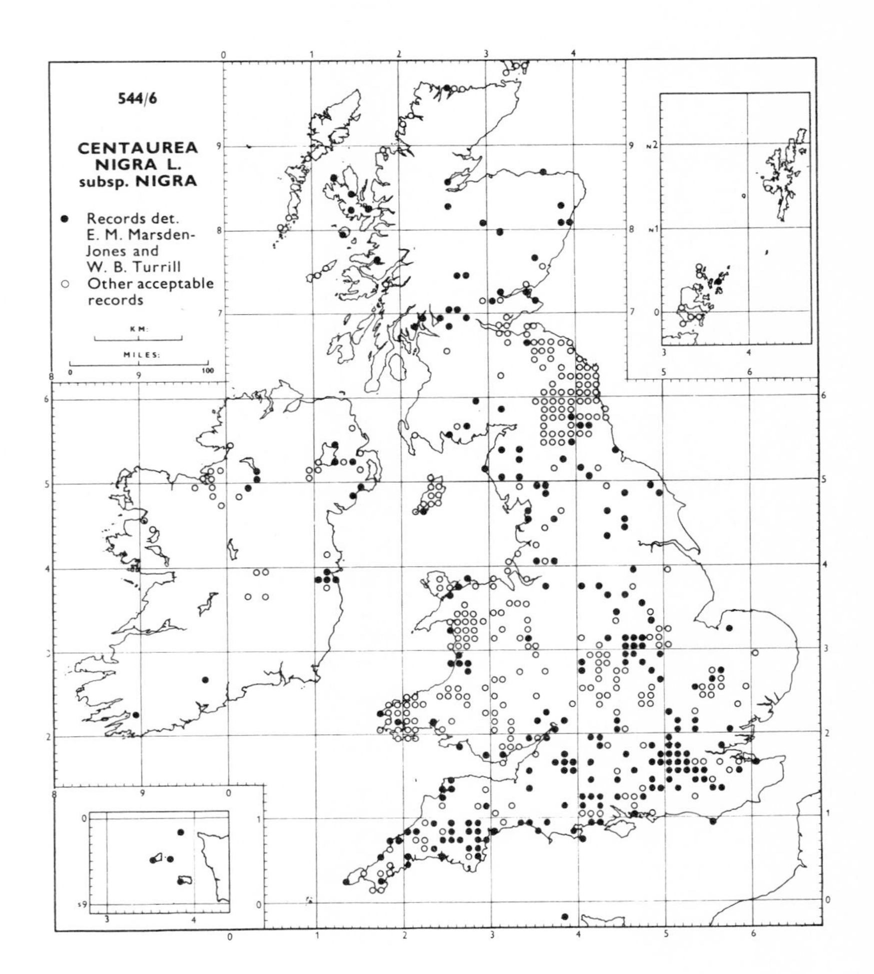

544/6 Centaurea nigra L.
subsp. **nigra**
544/7 subsp. **nemoralis** (Jord.) Gugler

The taxonomy of this group is that by A. R. Clapham in Clapham, Tutin and Warburg, 1962. The three maps are based mainly on the work of Marsden-Jones and Turrill (1954) who examined material from all parts of the British Isles and made statements about the presence or absence of the two taxa in numerous counties which they had visited. With this as a guide it has been possible to accept many of the field records received.

The map of subsp. *nemoralis* is probably reasonably accurate, and shows the concentration of this taxon on the chalk and limestone of southern England. The map of subsp. *nigra* is very incomplete. It seems almost certain that this taxon occurs throughout Scotland and Ireland as indicated by the map of the species in the *Atlas*, p. 291. The map of the intermediates is included to show that they occur outside the present range of subsp. *nemoralis*, notably in southern Ireland. In south-east England both subspecies form hybrid swarms with *C. jacea* L. Both occur in western Europe.

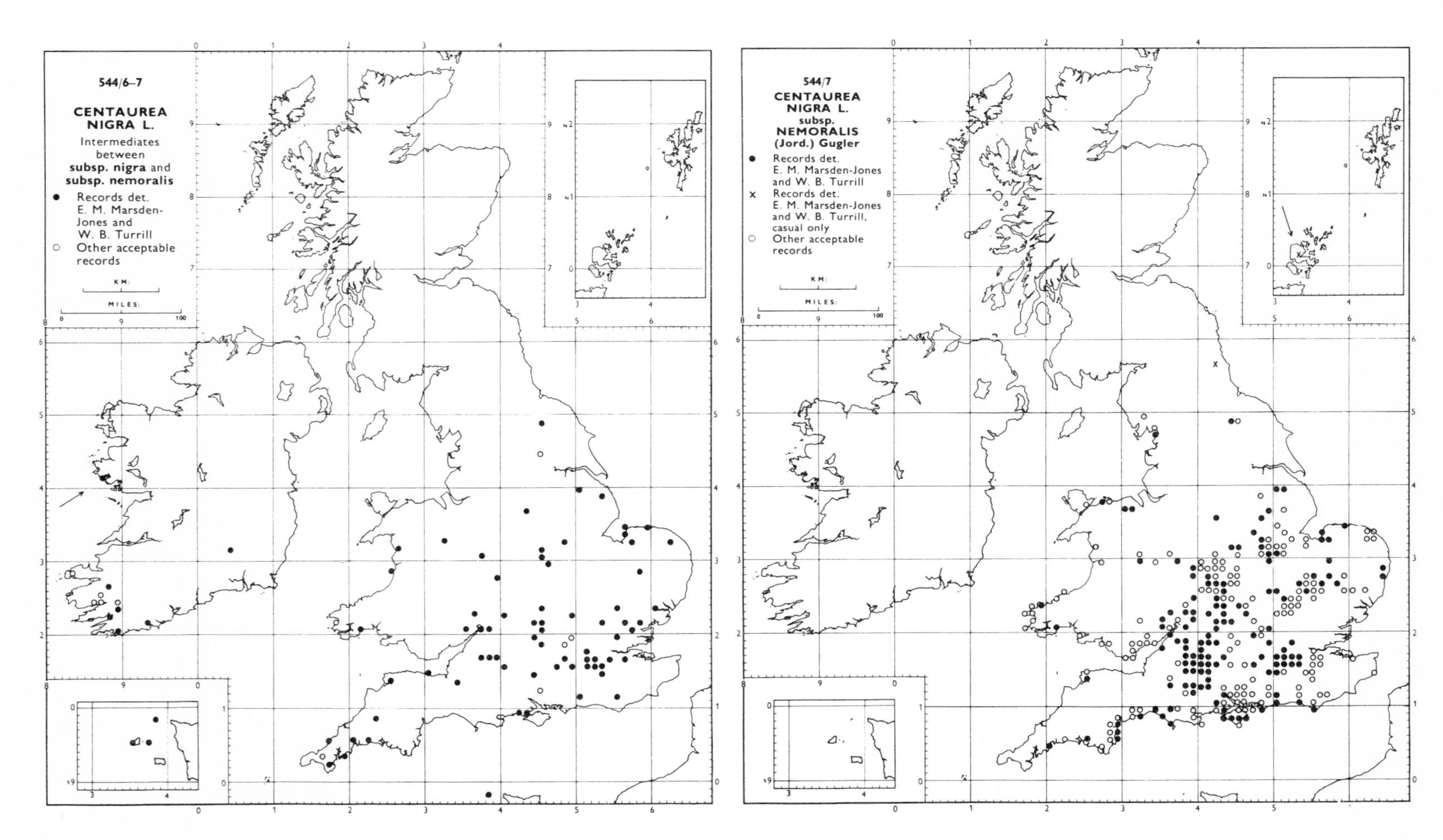

558/1 HIERACIUM L.*

The Genus *Hieracium* is one of the largest and most difficult in the British flora, and in the whole history of British botany only five men, J. Backhouse Jun. (1856), F. N. Williams (1902), F. J. Hanbury (1904), W. R. Linton (1905), and H. W. Pugsley (1948), have attempted to treat it critically as a whole, although a few others such as E. F. Linton, A. Ley and E. S. Marshall have made important contributions to our knowledge of the group. It is true that G. C. Druce (1928) enumerated many species in his *British Plant List*, and that J. Roffey (1925) provided a similar list in the 11th edition of *The London Catalogue of British Plants*, but both lists are compilations made from the work of earlier hieraciologists, plus more recent contributions from the continental botanists H. Dahlstedt and K. H. Zahn. The new species of Backhouse have all stood the test of time, but his records for them were few in number. Hanbury, W. R. Linton, E. F. Linton, Ley and Marshall made large collections, but their interpretation of species was rather wide, with the result that they described a large number of varieties. The account by F. N. Williams was apparently based on herbarium material, but we have come across no specimens annotated by him. The determinations by G. C. Druce and J. Roffey are far below the standard of those of the other authors mentioned above, and are on the whole unreliable. The revision of the genus by Pugsley is an excellent one which reduced chaos to order, but he was working during the difficult period of the Second World War and his account was consequently based mainly on material in the herbarium of the British Museum. Unlike Hanbury, E. F. and W. R. Linton, Marshall and Ley, who appear to have spent the best part of every summer in the field, Pugsley could not have seen many of the species in a fresh condition at the time he was describing them. For this reason there is much that is artificial in his account, which, in this respect, falls below the high standard of his important monographic works on *Euphrasia* and *Fumaria*; genera he knew in the field as well as in the herbarium. With the exception of Williams, all were amateurs who did their botanical work in their spare time. In marked contrast, on the Continent, the professional botanists C. J. M. Arvet-Touvet, H. G. A. Dahlstedt, K. Johansson, G. Samuelsson, S. O. F. Omang and K. H. Zahn spent much of their lives working on the *Hieracium* flora. To them we owe what little is known of the relationships of the British species to those of continental Europe. Since the publication of Pugsley's *Prodromus* we have described a number of species new to the British Isles (Sell and West 1955, 1962, 1965; and Sell, 1967) and in 1958 (Dandy, 1958) published a list of all the species then known to us.

To prepare maps of all the British species of *Hieracium* from all the published records would be unrealistic, because too many species have been interpreted in different ways by different authors. It was also thought advisable to give only those records which could be checked. Thus every dot on the maps is represented by one or more herbarium specimens. The main source of the records is some 50,000 specimens determined by us between 1950 and 1965. To these have been added records made by H. W. Pugsley (1948) in his *A Prodromus of the British Hieracia* and records from specimens, mainly of the more common species, determined by C. E. A. Andrews and J. N. Mills. Specimens supporting Pugsley's records are usually to be found in the British Museum. The very large collection at the University of Cambridge (CGE), containing the individual herbaria of A. Ley, E. S. Marshall, P. D. Sell and C. West, represents every species and most areas of distribution of the species themselves. Other important herbaria consulted, in part or in full are:

BEL, BM (includes the individual collections of F. J. Hanbury, E. F. Linton, H. W. Pugsley, J. Backhouse, J. Roffey, J. B. Syme), DBN, E, HWB (J. C. Melvill collection), K, LIVU (includes the W. R. Linton collection), MACO (includes the M. L. Wedgwood collection), OXF (includes the G. C. Druce collection), SHD, SLBI (includes the W. H. Beeby collection), TCD.

Many other private collections, too numerous to mention, have also been examined. We are most grateful to all those who sent us material, much of which was collected specially for this work. Special mention must be made, however, of the very important contributions made by A. G. Kenneth and A. McG. Stirling to our knowledge of the distribution of the species occurring in Scotland.

As far as is known, the species of the genus *Hieracium* as here defined (*Pilosella* being accepted as a distinct genus) are apomictic, with the exception of some forms of *H. umbellatum*. Species could have originated in one of at least three ways. First, they could have originated in the past, when some of the species were sexual, the variants resulting from hybridization reproducing apomictically, thus creating new stable species. In this way very diverse species could be created at the same time. Secondly, if a species which was once widespread died out in some areas and left a disjunct distribution, geographical isolation might lead to modifications in its characters which would be perpetuated by the mode of reproduction. One would assume such species to be very similar in characteristics, and to be species only because they reproduce apomictically. In sexual species such geographical variants, at least zoologically (and increasingly so botanically), are regarded as subspecies. In such widespread species as *H. vulgatum* geographical variation also occurs, but when the distribution is continuous it is difficult to provide distinguishing characters. Thirdly, every so often the offspring of a plant includes an individual that is aberrant in a few minor details. Once again the mode of reproduction allows this plant to reproduce itself, and if it should spread a new species would be formed. With native plants this may not happen very often. We have found single aberrant plants in populations, but they never seem to be numerous. With introduced plants of the Section *Vulgata*, which are common on roadsides and railway banks, it is different. Here there are a large number of variants, some widespread, some local; many of these are not included in this account as we have no names for them. N. Hylander (1943) reported a similar situation in Sweden, and described a large number of closely allied species. Some of these may be the same as those which occur in the British Isles. It is also possible that identical or very similar plants can arise from the same or an allied parent in widely separated areas.

We have assumed, perhaps quite wrongly, that the most widespread native species are the oldest and that closely related local species, especially if they grow with or replace them, have been derived from them. Our classification of the species within their major groups bears this in mind, and we have placed together species which we believe are likely to have had the same ancestor. Pugsley's classification was based entirely on one or two selected characters, with the result that species with the same probable ancestor were sometimes placed a long way apart.

Hieracium species are distributed throughout most of the British Isles, but they are much more common in some districts than in others. They are abundant on cliffs and ungrazed rocky areas throughout much of Wales, the Derbyshire and Staffordshire dales, the Yorkshire limestone, the English Lake District, and central and northern Scotland. Wherever there are large flocks of sheep,

* By P. D. Sell and C. West.

however, as in the Scottish lowlands and Shetland (112), they are usually restricted to cliffs and slopes inaccessible to these animals.

Between 1880 and 1920 the British hieraciologists E. F. and W. R. Linton, E. S. Marshall and A. Ley collected many thousands of specimens yearly, and there is some evidence that in those days *Hieracia* were more common in some areas than they are now. In the English lowlands *Hieracia* are usually only abundant on railway and roadside banks where they have been recently introduced. Certain species are probably native in open woodland and on sandy heaths, but some of these have probably been introduced. In Ireland, with the exception of the limestone areas of Burren and the north-east, where they are frequent, *Hieracia* are mainly restricted to river-sides, and cliffs in the coastal counties.

The distribution in Fennoscandia of some of our species can be seen in the work by G. Samuelsson (1954).

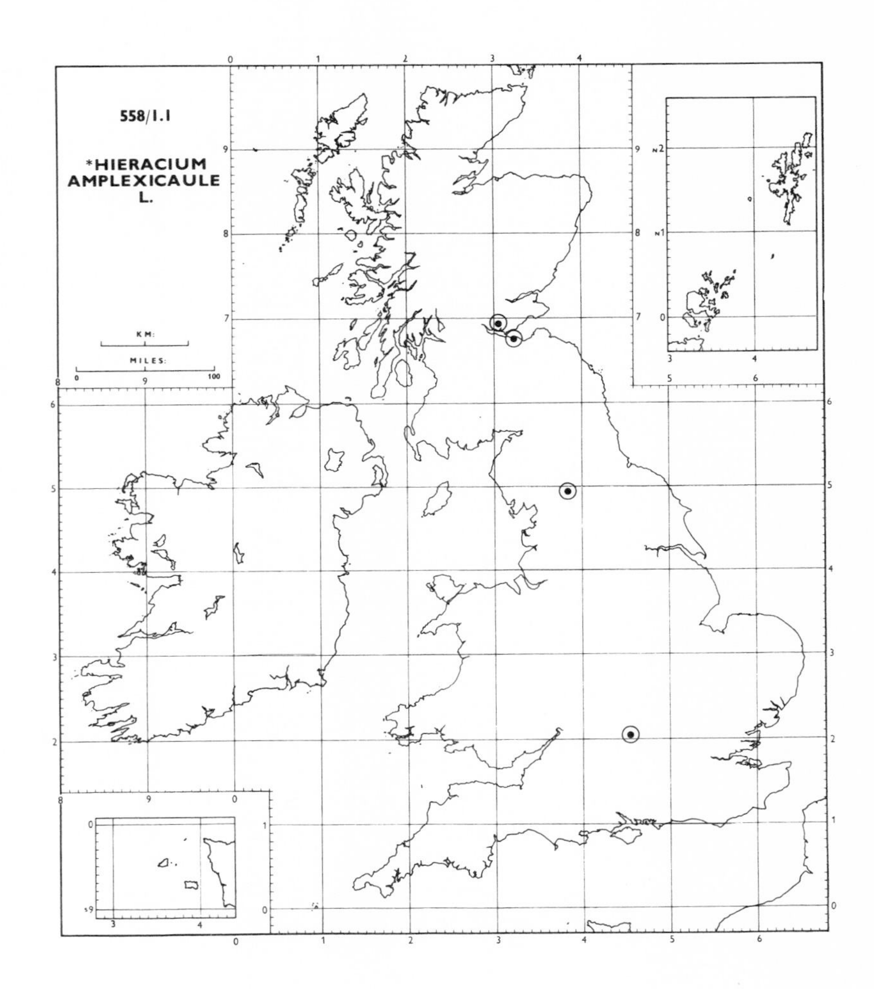

558/1.1–3 Section AMPLEXICAULIA Zahn

All three species, **H. amplexicaule** L., **H. pulmonarioides** Vill. and **H. speluncarum** Arv.-Touv. are natives of central Europe and the Pyrenees; they are naturalized in a few scattered localities in Great Britain, mainly on old walls. They show no signs of spreading from the localities where they were originally introduced.

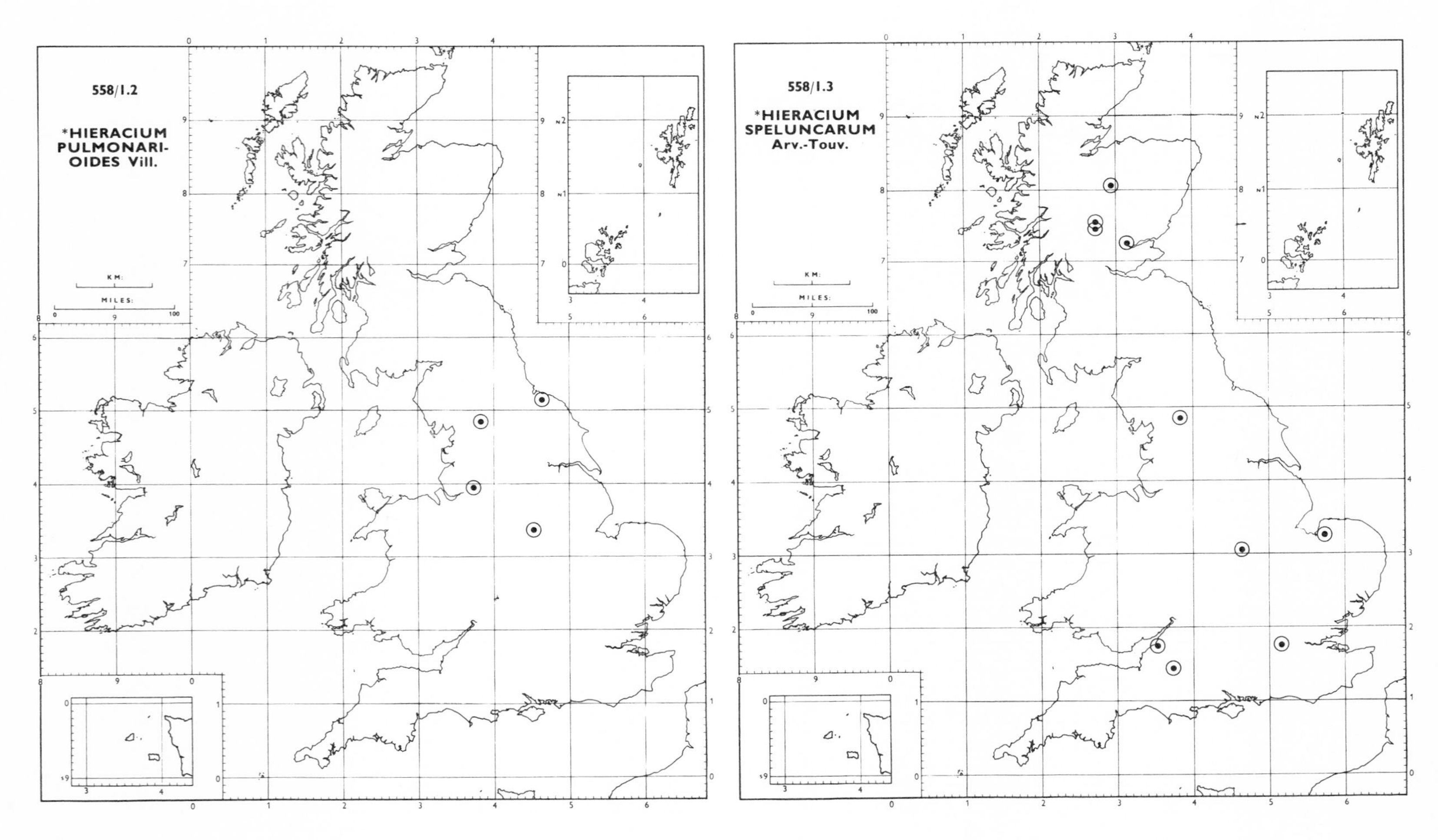

558/1.4–21 Section **ALPINA** F. N. Williams

This section, which is a natural, easily-defined group, is represented by 18 native species. They are plants of alpine rock ledges and bare, stony or grassy slopes, and screes on granite, mica-schist and slate, mostly at an altitude of over 2,000 ft (650 m.). The section inhabits arctic and subarctic Europe, Greenland and Asia, as well as the Alps of central Europe, and in the British Isles is mostly confined to mountainous regions in Scotland. Two species, however, are found in the English Lake District and one in north Wales. The flowering period of this section is usually in July and August. We are indebted to B. A. Miles for help with this section.

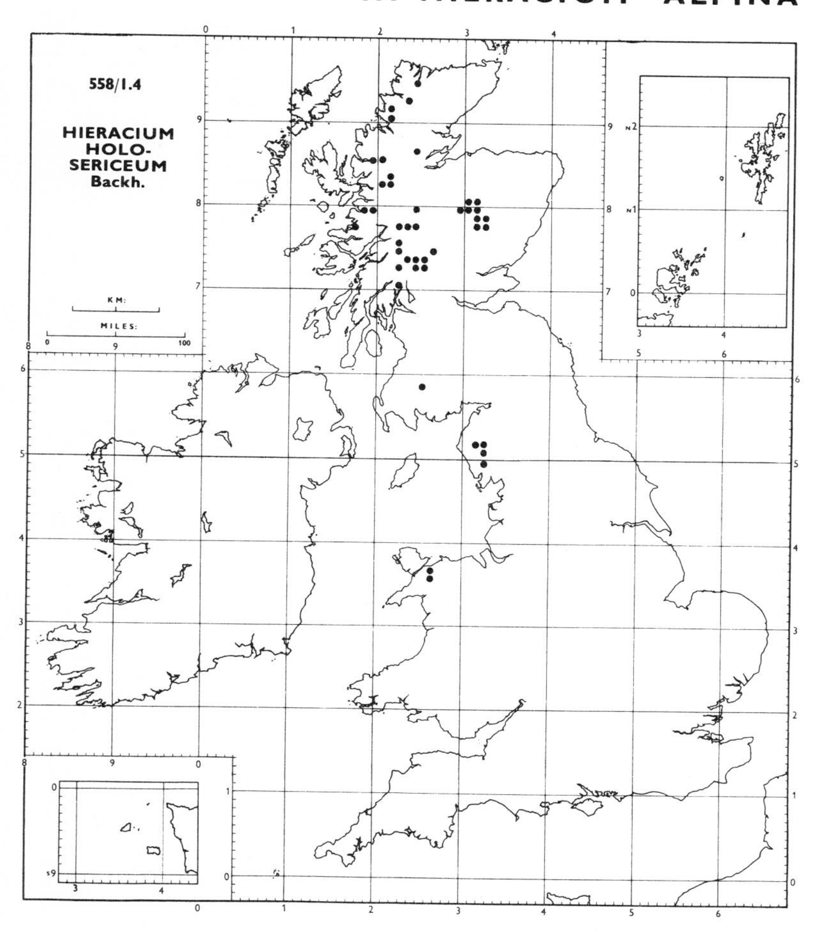

H. holosericeum Backh. is frequent in the Scottish Highlands, and occurs locally in Kirkcudbright (73), the English Lake District (69, 70) and north Wales. On the Continent it is found in the mountains of Scandinavia and central Europe. The closely allied **H. alpinum** L. is a much rarer species in Great Britain. Its head-quarters is the area occupied by the granitic mountains of the Cairngorm range and the mica-schist mountains of Angus (90). Isolated localities occur to the west in West Ross (105) and Inverness (97), and as far south as Ben Vorlich, Dunbarton (99). It has a wide distribution abroad, from Greenland and Iceland to Scandinavia, and in central Europe east to the Tatra Mountains. Both *H. holosericeum* and *H. alpinum* are clearly ancient, well-defined species. **H. alpinum** var. **insigne** Bab. is a rare plant recorded from Glen Callater (92) and Glen Feshie (96) which will probably have to be recognized as a distinct species.

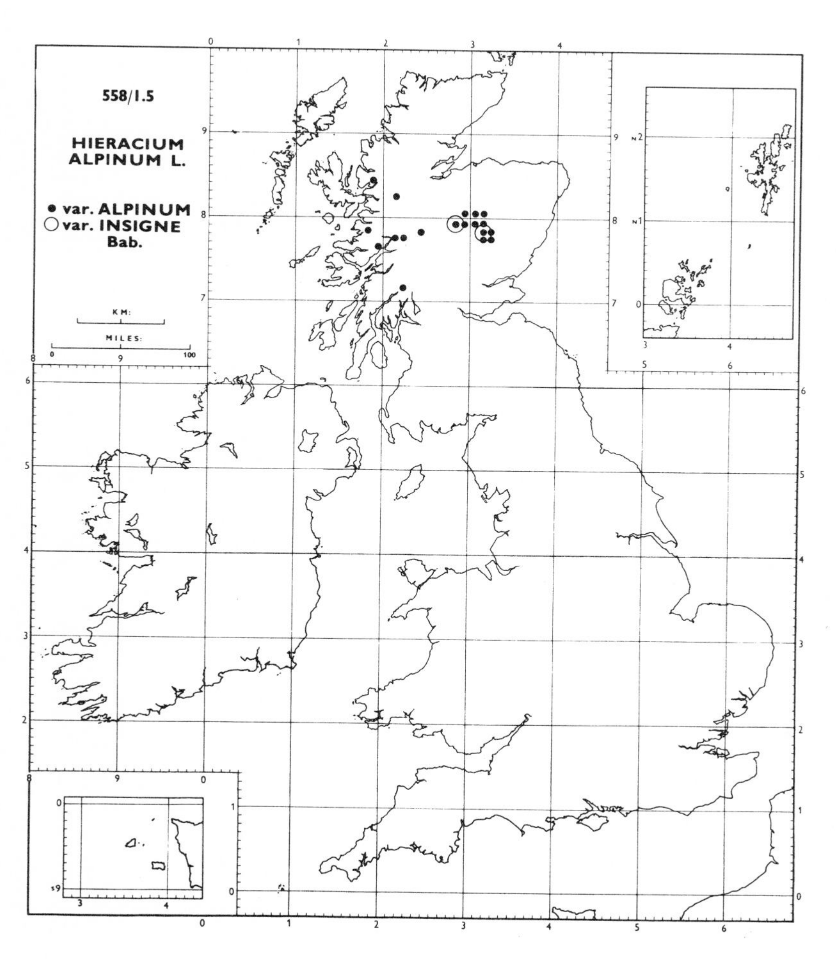

H. tenuifrons P. D. Sell & C. West (*H. gracilentum* Backh. (1856), non (Fries) Backh. (1853)) is closely allied to *H. alpinum* and in Great Britain it has a very similar distribution. The eastern populations differ from the western ones in several characters, and it is possible the eastern one will have to be described as a distinct species. On the Continent *H. tenuifrons* seems to occur in Norway. **H. pseudopetiolatum** (Zahn) Roffey is a very rare endemic confined to the Cairngorms. **H. subgracilentipes** (Zahn) Roffey is rather similar in appearance to *H. tenuifrons*. It is, however, endemic to the English Lake District (69, 70), and therefore geographically isolated from that species.

H. eximium Backh. is locally abundant in the central Scottish Highlands and has been recorded from Norway and Germany. It is closely allied to *H. halleri* Vill., a variable and widespread species in continental Europe. Dark-styled (var. *eximium*) and yellow-styled (var. *tenellum* Backh.) variants of *H. eximium* occur, but this slight difference does not seem to have any particular significance, although Backhouse and Hanbury say that the dark-styled type is commoner on the mica-slate and the yellow-styled more abundant on the granite. **H. notabile** P. D. Sell & C. West, known only from Ben Lawers and Ben More (88), is closely allied to *H. eximium*, and may well have been derived from it. **H. backhousei** F. J. Hanb., endemic to the Dubh Loch (92), appears at first sight to be a very distinct species, but **H. memorabile** P. D. Sell & C. West, which is endemic to Inverness (96, 97) and Aberdeen (92), is intermediate in characters between it and *H. eximium*. **H. marginatum** P. D. Sell & C. West (*H. globosiflorum* var. *lancifolium* Pugsl.), endemic to one mountain in West Ross (105) and two in West Sutherland, is closely allied to both *H. eximium* and *H. memorabile*.

H. grovesii Pugsl. is a very distinct endemic species which appears to have no close allies in the British Isles. It is confined to three localities in the Cairngorms and one in Argyll (98). The specimens from Argyll differ considerably from those of the Caingorms, and there is little doubt that they should be treated as a distinct species.

H. globosiflorum Pugsl. is a rather local plant of central and north Scotland, and is recorded for Norway and Lapland. The plants of the west and north have dark styles and a slightly different appearance and may have to be separated as a distinct species. Var. **larigense** Pugsl., which grows with var. **globosiflorum** in the Lairig Ghru (96), will also probably have to be treated as a distinct species. *H. globosiflorum* has no close allies in the British Isles, but it has certain characters in common with **H. graniticola** W. R. Linton, which is endemic to the Cairngorms.

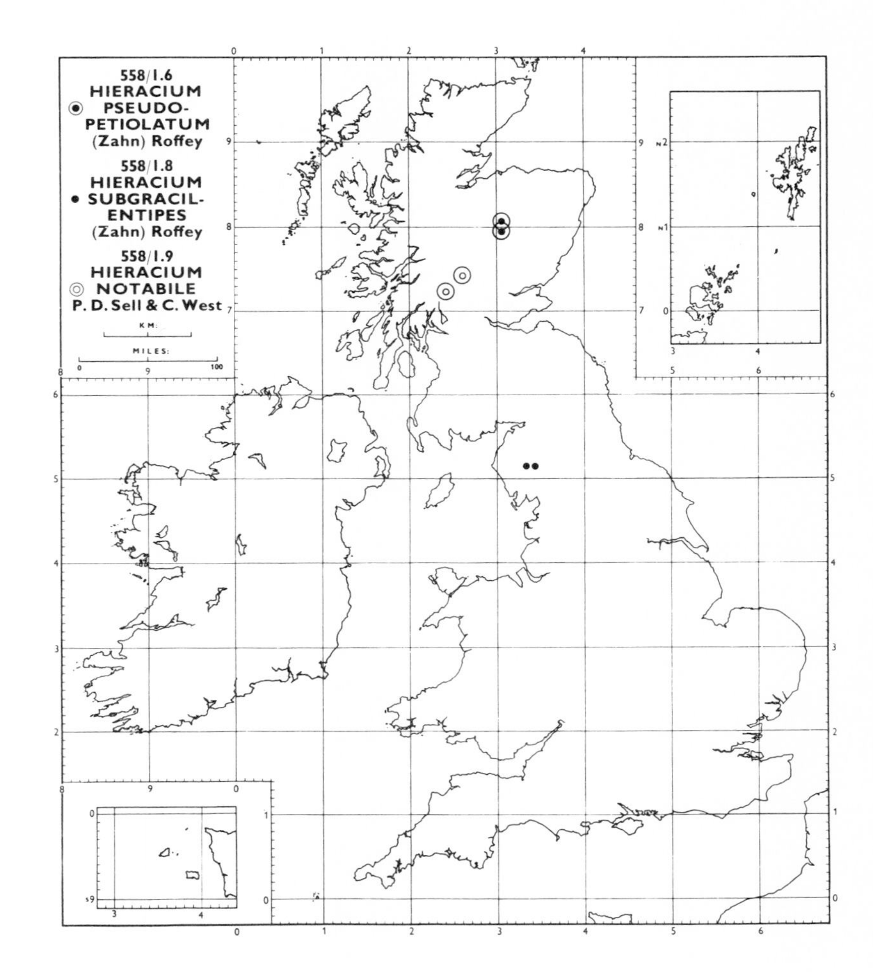

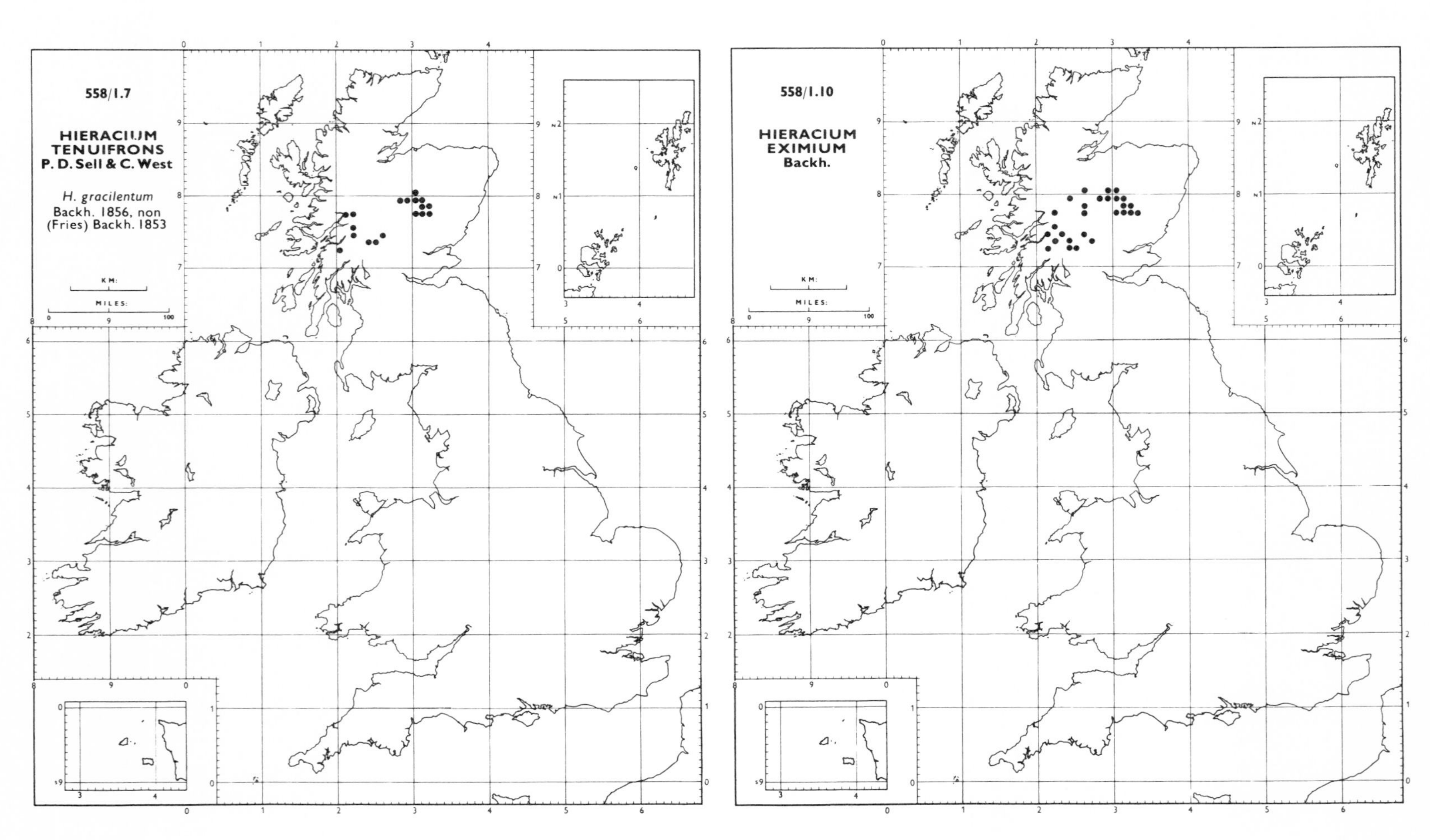

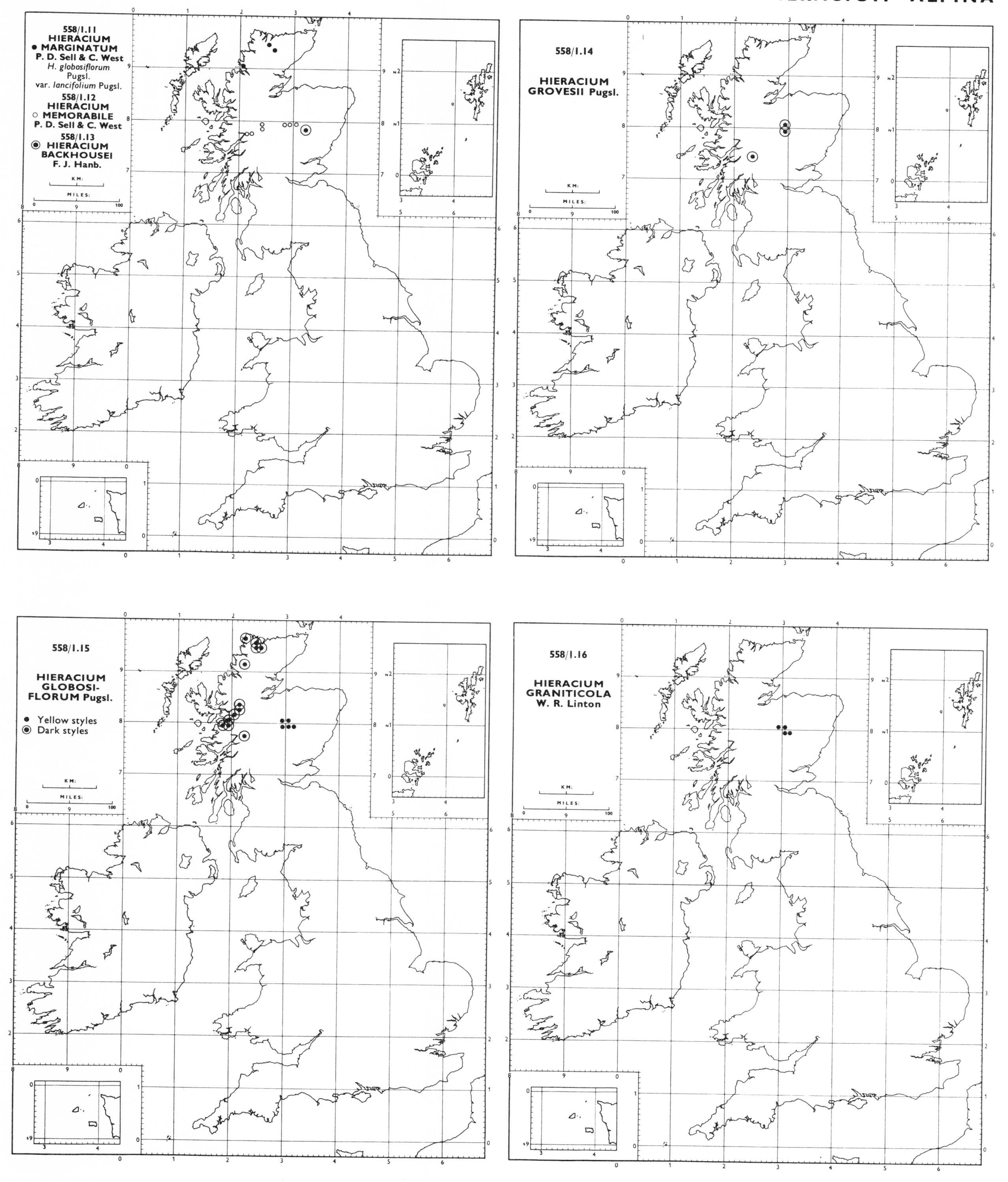
558/1.11
HIERACIUM
• MARGINATUM
P. D. Sell & C. West
H. globosiflorum
Pugsl.
var. lancifolium Pugsl.
558/1.12
HIERACIUM
○ MEMORABILE
P. D. Sell & C. West
558/1.13
◉ HIERACIUM
BACKHOUSEI
F. J. Hanb.
558/1.14
HIERACIUM
GROVESII Pugsl.
558/1.15
HIERACIUM
GLOBOSI-
FLORUM Pugsl.
• Yellow styles
◉ Dark styles
558/1.16
HIERACIUM
GRANITICOLA
W. R. Linton
KM:
MILES:

H. calenduliflorum Backh. and **H. pseudocurvatum** (Zahn) Pugsl. are both scattered over the mountains of central Scotland, and are closely allied to each other. In both species the western forms can be distinguished from the eastern and they may have to be recognized as distinct species. **H. macrocarpum** Pugsl. is a very rare endemic of the Cairngorms.

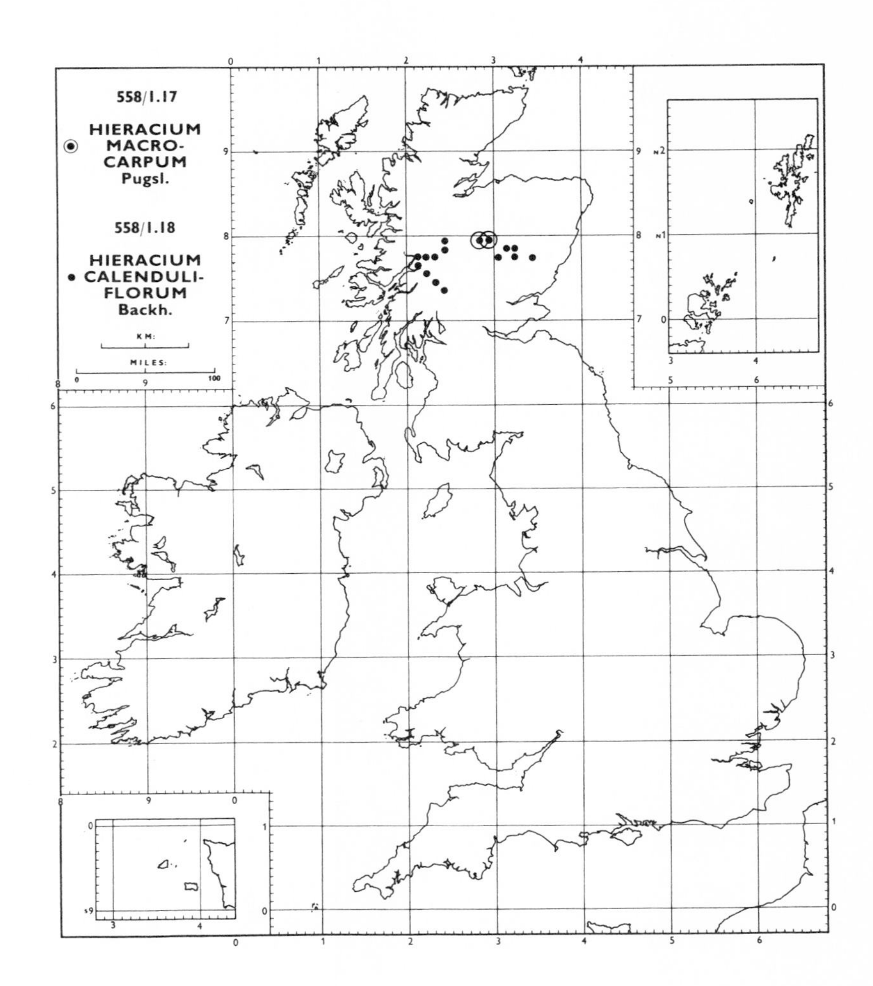

H. hanburyi Pugsl., endemic to central Scotland, is not uncommon in many of the higher alpine glens. It is a most variable species, but the variants appear to have neither geographical nor ecological significance, more than one variant often being found in the same population. **H. atraticeps** (Pugsl.) P. D. Sell & C. West (*H. hanburyi* var. *atraticeps* Pugsl.), which is very closely allied to *H. hanburyi*, occurs in a few scattered localities within the range of the latter, and is almost certainly derived from it.

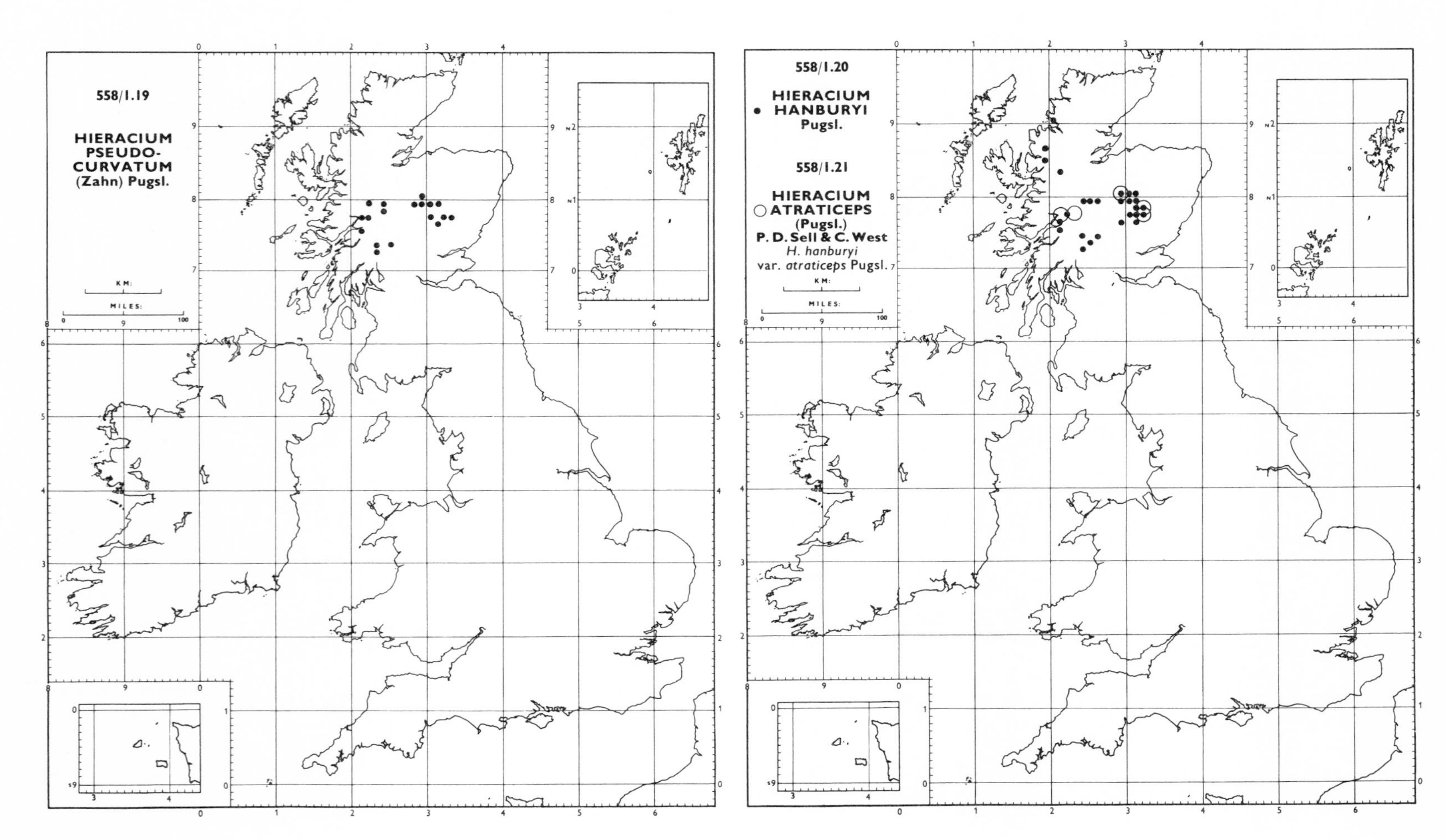

558/1.22–52 Section **SUBALPINA** Pugsl.

This section contains 30 native species. They do not form a natural group and are often not easy to define. Their probable origin was from crosses between the true *Alpina* and the Sections *Vulgata* and *Cerinthoidea*. Like the *Alpina*, they are mainly confined to the mountains of Scotland: only *H. cumbriense* of the English Lake District, and *H. vennicontium* with a range extending into Yorkshire, occur outside the main area. The *Subalpina* are plants of rock ledges and rocky stream-sides usually above 1,500 ft (450 m.). The flowering period of these plants is from mid June to the beginning of August.

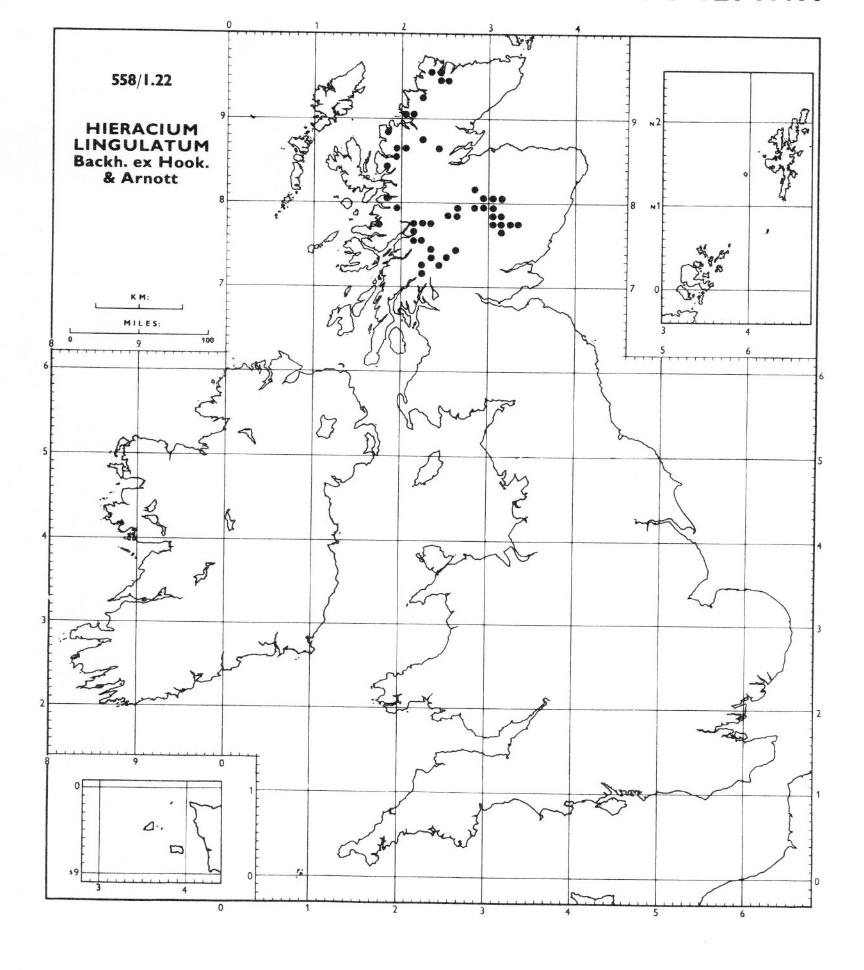

H. lingulatum Backh. ex Hook. & Arnott, which is one of the commoner species of the higher glens, is unlike other species in this section and is closely related to the true *Alpina*.

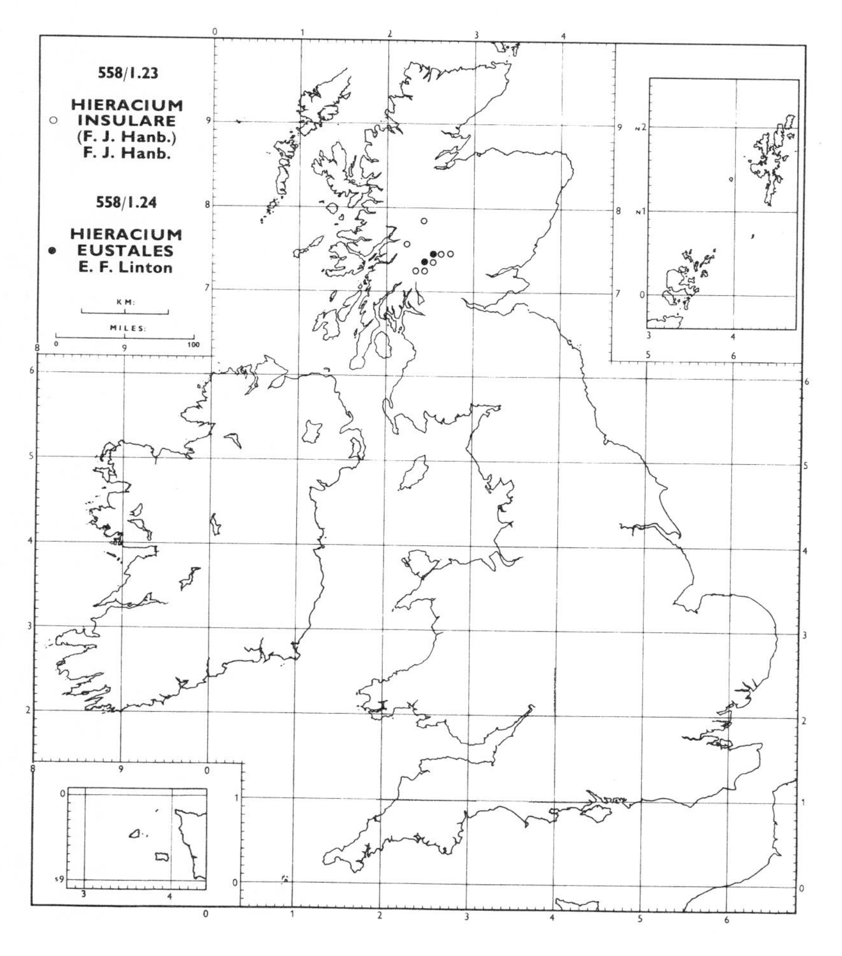

H. eustales E. F. Linton and **H. insulare** (F. J. Hanb.) F. J. Hanb. are closely allied, very local endemics of the Perthshire hills (88), the latter species extending into Inverness (97) and Argyll (98).

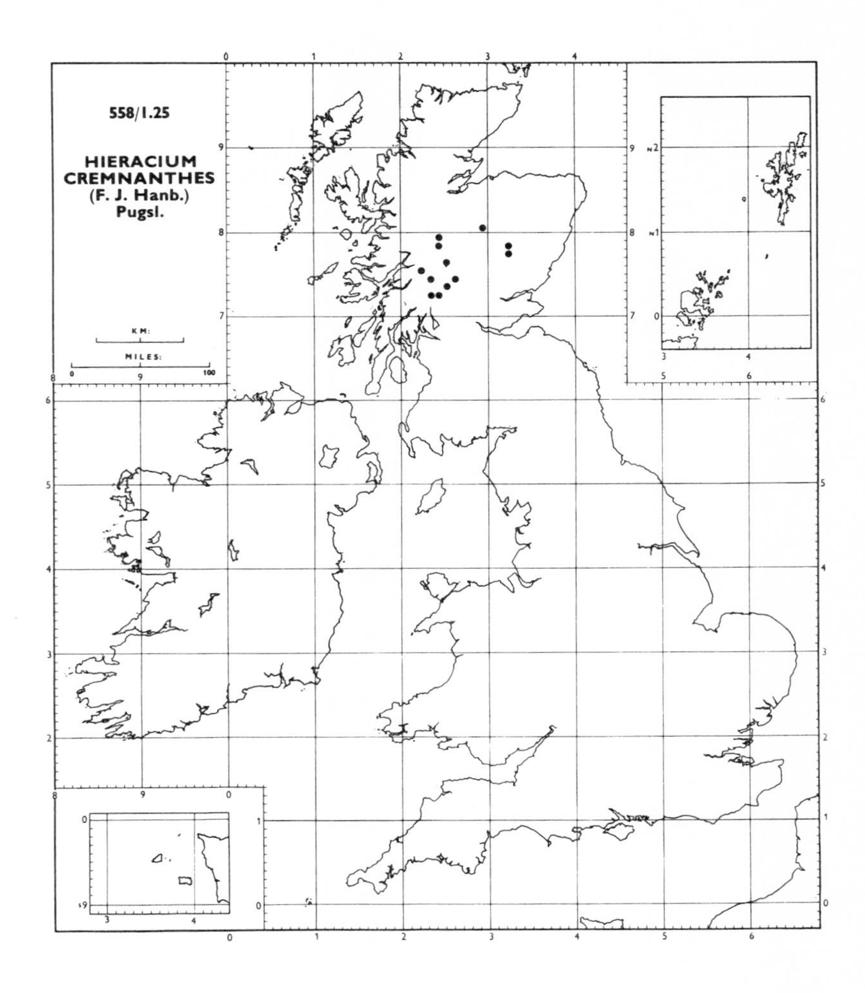

H. cremnanthes (F. J. Hanb.) Pugsl. is an endemic species scattered over central Scotland. **H. sinuans** F. J. Hanb. (*H. pulmonarium* auct.) is a very distinct endemic species of rocky burns and moist ledges of the ravines of west-central Scotland. **H. senescens** Backh., one of the most widespread and variable species of the section, is endemic to Scotland, where it is found on grassy mountain ledges and rocky crevices. **H. molybdochroum** (Dahlst.) Omang is a local species of the mountains of south Aberdeen (92) and Glen Shee (89). It also occurs in Scandinavia. The allied **H. marshallii** E. F. Linton is a widespread, but not common, endemic of central Scotland. The last five species mentioned, although having certain points in common, are not closely allied, and would appear to have had independent origins.

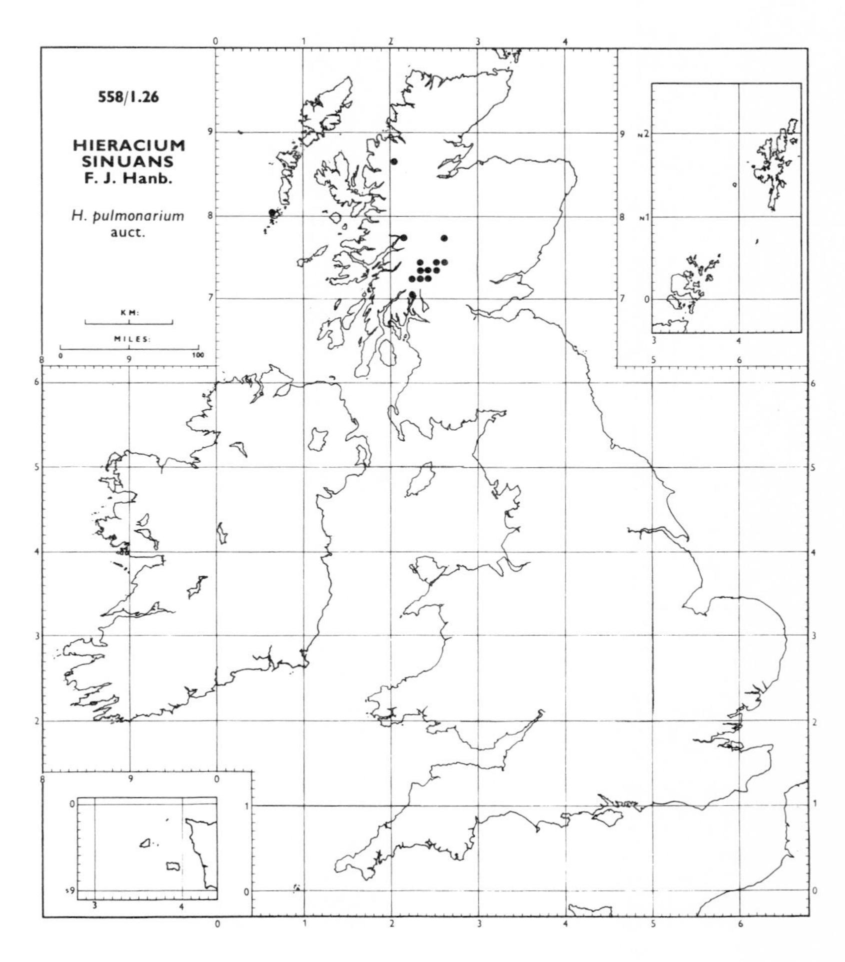

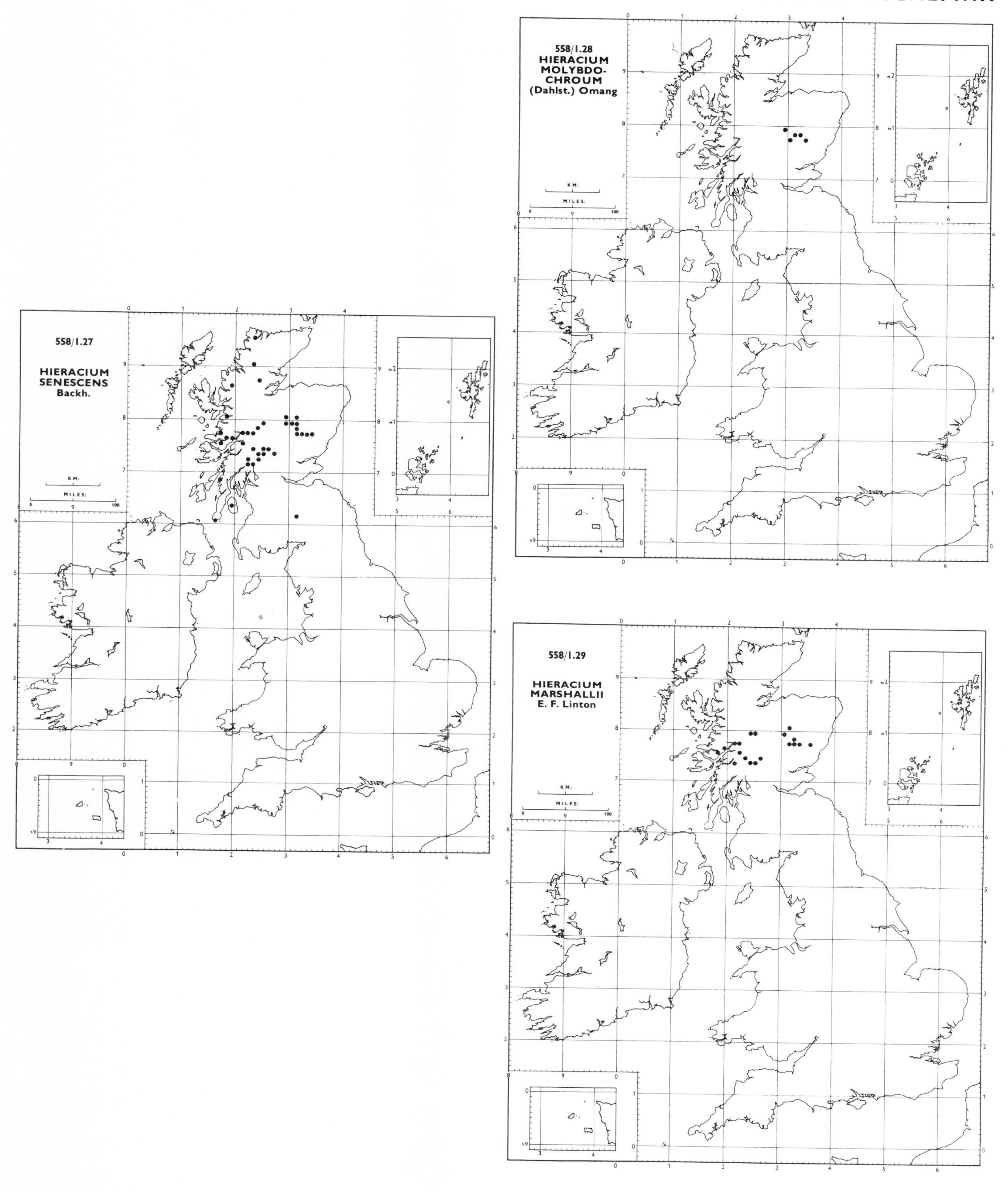
558/1.28
HIERACIUM MOLYBDOCHROUM (Dahlst.) Omang
558/1.27
HIERACIUM SENESCENS Backh.
558/1.29
HIERACIUM MARSHALLII E. F. Linton
KM:
MILES:

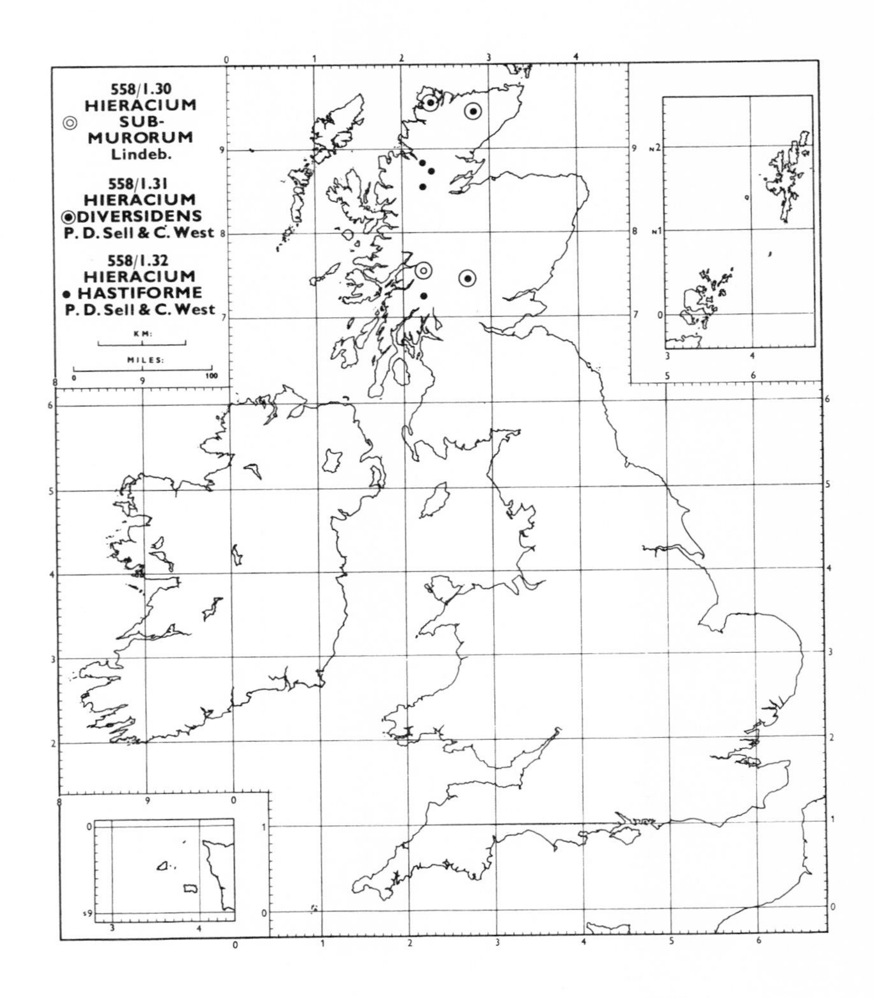

H. submurorum Lindeb., in Britain confined to a few localities around Kingshouse, Argyll (98), occurs also in alpine and arctic regions of Norway. The closely allied endemic **H. diversidens** P. D. Sell & C. West has been extensively collected by the Allt Odhar, near Fortingal, Perth (88), while single sheets of what appears to be the same species were collected on Ben Griam Beg, East Sutherland (107), and Beinn Spionnaidh (108). **H. hastiforme** P. D. Sell & C. West is an endemic species occurring in three localities in Ross (105, 106) and one in Argyll (98). The inclusion of **H. dissimile** (Lindeb.) Johans. in the British List rests solely on a specimen collected by A. G. Kenneth near Loch an Dughaill, Kintyre (101), in 1962, which exactly matches Scandinavian type material. **H. cuspidens** P. D. Sell & C. West (*H. dissimile* var. *majus* Pugsl.) is an endemic species, which appears to be of frequent occurrence in Perthshire (88, 89), with a single record from Argyll (98). **H. anfractiforme** E. S. Marshall is an endemic, which is frequent in Argyll (98), Perth (88) and Inverness (97), with a single locality on Ben Vorlich, Dunbarton (99). All the species so far mentioned in this paragraph have much in common, and may well have had the same ancestral origin. The endemic **H. centripetale** F. J. Hanb., which occurs in a number of widely scattered localities in Scotland, from Dumfries (72) to Sutherland (108), is perhaps slightly more distinct, but its affinities are clearly with this group. The allied **H. longilobum** (Dahlst. ex Zahn) Roffey is native in the British Isles at the Midlaw Burn, near Moffat, Dumfries (72). This species also occurs in Sweden.

The endemic **H. hyparcticoides** Pugsl. occurs in several localities in Sutherland (108), Inverness (96, 97) and Ross (106), with an outlying record from Aberdeen (92). The very closely allied **H. glandulidens** P. D. Sell & C. West is endemic to Sutherland (108), Ross (105, 106) and Argyll (98).

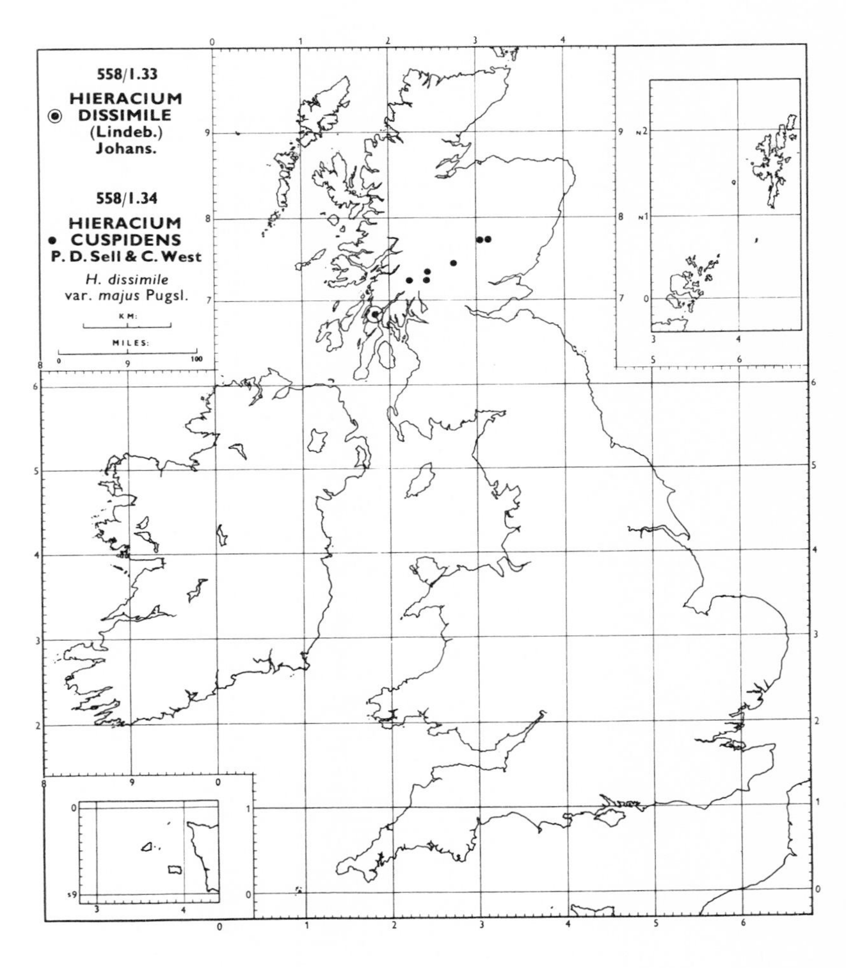

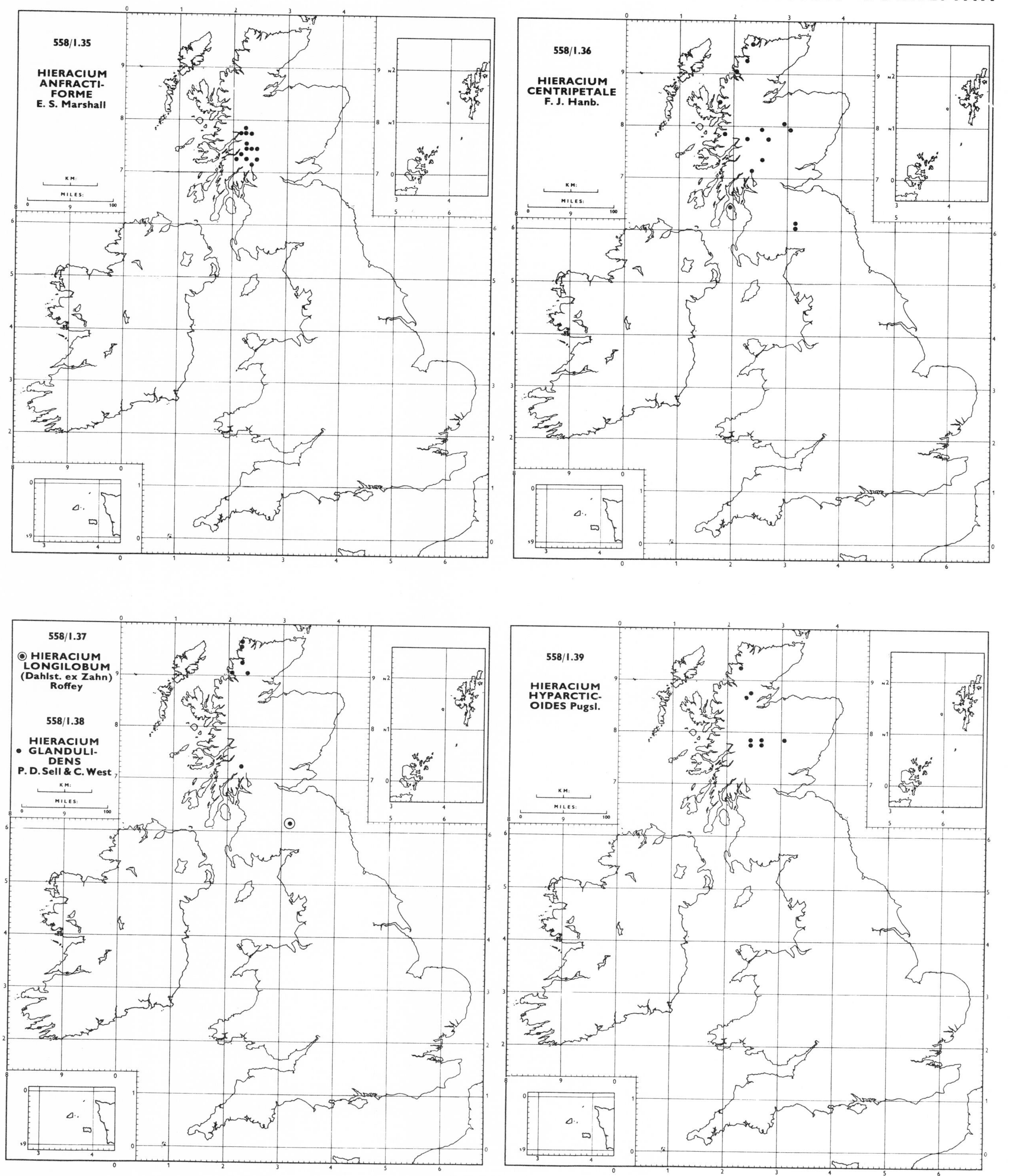
558/1.35
HIERACIUM ANFRACTI-FORME E. S. Marshall
558/1.36
HIERACIUM CENTRIPETALE F. J. Hanb.
558/1.37
◉ HIERACIUM LONGILOBUM (Dahlst. ex Zahn) Roffey
558/1.38
HIERACIUM
• GLANDULI-DENS P. D. Sell & C. West
558/1.39
HIERACIUM HYPARCTIC-OIDES Pugsl.
KM:
MILES:

H. callistophyllum F. J. Hanb. is an endemic species, locally frequent in west-central Scotland, with isolated localities in Ross (106), Aberdeen (92), Angus (90) and Dumfries (72). Although somewhat variable, it has no close allies.

H. isabellae E. S. Marshall, endemic to a few localities in central Scotland, is a very distinct species, which has several characters in common with *H. duriceps* of the Section *Vulgata*.

H. vennicontium Pugsl. is an endemic species, common in central Scotland, with isolated localities in north England and the Outer Hebrides (110). The closely allied endemic **H. melanochlorocephalum** Pugsl. is restricted to the Cairngorm Mountains. These two species also have strong affinities with the Section *Vulgata*.

H. clovense E. F. Linton has been variously placed in Sections *Cerinthoidea*, *Oreadea* and *Vulgata*, but it is here put in the position originally given to it by E. F. Linton. It is endemic to the mountains of a small area of east-central Scotland and of the English Lake District (69, 70), with an isolated record from Argyll (98). The closely allied **H. chrysolorum** P. D. Sell & C. West is endemic to the Midlaw Burn, near Moffat, Dumfries (72), and Hen Hole, Northumberland (68).

H. gracilifolium (F. J. Hanb.) Pugsl. has strong affinities with some of the glandular-headed species of *Vulgata*, but the presence of glandular hairs on the leaves suggests that it is best placed in the Section *Subalpina*. A common species of central Scotland with isolated records from Kirkcudbright (73) and West Ross (105), it also occurs in Scandinavia and central Europe. **H. nigrisquamum** P. D. Sell & C. West, endemic to a few localities in central Scotland, is intermediate in character between *H. gracilifolium* and **H. dasythrix** (E. F. Linton) Pugsl., which is endemic to west and central Scotland.

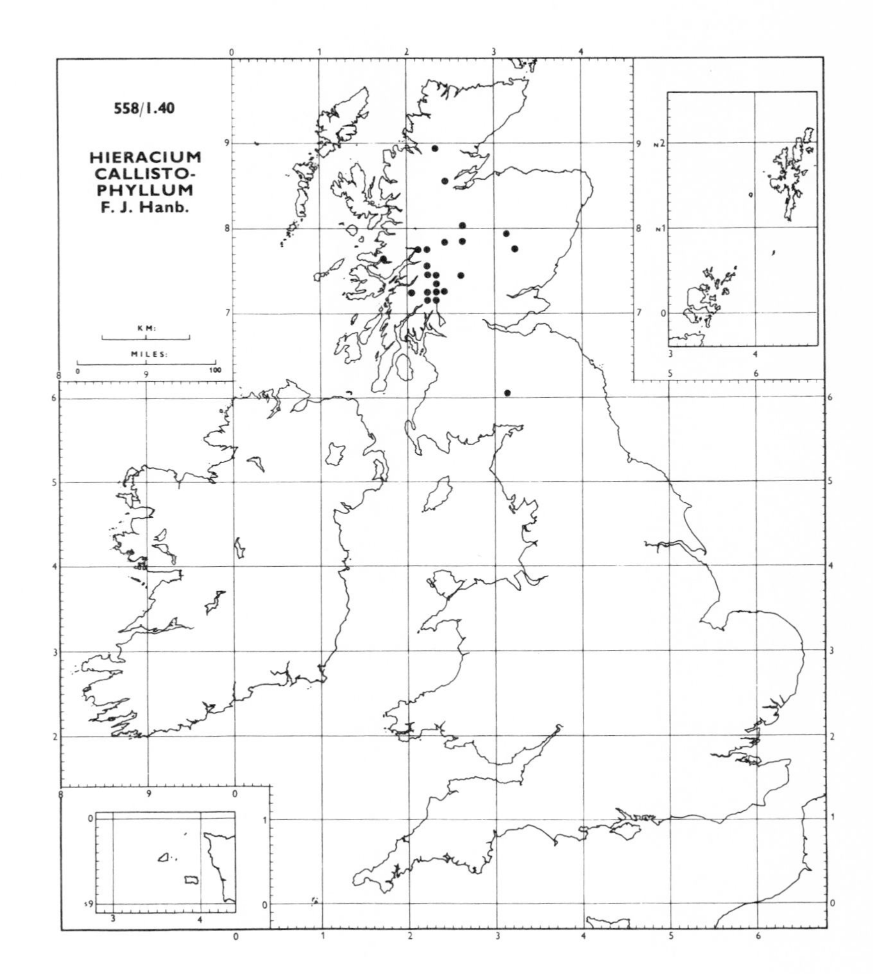

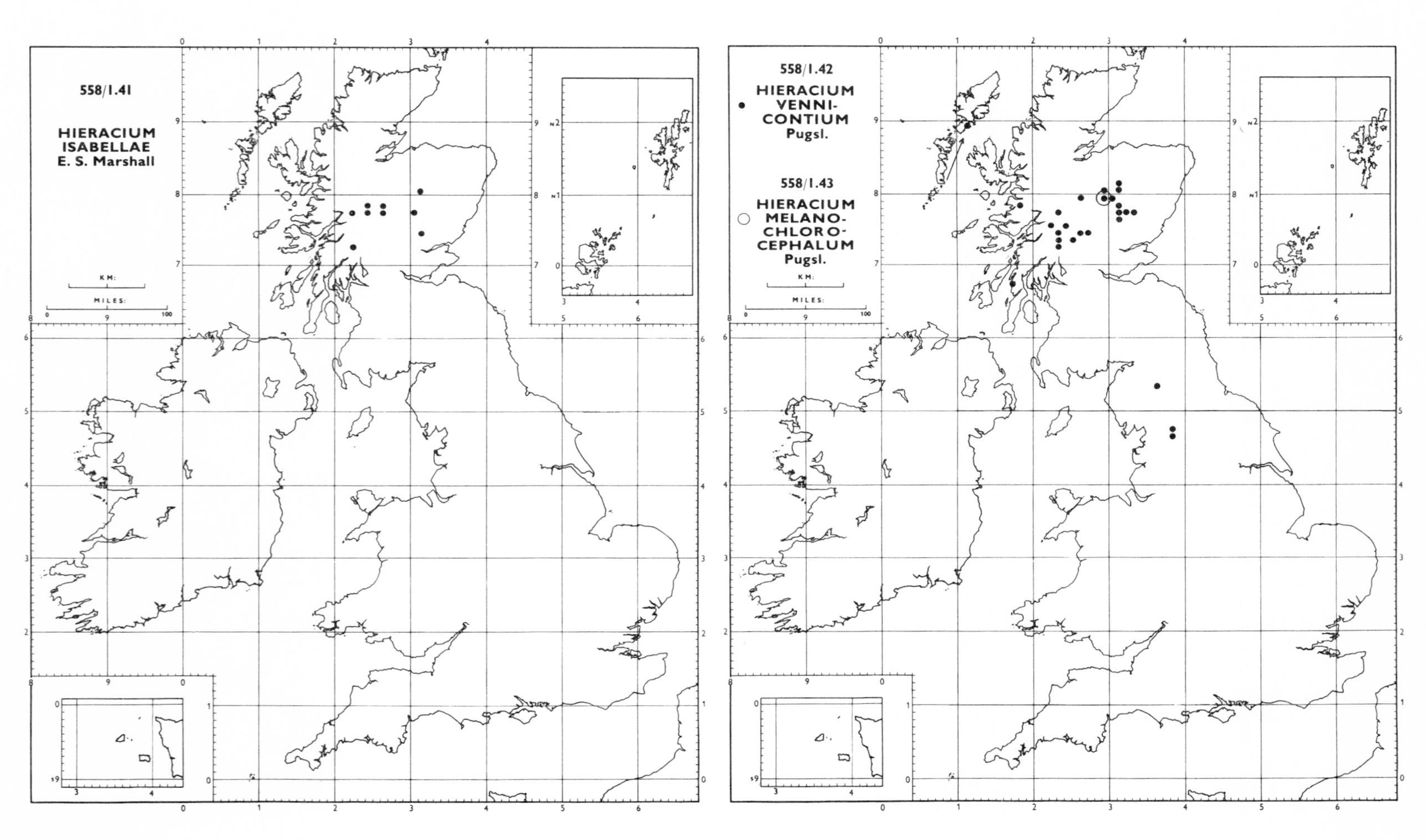

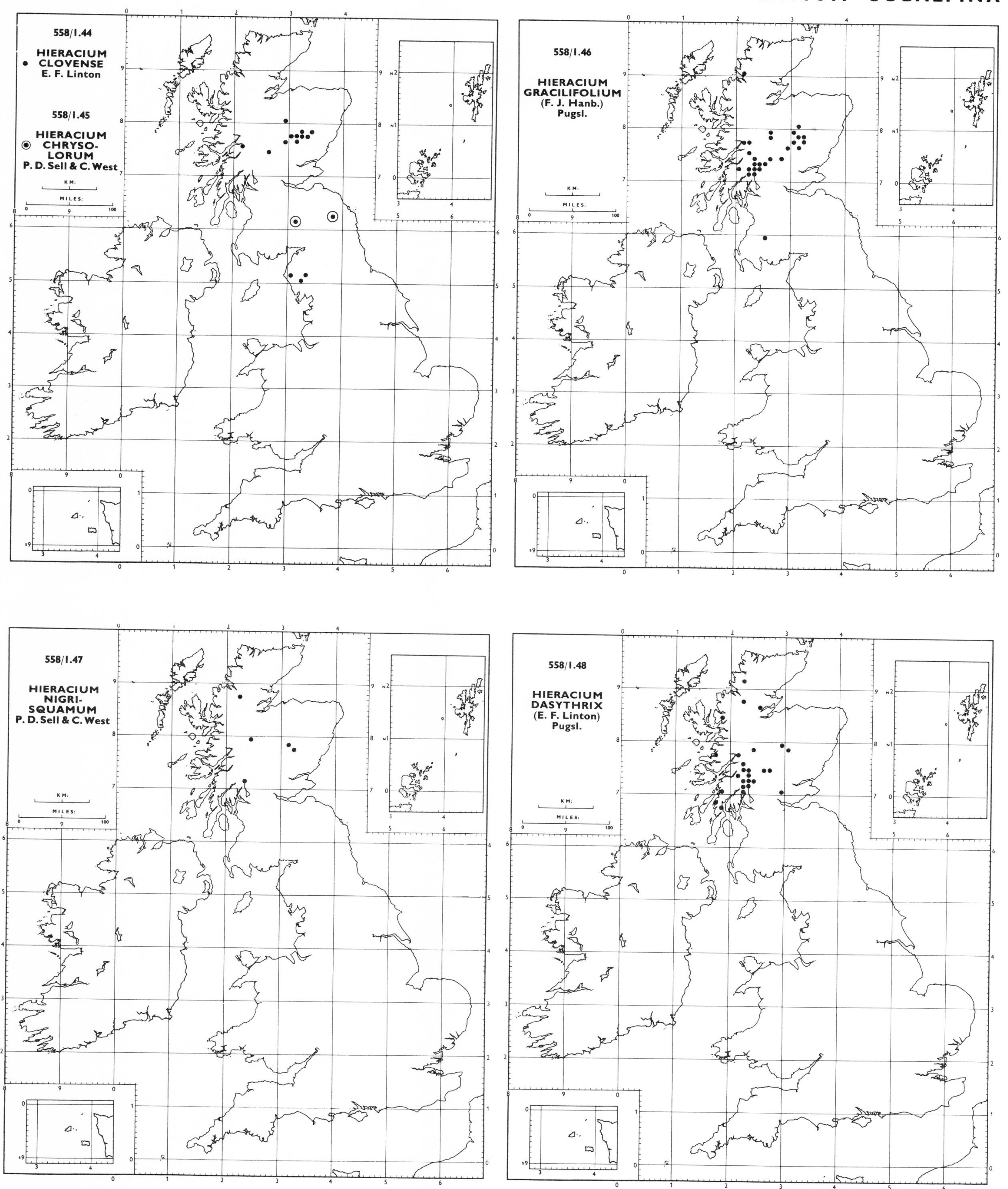
558/1.44
HIERACIUM
• CLOVENSE
E. F. Linton
558/1.45
HIERACIUM
◉ CHRYSO-
LORUM
P. D. Sell & C. West
558/1.46
HIERACIUM
GRACILIFOLIUM
(F. J. Hanb.)
Pugsl.
558/1.47
HIERACIUM
NIGRI-
SQUAMUM
P. D. Sell & C. West
558/1.48
HIERACIUM
DASYTHRIX
(E. F. Linton)
Pugsl.
KM:
MILES:

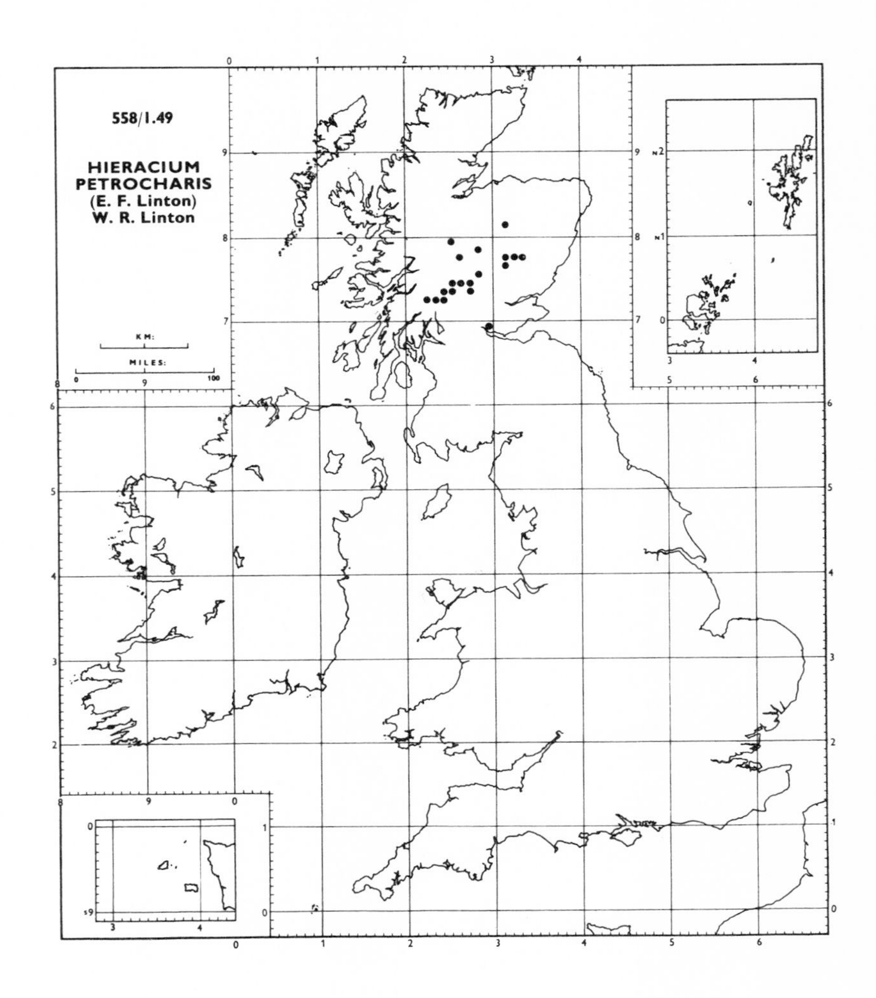

The remaining four species of the Section *Subalpina* show strong affinities with the Section *Cerinthoidea*, with which they connect the Section *Subalpina*. **H. petrocharis** (E. F. Linton) W. R. Linton is endemic to central Scotland. **H. pseudanglicum** Pugsl. is an endemic species, frequent in central Scotland, with isolated records from Ross (105), Rhum (104) and Antrim (H. 39). **H. cumbriense** F. J. Hanb. is endemic to Westmorland (69) and Cumberland (70). **H. pseudanglicoides** J. E. Raven, P. D. Sell & C. West, endemic to central Scotland, is sometimes difficult to distinguish from some forms of *H. flocculosum*. In general appearance *H. pseudanglicum* suggests an alpine equivalent of *H. anglicum*, while *H. cumbriense* suggests a similar equivalent of *H. ampliatum*.

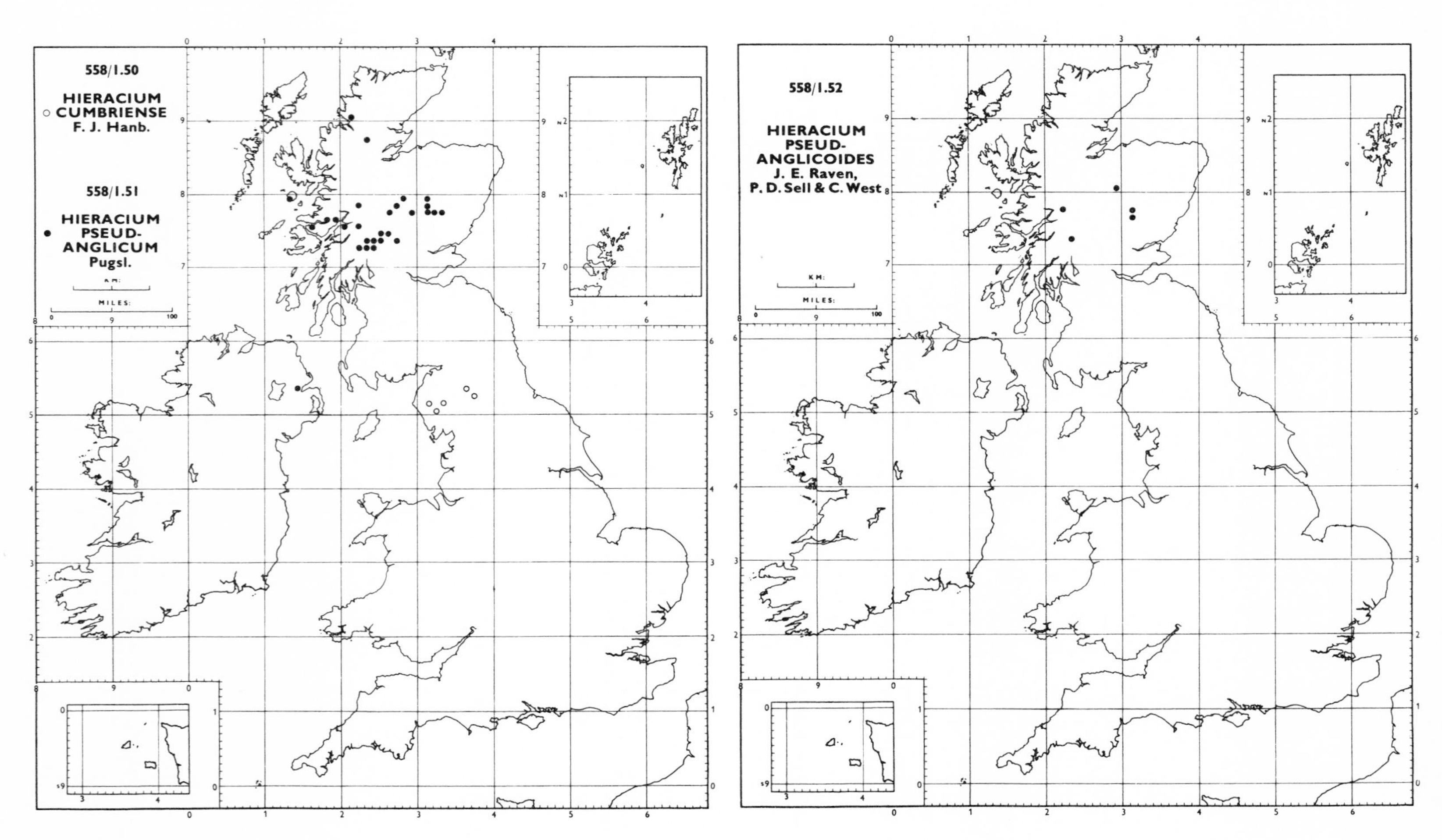

558/1.53–62 Section **CERINTHOIDEA** Koch

This section forms a natural group of ten closely allied native species. They are abundant on cliff ledges and by rocky streams in coastal and upland regions of north England, Scotland and the north and west perimeter of Ireland. In some localities, such as the Traligill Burn above Inchnadamph (108), as many as six of these species occur, and it is highly probable that all ten species have the same ancestral origin. They are certainly most abundant on basic rock, although not confined to it. The flowering period in a warm spring starts in late May, and may continue until September or even October. In this group the characters of the species seem to remain constant throughout. It is an interesting fact that whereas species of this group occur in France and the Pyrenees, and are also found in Iceland and the Faeroes, not one of them has been recorded from Shetland or Scandinavia.

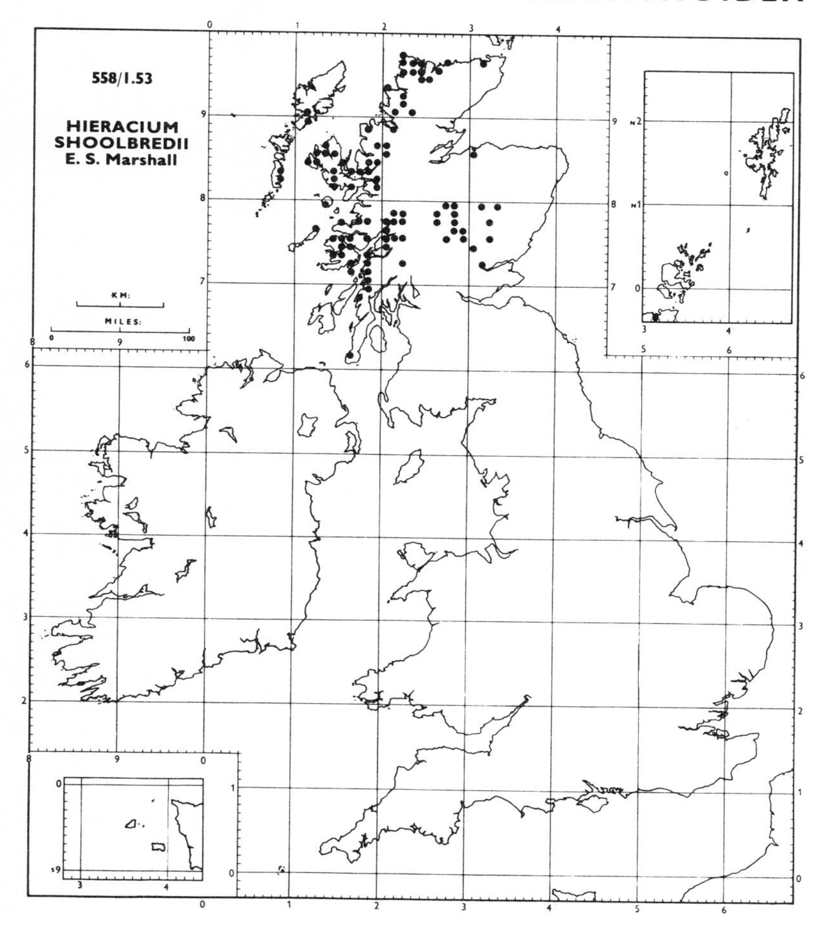

H. shoolbredii E. S. Marshall is a common, endemic, uniform species of central and northern Scotland, with an isolated locality on the Mull of Kintyre (101). The closely allied, rather variable, endemic **H. flocculosum** Backh. is less frequent, having its main distribution in central Scotland, with isolated localities in northern Scotland, the Hebrides (103, 110), Sligo (H. 28) and Down (H. 38).

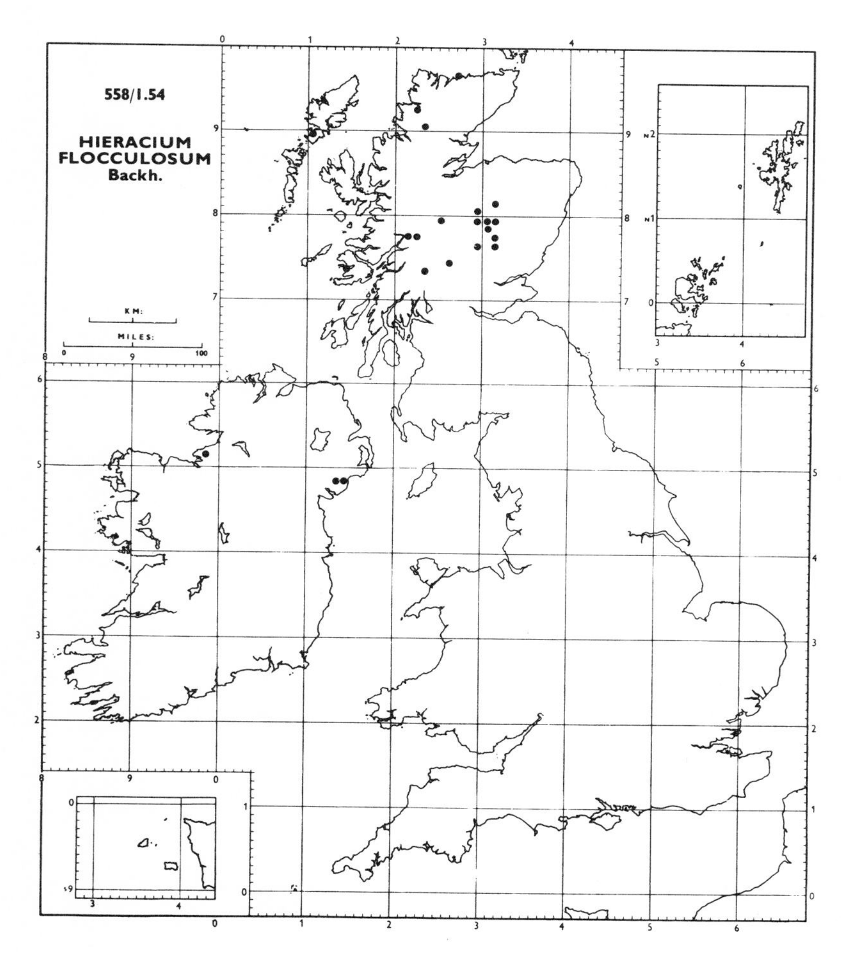

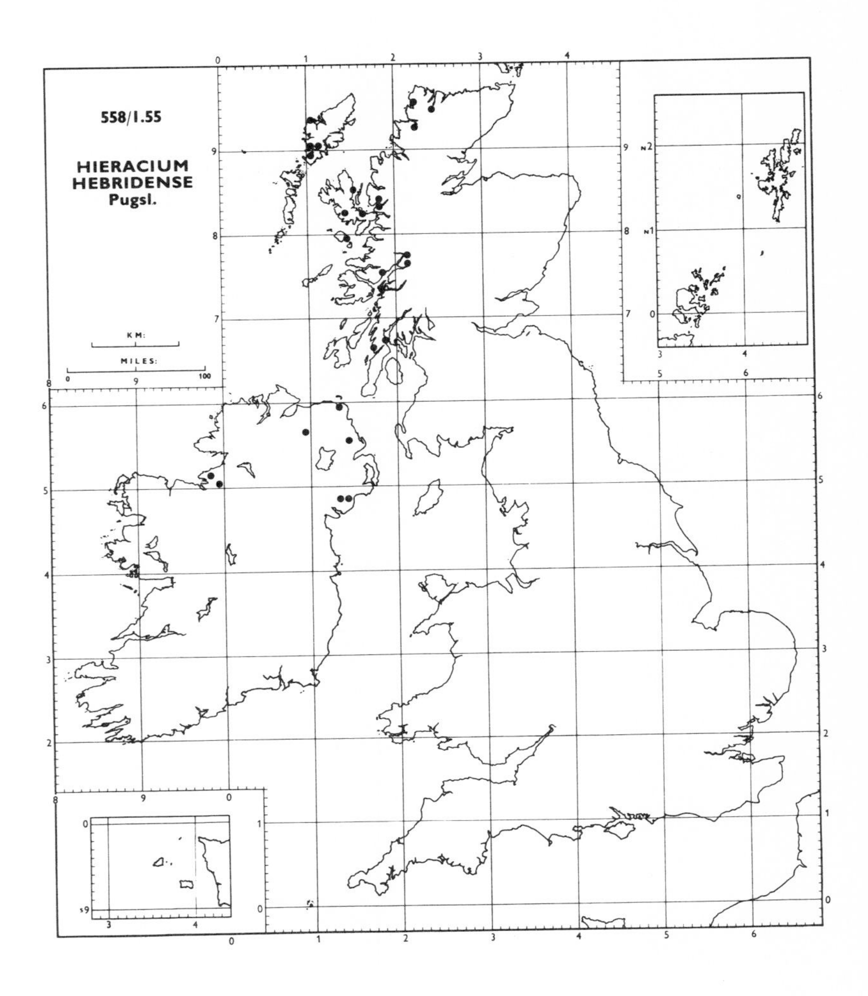

H. hebridense Pugsl. is a uniform, rather local endemic species of western Scotland and northern Ireland. The variable endemic **H. ampliatum** (W. R. Linton) A. Ley is an abundant, characteristic species of basic rocks in north-west England. It is scattered throughout western Scotland and north-eastern Ireland, with an isolated locality at Doon Lough in County Leitrim (H. 29). Herbarium specimens of this species are sometimes difficult to distinguish from *H. anglicum*, with which it often grows, but in the field the two species are unmistakable. The uniform endemic **H. langwellense** F. J. Hanb., which is widespread in central and northern Scotland, with an isolated locality in Galway (H. 16), is easily distinguished in the field from all other species of the section by its clear green leaves.

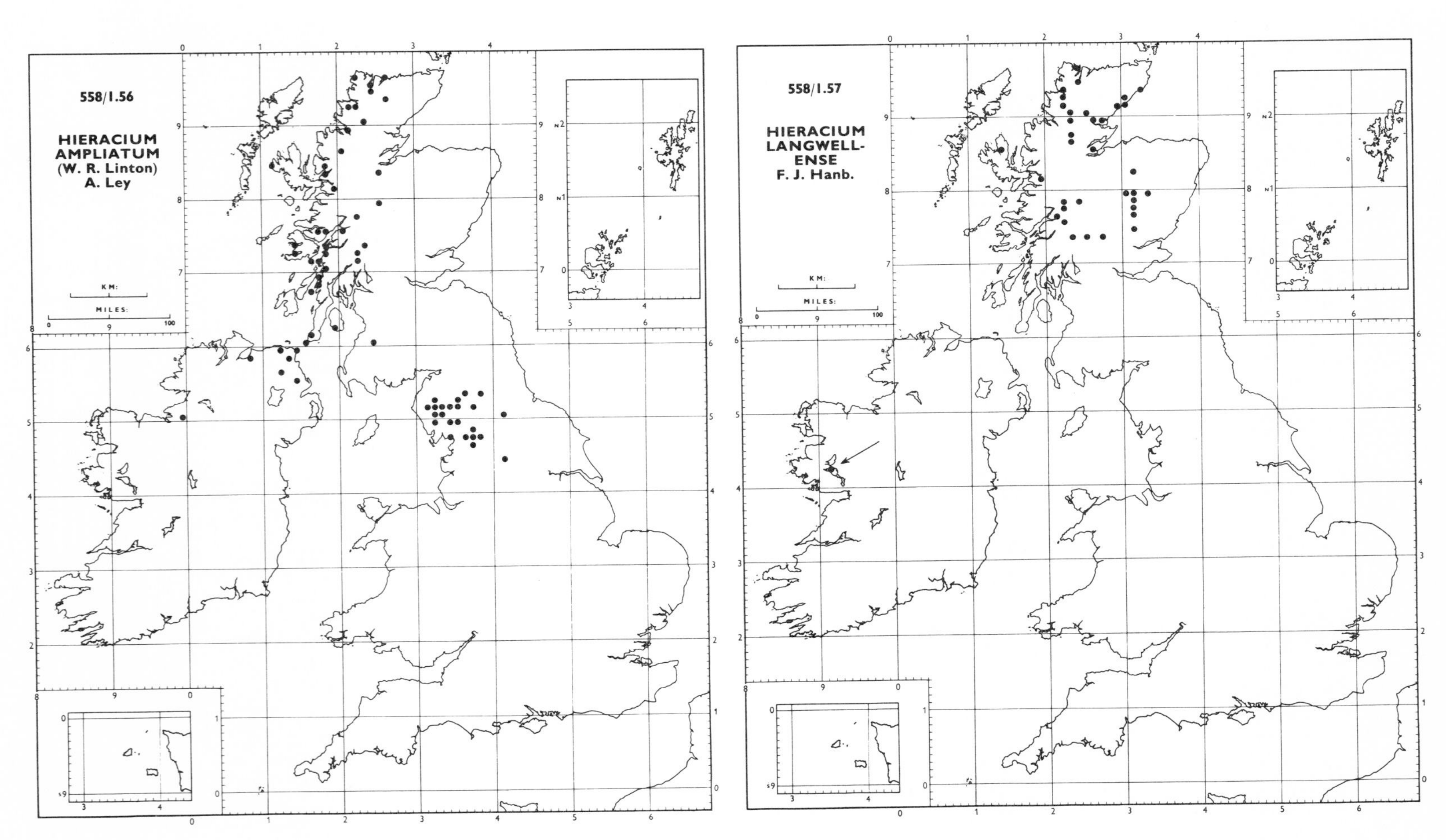

H. hartii (F. J. Hanb.) P. D. Sell & C. West is a distinct, local endemic of north-western Ireland; it was said to be the *Hieracium* of Slieve League (H. 27) by F. J. Hanbury and H. C. Hart in the 1880s. Although it has been searched for in all its localities, it has not been found during the present century. **H. anglicum** Fries is the commonest species of the section, being widely distributed and very variable. In the second half of June in northern England, central and northern Scotland and the northern and western perimeter of Ireland the rather glaucous leaves, dark involucres and handsome flowers of this plant are characteristic of many areas. Dwarf forms of the Scottish hills, with very dark heads, strongly approach *H. pseudanglicum* of the Section *Subalpina*. In western Ireland very luxuriant forms of low stature with large spreading leaves and capitula of the largest size are to be found. Plants with two to four cauline leaves (var. *amplexicaule* Backh. ex Bab.) are sometimes difficult to distinguish from *H. iricum*. The only record for Wales, which is also the only record for a Cerinthoidean species in that country, is based on a specimen gathered by C. West at Craig Breidden, Montgomery (47). *H. anglicum* is also found in Iceland and the Faeroes. The closely allied **H. iricum** Fries has much the same general distribution as *H. anglicum*, but is more local. *H. iricum* also occurs in the Faeroes. **H. magniceps** P. D. Sell & C. West, endemic to Glen Shee, Perthshire (89), and **H. scarpicum** Pugsl., endemic to Lewis and Scarp in the Outer Hebrides (110), are closely allied to *H. iricum*.

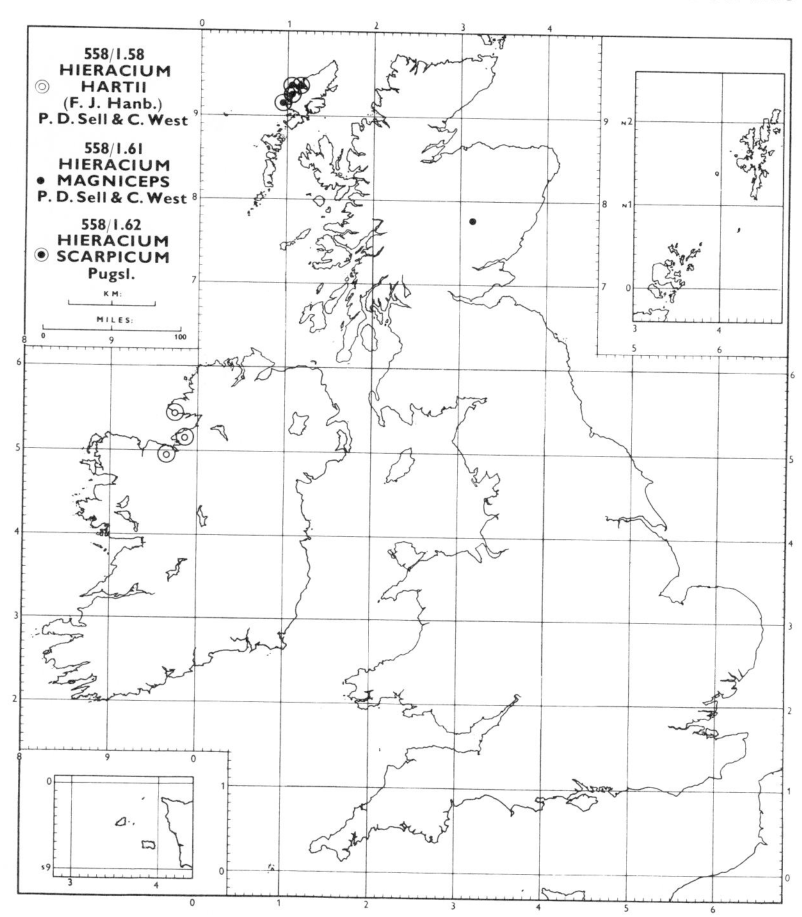

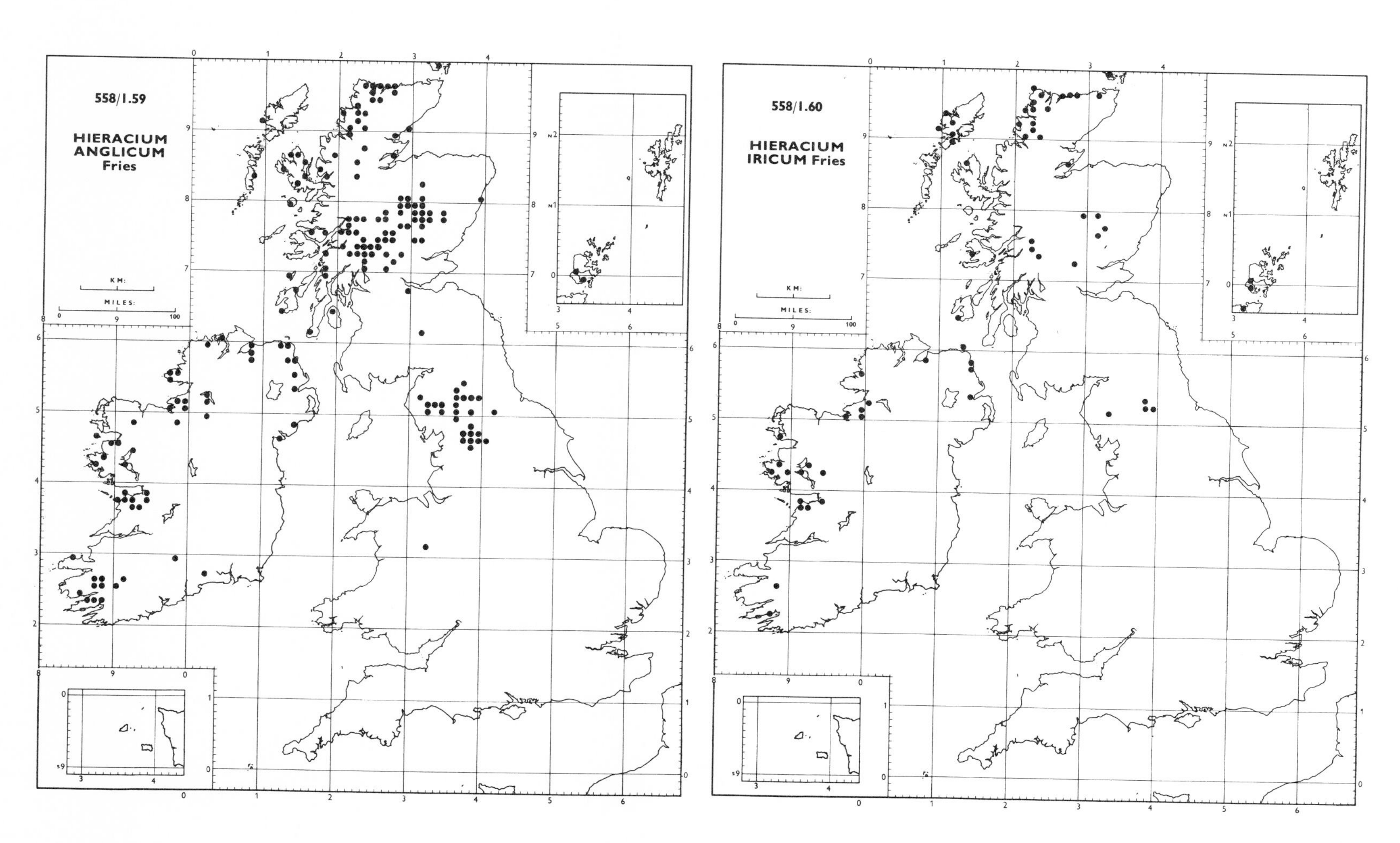

558/1.63–106 Section **OREADEA** Zahn

This section consists of a very diverse group of 44 native species which are, however, unlikely to be confused with the species of any other section. They flower in great abundance, usually on limestone and other basic cliffs, in late May and June, from Somerset and Devon, through Wales, central and northern England and Scotland to the Shetlands. In Ireland they may be found wherever there are suitable habitats. They are endemic unless otherwise stated. Although flowering specimens may be found until September, these do not usually have the characteristic appearance of the earlier flowering ones. Species of this section apparently occur throughout Europe and the Near East.

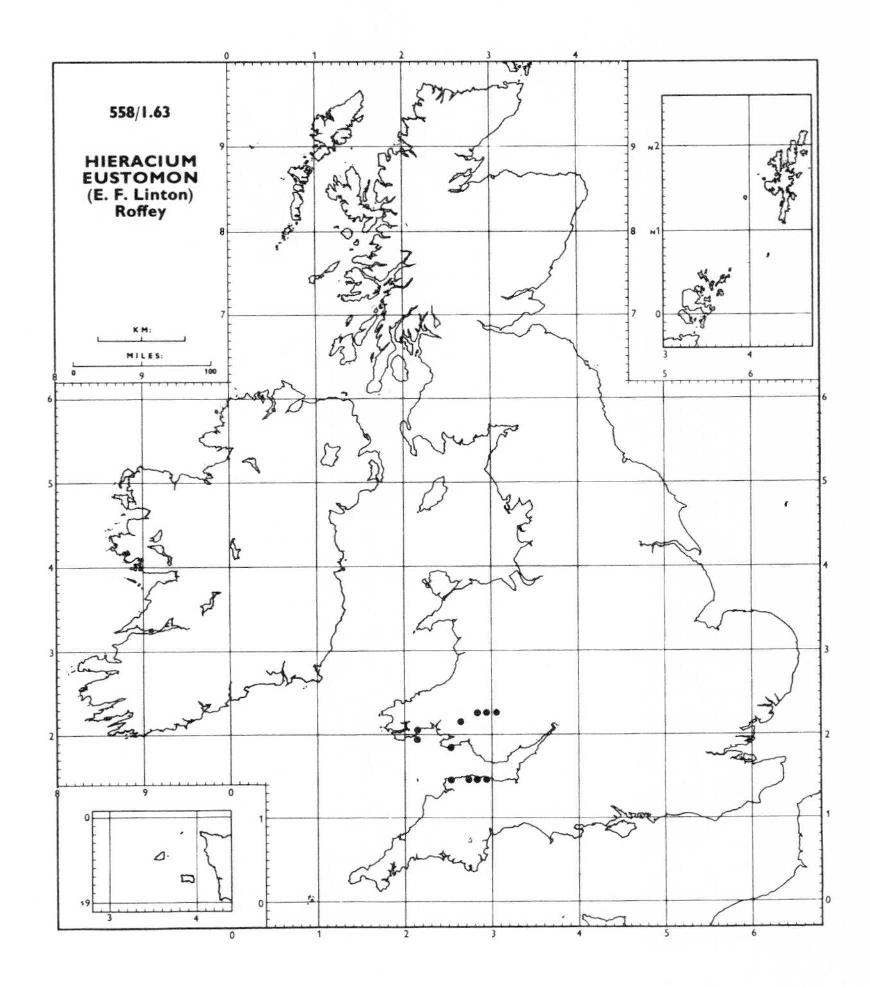

H. eustomon (E. F. Linton) Roffey, a local species of north Devon (4) and south Wales, does not appear to be closely allied to any other British species. *H. eustomon* is, however, doubtfully distinct from *H. veterascens* Dahlst. (from the Faeroes), though they must surely have had an independent origin.

H. stenopholidium (Dahlst.) Omang has a very disjunct distribution over the western half of Britain from Devon to the Hebrides. It is also found in Iceland. In general appearance it bears a strong resemblance to *H. stenolepiforme*.

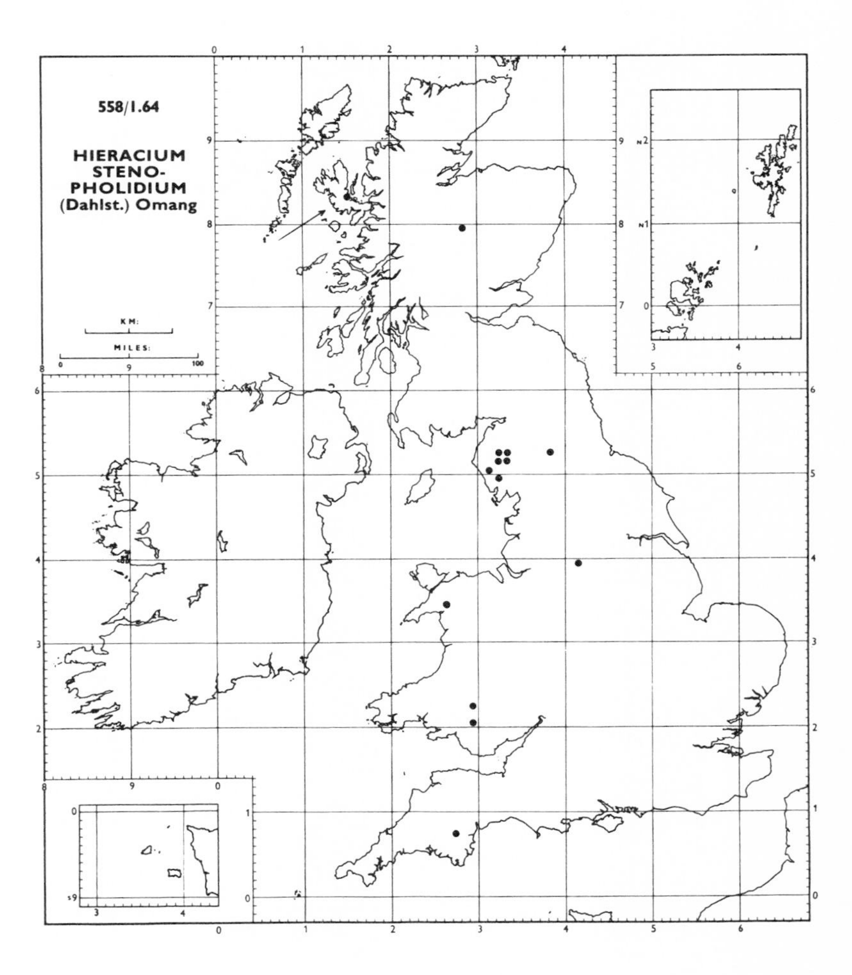

H. stenolepiforme (Pugsl.) P. D. Sell & C. West, **H. britannicum** F. J. Hanb., **H. subbritannicum** (A. Ley) P. D. Sell & C. West, **H. britanniciforme** Pugsl. and **H. dicella** P. D. Sell & C. West, form a group of closely allied species. The most widespread of these, *H. dicella*, grows in at least some of the localities of all the others, except *H. stenolepiforme* (which is endemic to Cheddar Gorge, Somerset (6)), and may well be the species from which they were derived. On the limestone cliffs of south Wales, where *H. subbritannicum* and *H. dicella* are frequent, both species are quite distinct. On the limestone on and around Great Ormes Head (49), where *H. britanniciforme* is a very local endemic, *H. dicella* is rare. *H. britanniciforme* also occurs at Ballygroggan, Kintyre (101). *H. britannicum* is an abundant and characteristic species of the Derbyshire (57) and Staffordshire (39) dales, where its handsome flowers form sheets of yellow. In these localities *H. dicella*, the much scarcer plant, so closely resembles *H. britannicum* in appearance that one has difficulty in finding it. On the Durness limestone (108) the very similar **H. sarcophylloides** Dahlst., which has a few relict stations in west Scotland, Orkney (111) and the Faeroes, also occurs. Throughout the remainder of its range *H. dicella* is not associated with any closely allied species. However, **H. fratrum** Pugsl., from disjunct localities in Dumfries (72), Argyll (98) (near where *H. dicella* occurs), Aberdeen (92) and Inverness, is also a closely related species. In Ireland *H. dicella* is replaced by **H. basalticola** Pugsl., a species very similar in appearance, which may have become modified as a result of geographical isolation. **H. subplanifolium** Pugsl. is also perhaps best included in this group. Its disjunct distribution defies explanation. It matches exactly three original specimens of *H. riukanense* Omang from Scandinavia, but it is thought inadvisable to unite the two species on such slender evidence.

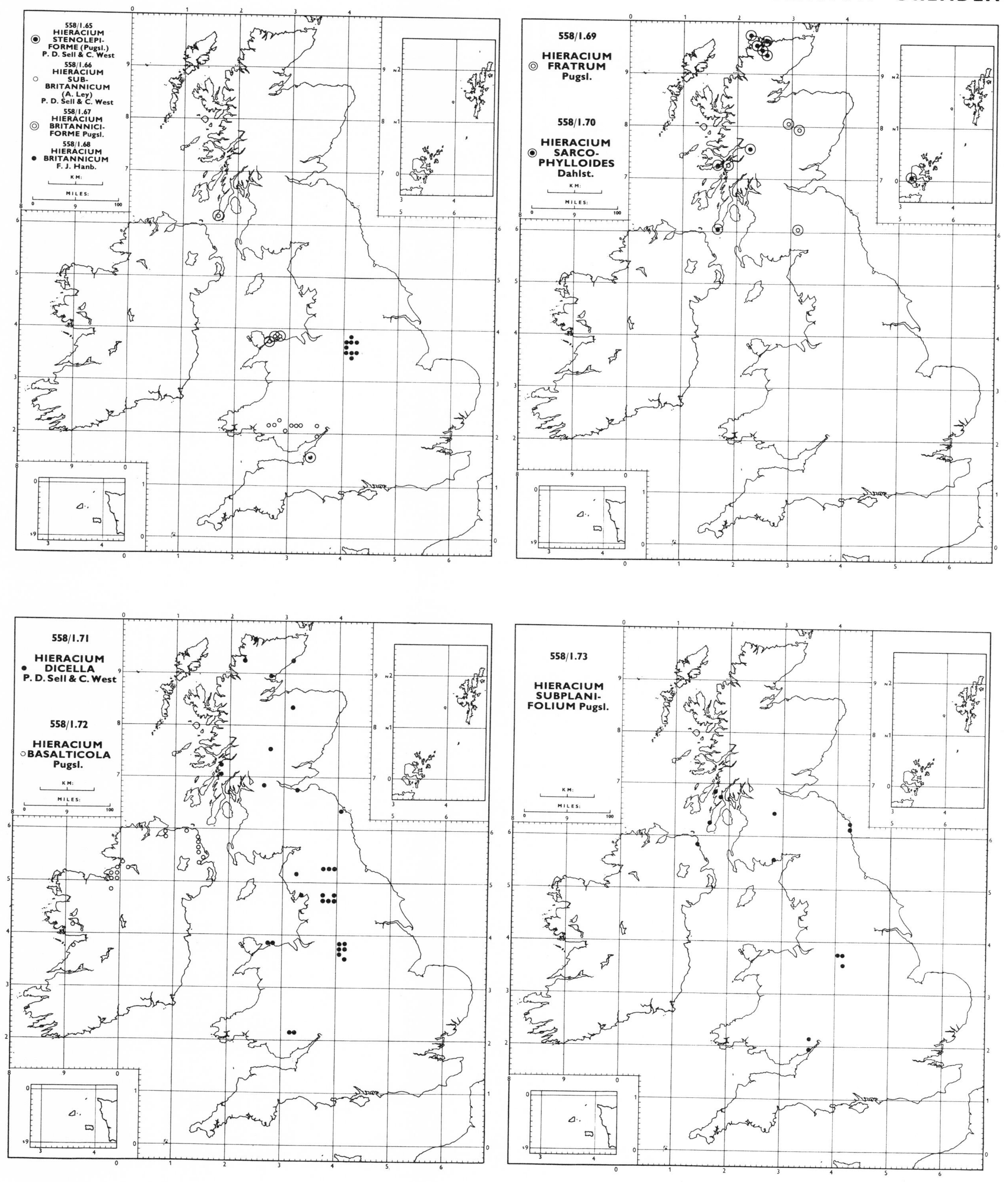
558/1.65
HIERACIUM STENOLEPIFORME (Pugsl.) P. D. Sell & C. West
558/1.66
HIERACIUM SUBBRITANNICUM (A. Ley) P. D. Sell & C. West
558/1.67
HIERACIUM BRITANNICIFORME Pugsl.
558/1.68
HIERACIUM BRITANNICUM F. J. Hanb.
KM:
MILES:
558/1.69
HIERACIUM FRATRUM Pugsl.
558/1.70
HIERACIUM SARCOPHYLLOIDES Dahlst.
558/1.71
HIERACIUM DICELLA P. D. Sell & C. West
558/1.72
HIERACIUM BASALTICOLA Pugsl.
558/1.73
HIERACIUM SUBPLANIFOLIUM Pugsl.

H. saxorum (F. J. Hanb.) P. D. Sell & C. West is a frequent endemic species of cliffs and rocky places in north and central Scotland, with scattered localities in central and northern England and Wales.

H. praetermissum P. D. Sell & C. West is endemic to the north coast of Sutherland (108). **H. ebudicum** Pugsl. occurs only in a few localities in Harris and Lewis in the Outer Hebrides (110). *H. praetermissum* and *H. ebudicum* have no closely related species.

H. riddelsdellii Pugsl. and **H. repandulare** Druce are closely related species with no other very close allies. On the Breconshire limestone (42) they grow adjacent to each other, and are almost certainly derived from the same ancestor. *H. repandulare* has an isolated locality on Carnedd Dafydd, Caernarvon (49). Plants at Burnmouth, Berwickshire (81), match exactly the Breconshire specimens of *H. riddelsdellii* and must be regarded as conspecific, though they may well be of different origin.

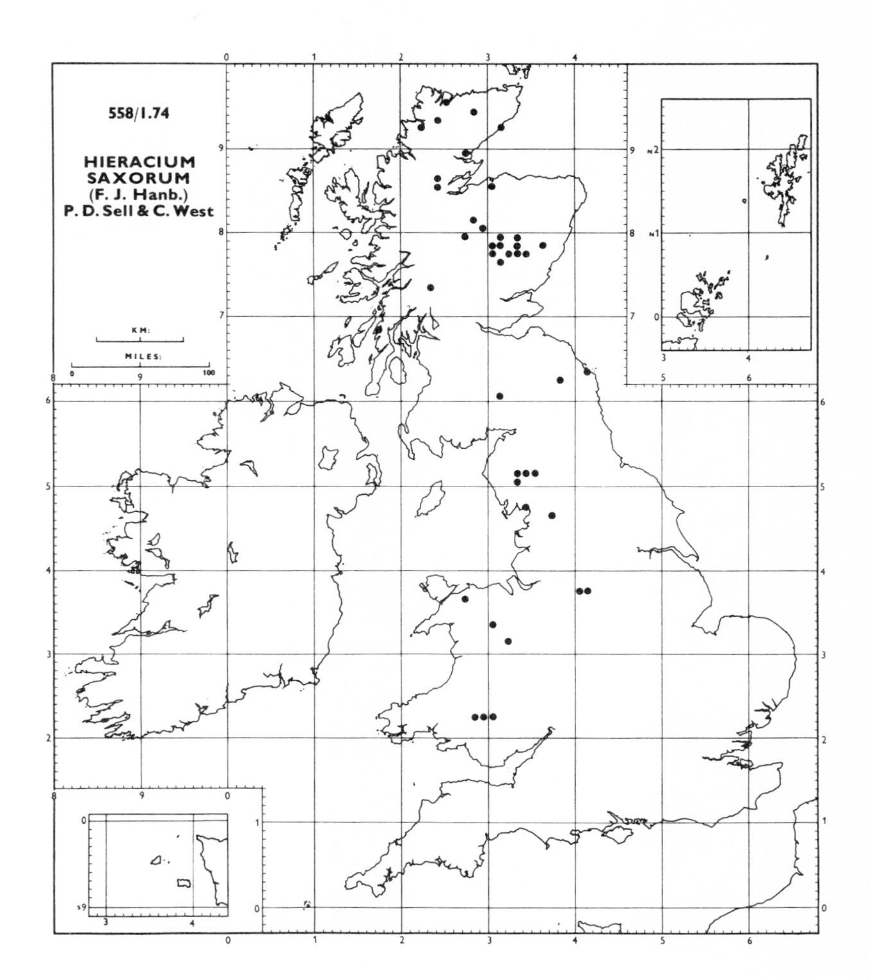

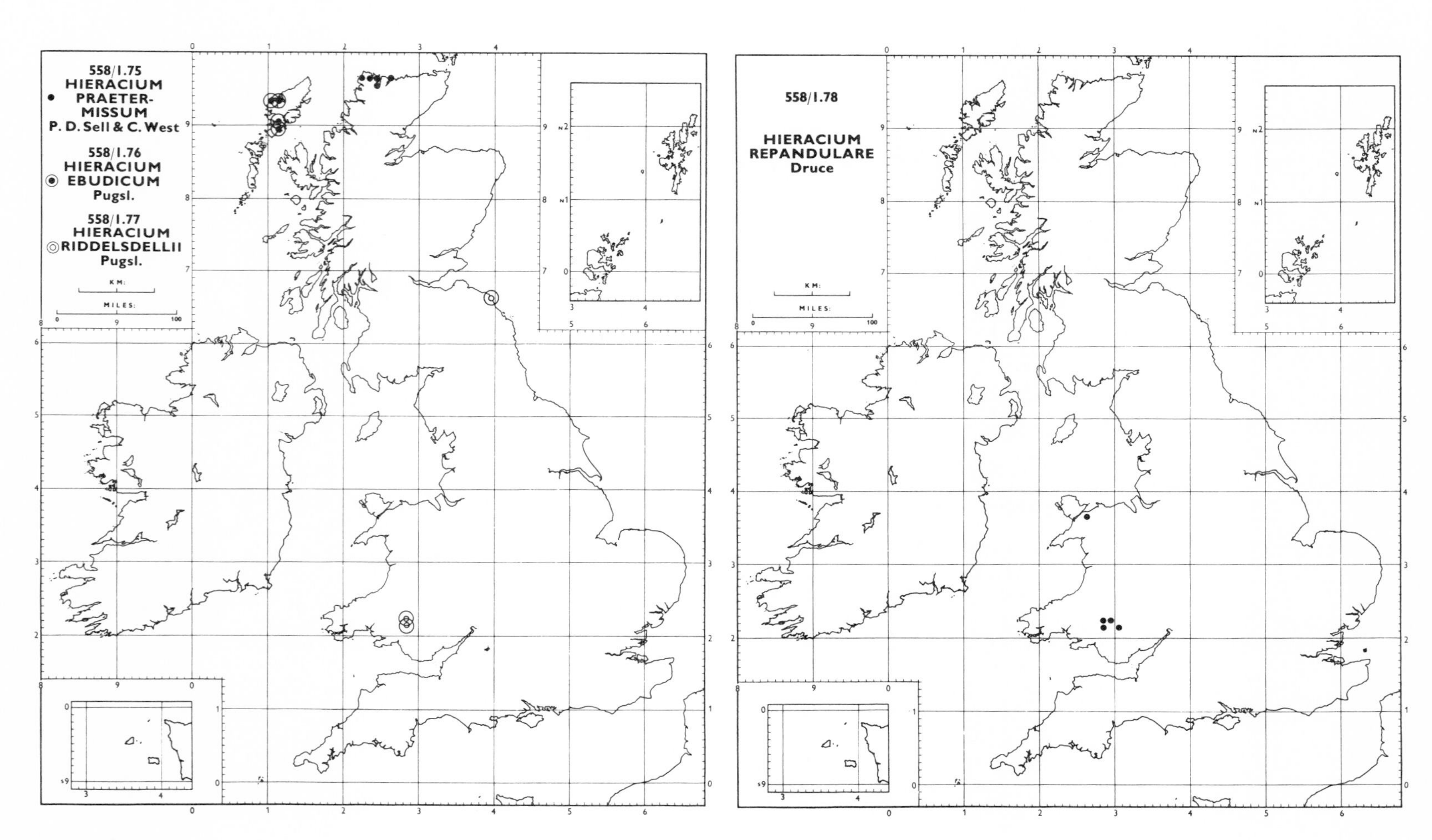

H. hypochoeroides Gibson is a beautiful, distinct species of the limestone in Dyffryn Crawnon, Brecon (42), Tutshill, Gloucester (34), near Llangollen, Denbigh (50), mid and north-west Yorks (64 and 65), the English Lake District, Antrim (H. 39) and the Burren, Clare (H. 9).

H. schmidtii Tausch is scattered throughout western Britain from Devon to Sutherland, with its headquarters in the eastern Highlands of Scotland. It has isolated localities in northern and western Ireland. Abroad, the distribution of this most variable species ranges from Iceland and Scandinavia through Germany, Switzerland and France to Spain. **H. brigantum** (F. J. Hanb.) Roffey, confined to the limestone of Yorkshire, Westmorland (69) and Cumberland (70), has been placed in the Section *Cerinthoidea*, but its rounded, strongly setose leaves and yellow styles clearly indicate its affinity with *H. schmidtii* of the Section *Oreadea*, where it was originally placed by Hanbury.

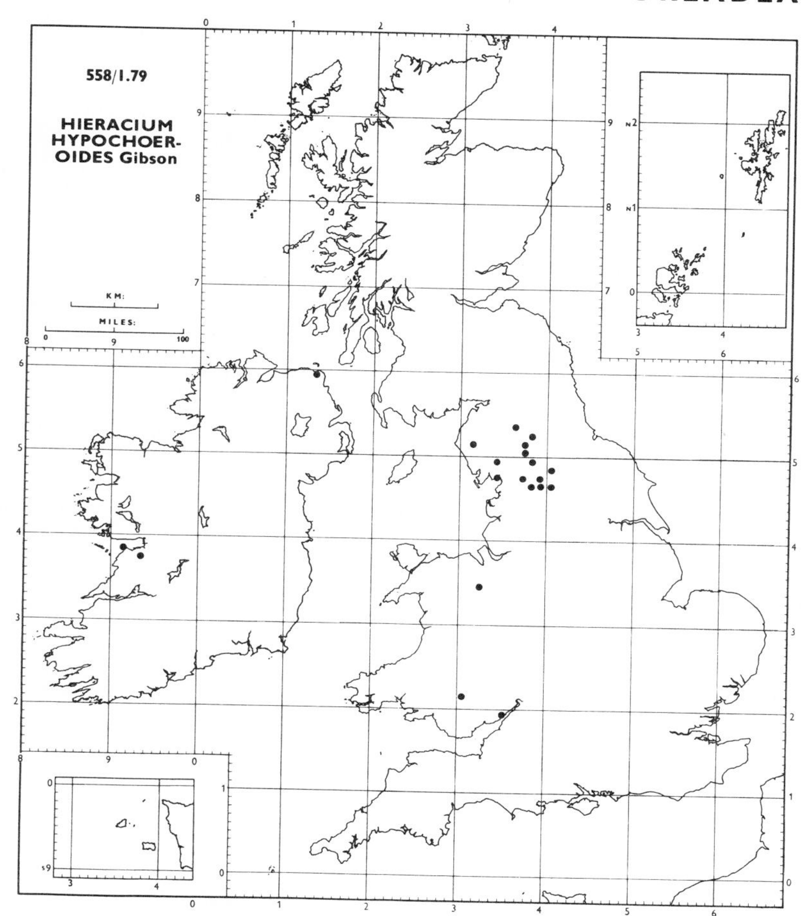

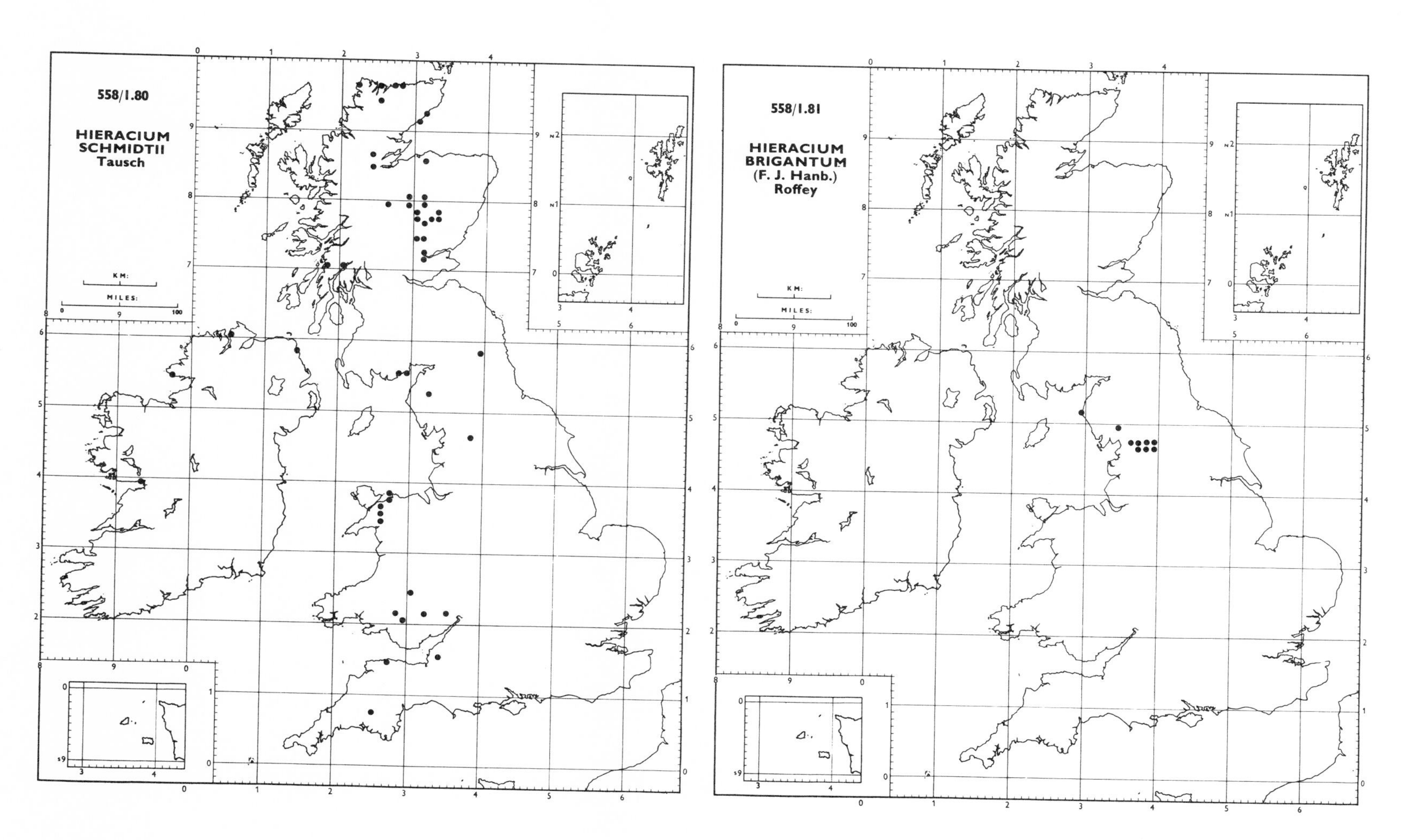

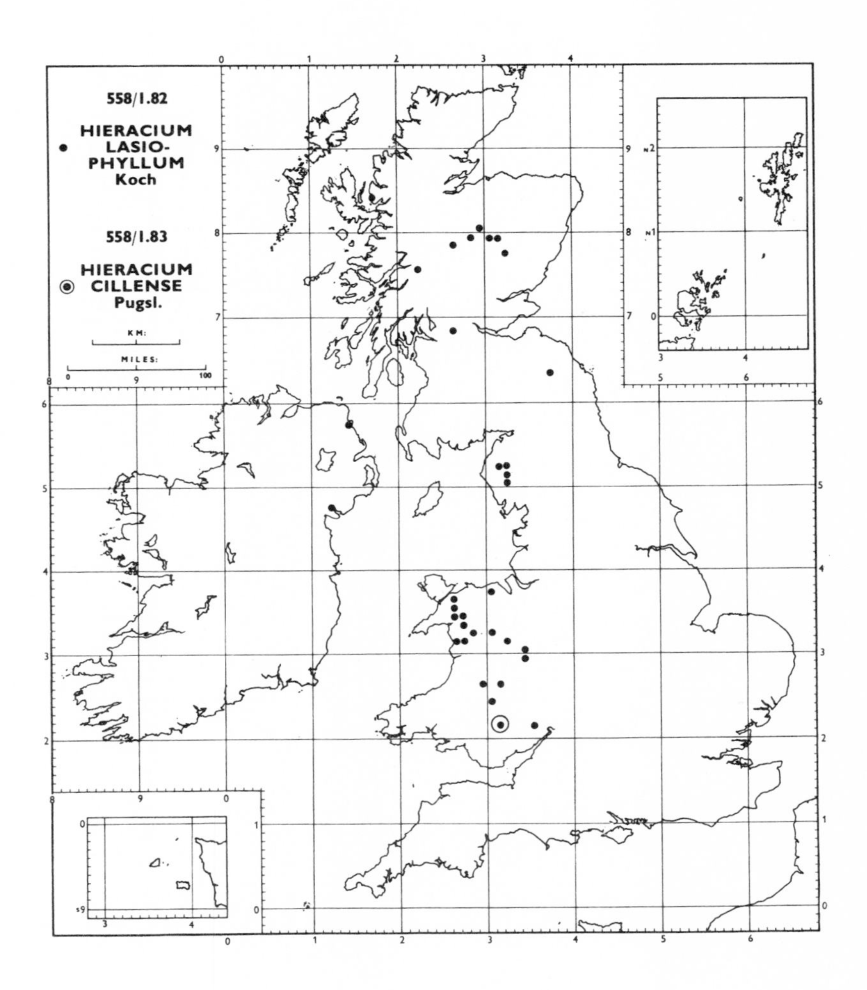

H. lasiophyllum Koch is a very distinct species, scattered over Wales, the English Lake District and Scotland, with two localities in north-eastern Ireland. On the Continent it is recorded in central Europe from France to Romania, but some of these plants may perhaps better be referred to *H. schmidtii*. It is probably always on limestone or basic rocks. **H. cillense** Pugsl., confined to the cliffs of the Llangattock Quarries, near Crickhowell, Brecon (42), is very similar to *H. lasiophyllum* and may have been derived from it.

H. nitidum Backh., **H. jovimontis** (Zahn) Roffey and **H. leyi** F. J. Hanb. have many points in common. *H. nitidum* is a widespread variable plant of cliff ledges in central and northern Scotland, with an isolated record from Moffat, Dumfries (72). *H. jovimontis* is a locally frequent plant in central Scotland, with isolated records from Lewis in the Outer Hebrides (110) and Sutherland (108). It also occurs in Germany, and is recorded from Spain. *H. leyi* is a characteristic species of rocky places in the mountains of north Wales, from which locality the species was originally described. It also occurs in isolated localities in Brecon (42) and Glamorgan (41). The plants of north England and Scotland have a slightly different facies.

H. cyathis (A. Ley) W. R. Linton is confined to a few limestone cliffs in south Wales and the Mendips (6). The general facies of the Mendip plants is slightly different from that of the plants from south Wales. *H. cyathis* does not appear to be closely related to any other British species.

H. decolor (W. R. Linton) A. Ley (*H. subcyaneum* (W. R. Linton) Pugsl.) is an abundant species on the limestone of the Derbyshire dales (57) and of north-west England. It had previously been placed in the Section *Vulgata*, but its furcate-corymbose inflorescence of large capitula and its more or less yellow styles suggest a stronger affinity with the *Oreadea*. It is an extremely variable species, especially in its leaves, which may vary in shape from oblong-ovate to broadly-ovate, and may be shallowly to deeply dentate, slightly hairy to densely hairy and light green to glaucous green or even spotted. *H. decolor* flowers from May to September, and it is possible that different types of variation can be correlated with time of flowering, especially as different types at a different stage can often be found in the same colony. The closely allied **H. pseudoleyi** (Zahn) Roffey (*H. decolor* auct.) is endemic to the limestone of north Wales.

H. sommerfeltii Lindeb. is a very local plant of central Scotland and the English Lake District. It also occurs in Norway. The endemic **H. carneddorum** Pugsl., which is frequent in rocky places on the cliffs of north Wales, and has been recorded from Fleetwith Pike, Cumberland (70), closely resembles *H. sommerfeltii* in appearance.

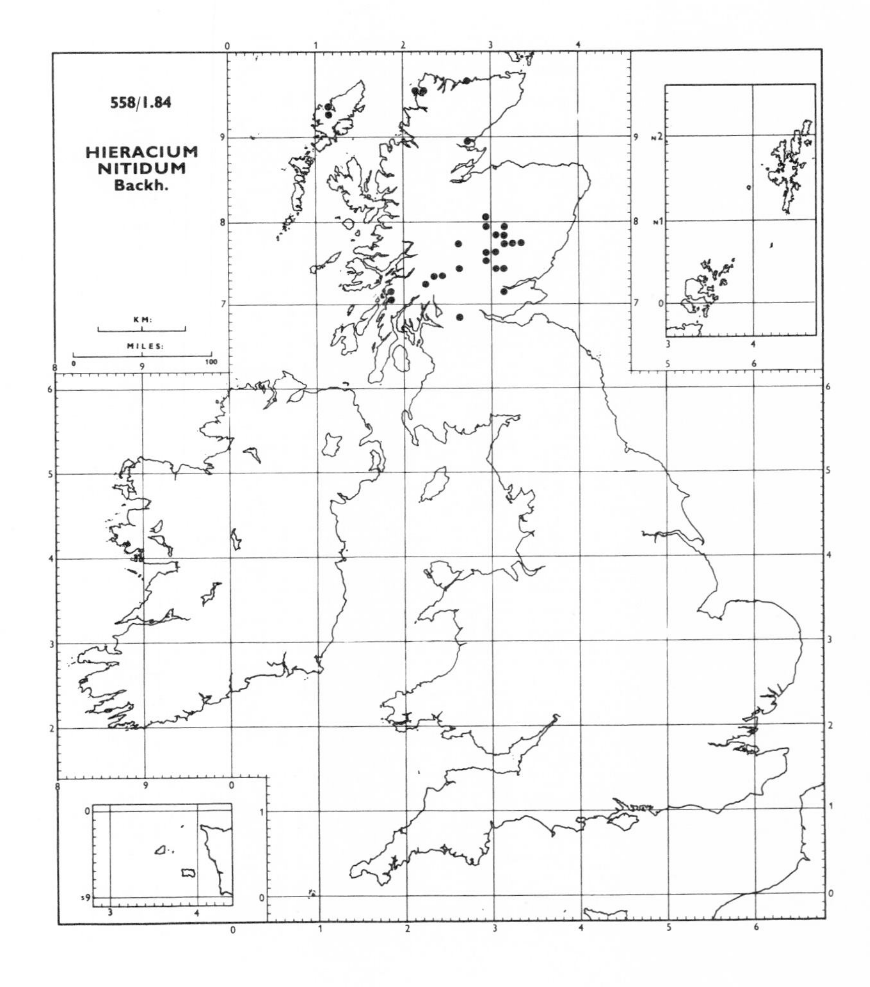

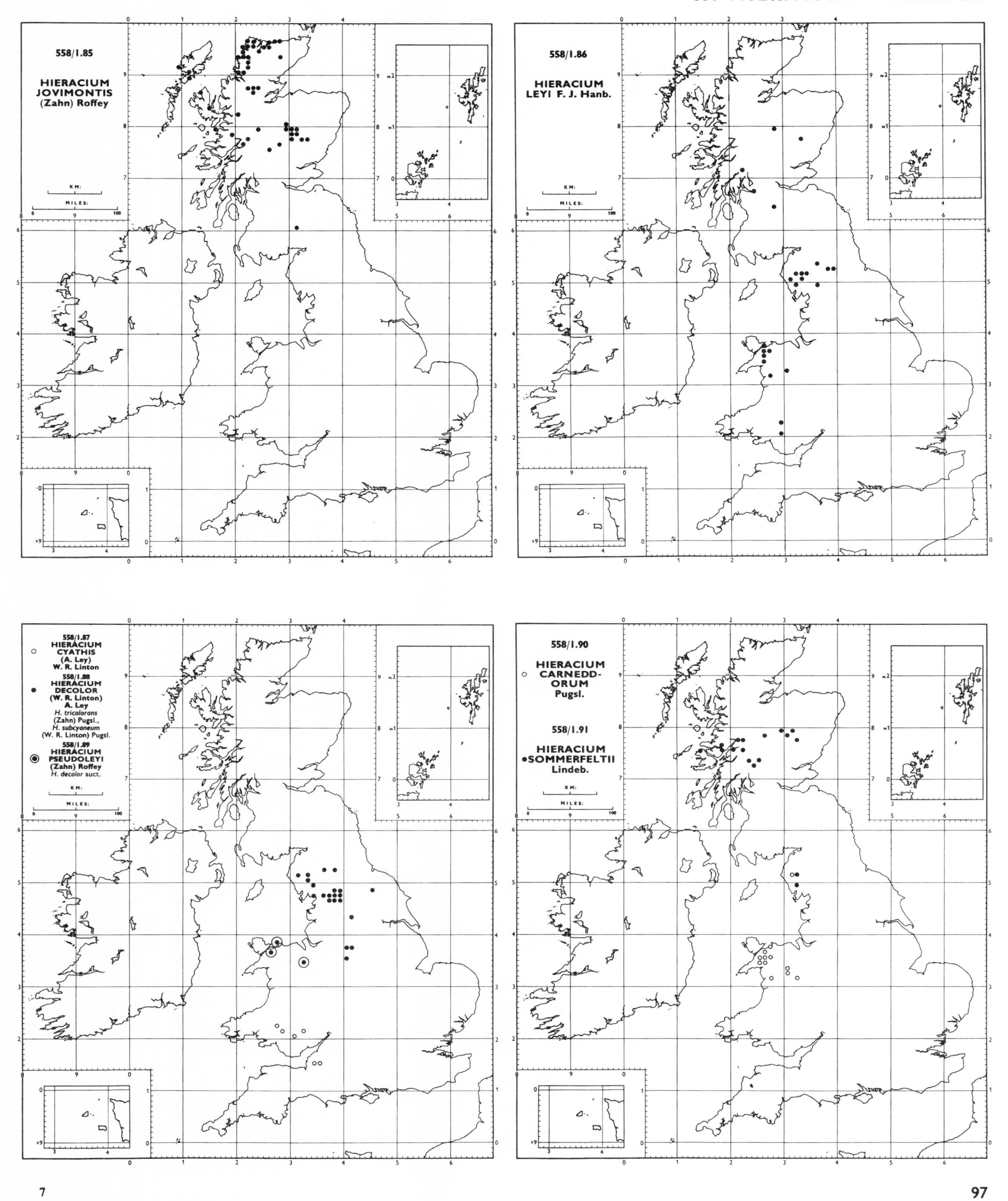
558/1.85
HIERACIUM
JOVIMONTIS
(Zahn) Roffey
558/1.86
HIERACIUM
LEYI F. J. Hanb.
558/1.87
HIERACIUM
CYATHIS
(A. Ley)
W. R. Linton
558/1.88
HIERACIUM
DECOLOR
(W. R. Linton)
A. Ley
H. tricolorans
(Zahn) Pugsl.,
H. subcyaneum
(W. R. Linton) Pugsl.
558/1.89
HIERACIUM
PSEUDOLEYI
(Zahn) Roffey
H. decolor auct.
558/1.90
HIERACIUM
CARNEDD-
ORUM
Pugsl.
558/1.91
HIERACIUM
SOMMERFELTII
Lindeb.
KM:
MILES:

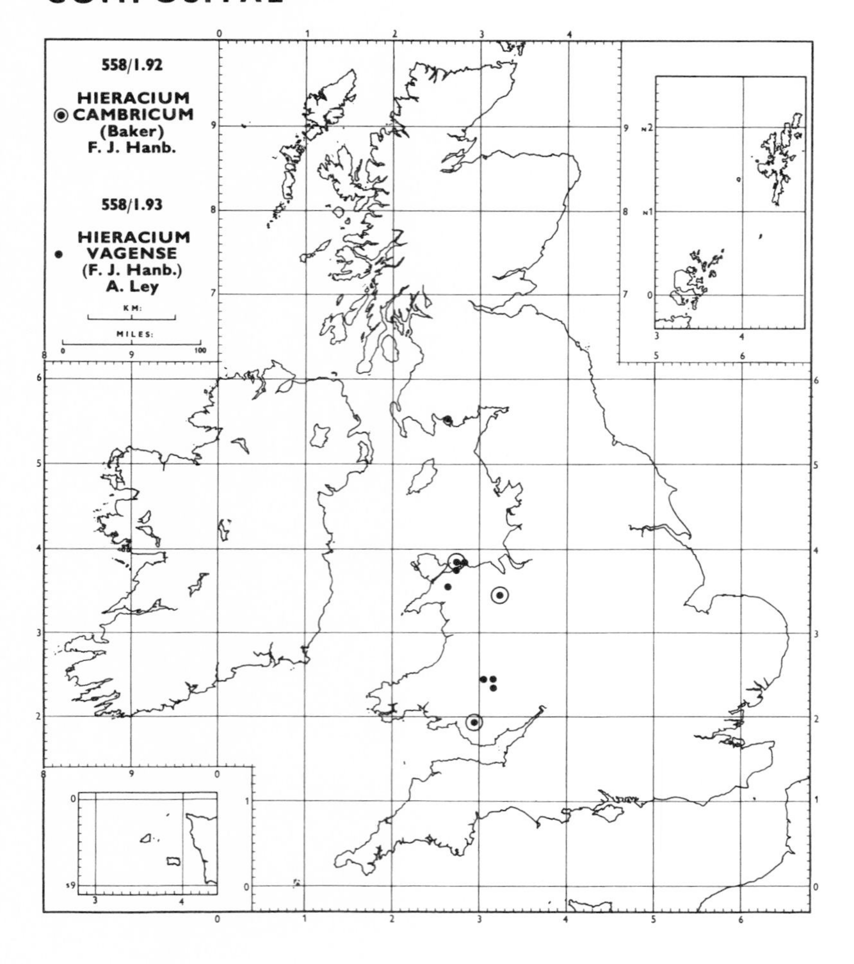

H. cambricum (Baker) F. J. Hanb. and **H. vagense** (F. J. Hanb.) A. Ley are closely related species with no near allies. *H. cambricum* is a local species of limestone cliffs. It is frequent on the Great Ormes Head in Caernarvon (49), but has not been seen in its other two localities, Treorchy, Glamorgan (41), and Llangollen, Denbigh (50), for over 60 years. In its glaucous leaves, furcate inflorescence and yellow styles, *H. cambricum* resembles the other members of Section *Oreadea*, but differs from them in having glabrous leaves. In this respect it approaches the Section *Glauca* of central Europe. *H. vagense* is a very local species of the limestone of Caernarvon, and of the low limestone banks of the River Wye in Brecon (42) and Radnor (43). A population of plants found on the banks of the River Dee, Tongland, Kirkcudbright (73) matches exactly the plants of this very characteristic species from Wales. The characteristics of the Kirkcudbright plants, like those from Wales, are retained in cultivation.

H. argenteum Fries is a common, variable species of rocky ledges, stream-sides and grassy banks (from sea-level to more than 2,000 ft (650 m.)) in central and northern Scotland. It is less frequent in northern England and Wales, and has a few scattered localities in Ireland. It is found on all types of rock, but shows a preference for those which are basic. Outside the British Isles it is a common plant in Iceland and Scandinavia. The usual form throughout its range has lanceolate leaves. In Scotland a form with elliptic-lanceolate to ovate leaves (var. *septentrionale* F. J. Hanb.) is sometimes found in the same colony as the narrow-leaved plant, to which it is connected by intermediates. The plant described by Pugsley (1948) as *H. orimeles* var. *argentatum* has more characters in common with *H. argenteum* than with *H. orimeles*, and is better regarded as a variant of that species. *H. argenteum* var. *subglabratum* (F. J. Hanb.) Pugsl., which occurs in a few localities in Scotland, has broader, more sharply dentate leaves, larger heads and pilose-tipped ligules,

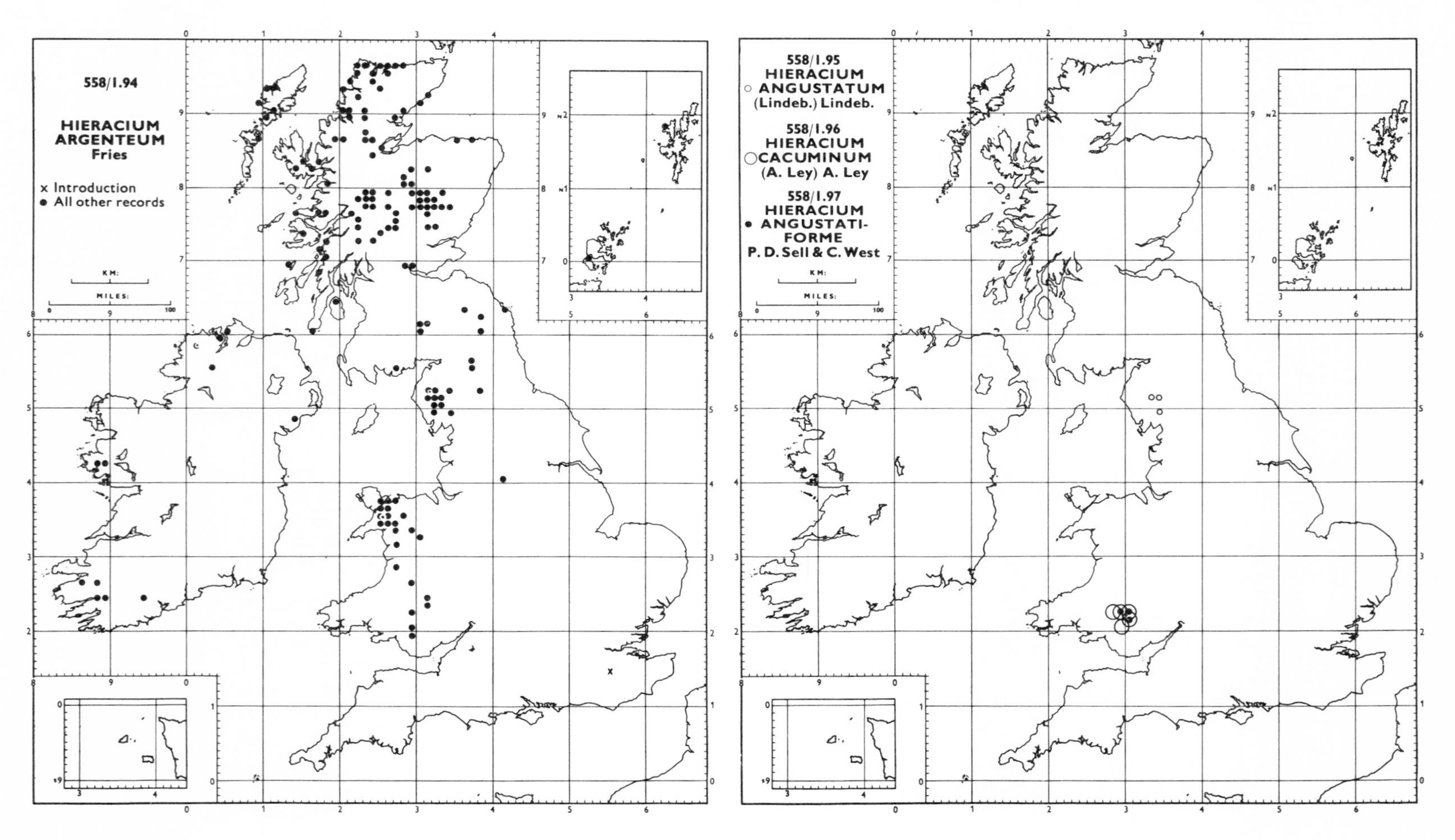

and should perhaps be regarded as a distinct species.

H. cacuminum (A. Ley) A. Ley, **H. angustatiforme** P. D. Sell & C. West and **H. angustatum** (Lindeb.) Lindeb. are closely allied species. *H. cacuminum* is a locally common plant of stream-sides and cliffs in Glamorgan and Brecon (41, 42). *H. angustatiforme* occurs in similar situations within the same geographical areas as *H. cacuminum*. *H. angustatum*, a local species of the English Lake District, also occurs in Scandinavia. It has not been seen in its British localities for nearly 60 years.

H. leyanum (Zahn) Roffey, **H. holophyllum** W. R. Linton, **H. caledonicum** F. J. Hanb. and **H. angustisquamum** (Pugsl.) Pugsl. have many characters in common. *H. leyanum* is a locally common plant of cliff ledges in south Wales. Plants collected from Portland, Dorset (9), differ slightly, but not enough to merit specific rank. The species has not been seen at this locality in recent years. *H. holophyllum* is a local species of the limestone of north Wales and the Derbyshire dales (57). *H. angustisquamum* has a very strange distribution. The localities in Somerset (6), Sutherland (108), east Ireland and Loch Conn in West Mayo (H. 27) are on limestone, while those in Kintyre (101) are on basic coastal cliffs. At Dunston Hill near Petworth, Sussex (13), the plant occurs in old chalk woodland. At Grays, Essex (18), it occurs in an old chalk quarry, where it may have been introduced. *H. caledonicum* is a widely distributed species in Wales, northern and central England, Scotland and northern Ireland, with outlying localities in Wicklow (H. 20). Throughout most of its range it is a uniform characteristic species, but on the north coast of Scotland and in the Faeroes it shows much variation in its leaves. In view of the wide range of this species, and its close affinity with the other three species mentioned in this paragraph, it may well be the parent from which they have been derived.

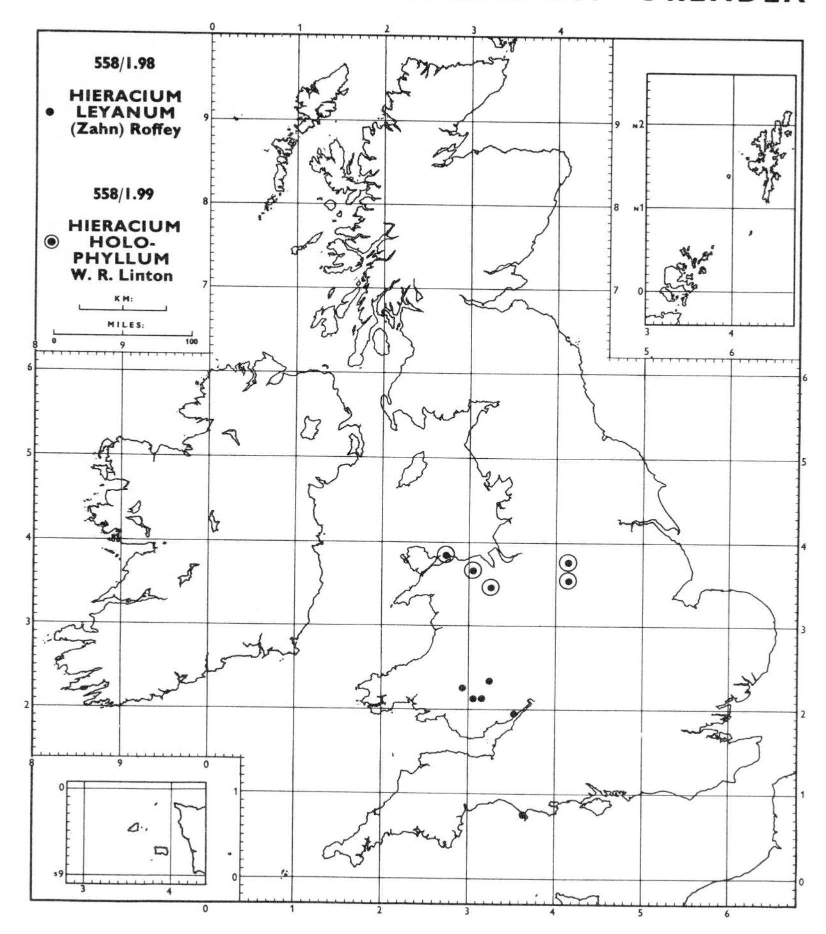

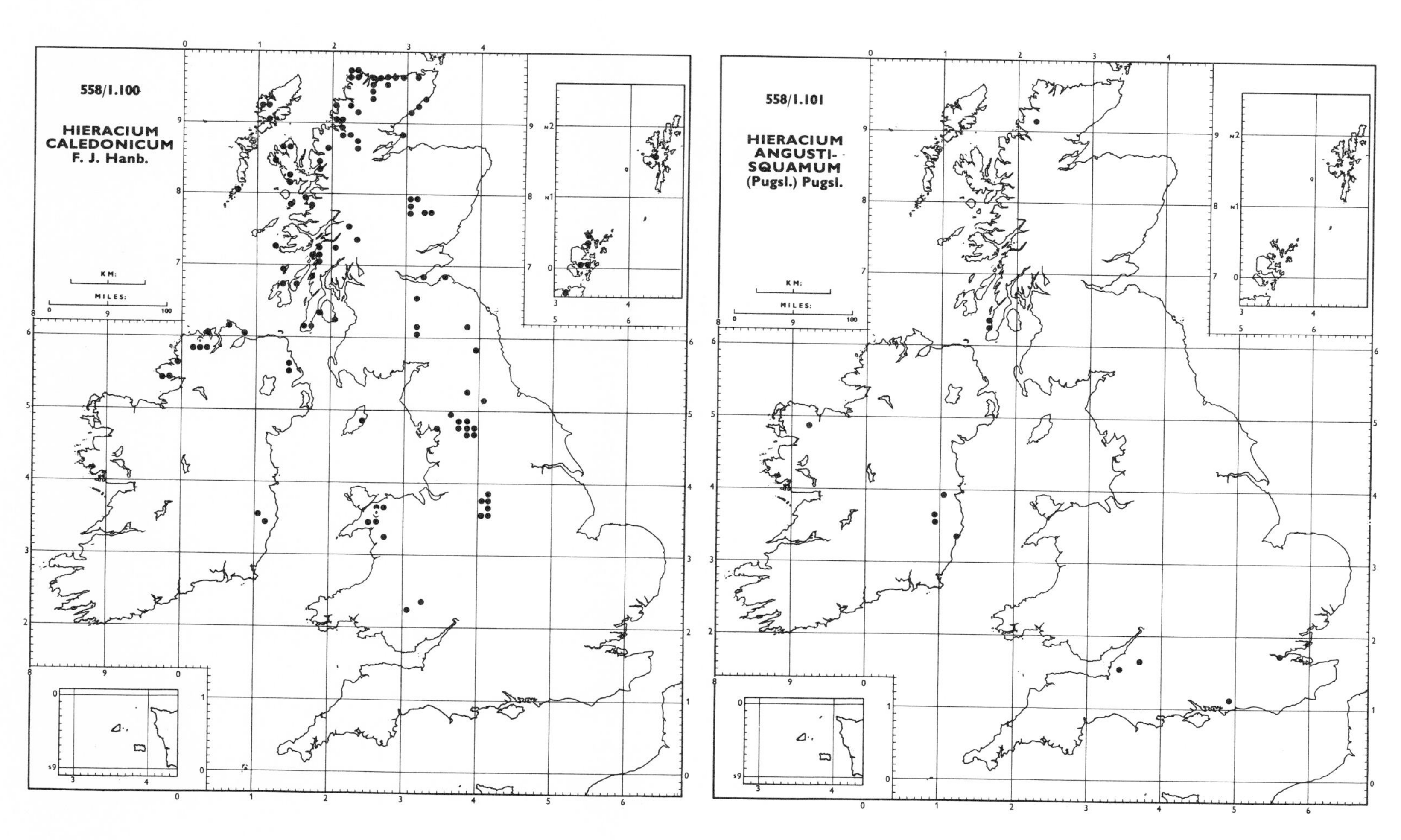

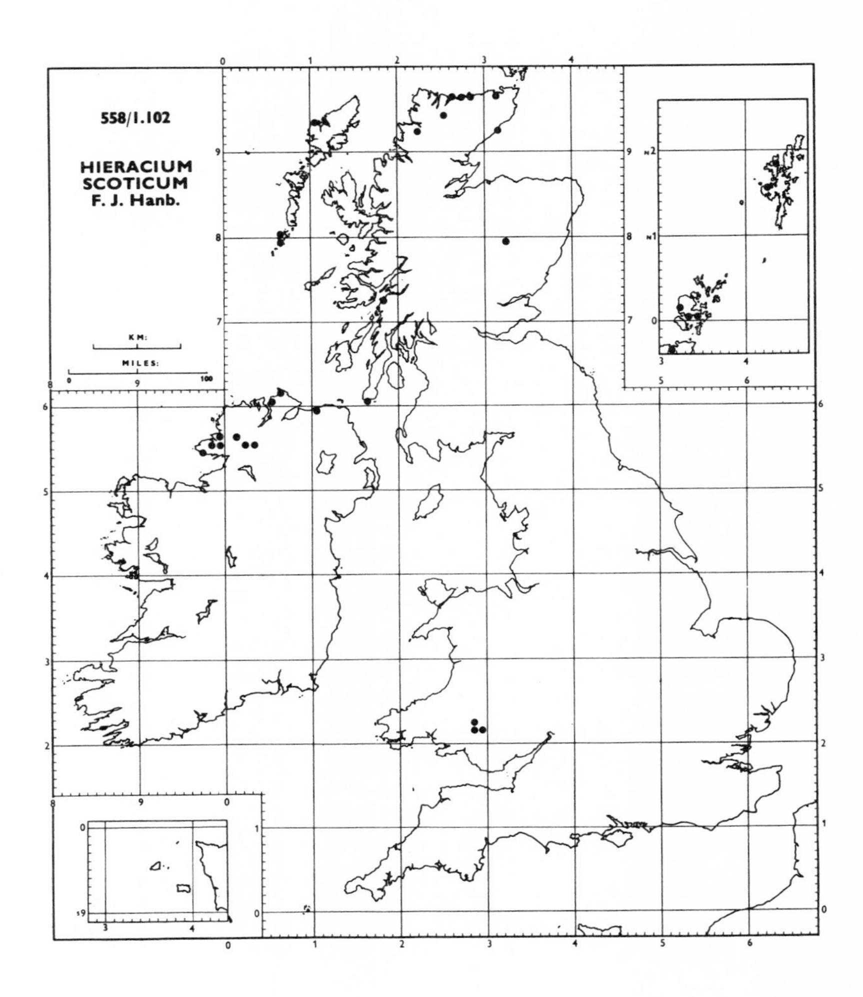

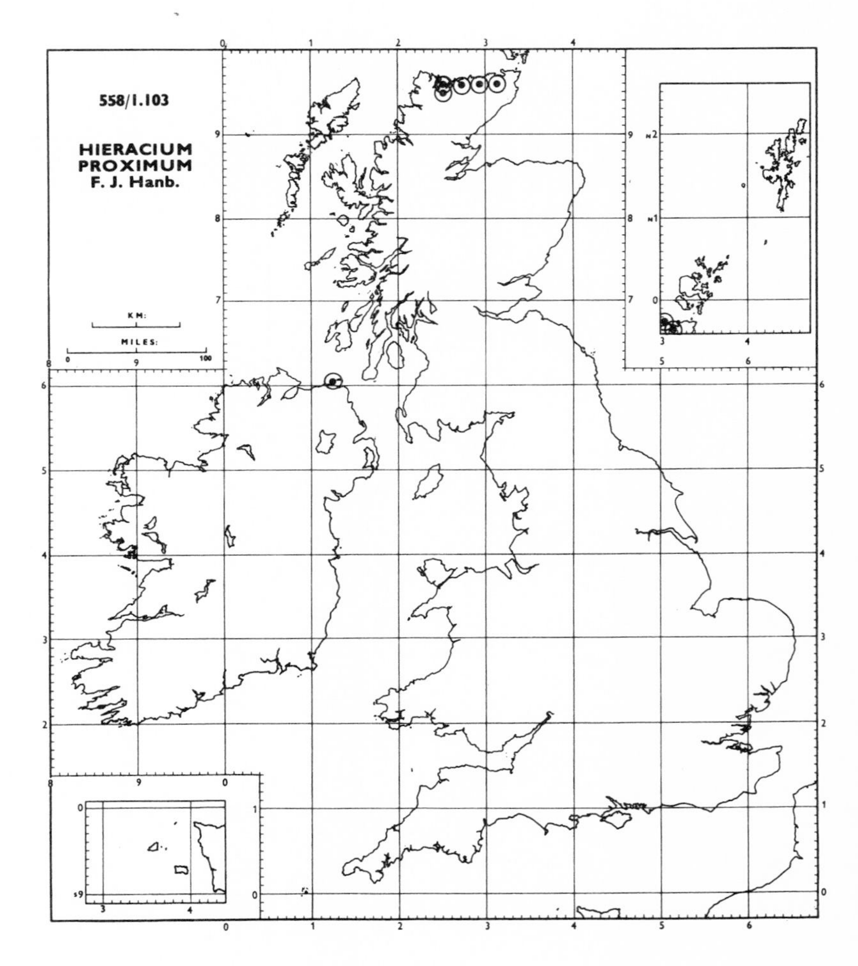

H. scoticum F. J. Hanb. is a local species of low cliffs and grassy headlands on the coasts of northern and western Scotland, Orkney (111), Shetland (112) and northern Ireland. It also occurs in a few localities inland in northern Ireland, south Wales and central and north Scotland. The plants from Brecon (42) differ from the others in having smaller capitula. Outside the British Isles it is known only from the Faeroes. **H. proximum** F. J. Hanb. is a very local species of low cliffs and grassy slopes on the north coast of Scotland and at Ballintoy in Antrim (H. 39). It is so closely allied to *H. scoticum*, and has such a similar distribution, that it must surely have been derived from the same ancestor.

H. subrude (Arv.-Touv.) Arv.-Touv., **H. orimeles** F. J. Hanb. ex W. R. Linton and **H. chloranthum** Pugsl. form a closely allied group of species, all of which have a rather wide distribution in the British Isles. *H. orimeles* is an abundant species of cliff ledges and rocky stream-sides in north Wales, the Outer Hebrides (110) and Shetland (112), with scattered localities in south Wales, northern England, Scotland, Orkney and northern Ireland. *H. subrude* is a local plant with a range somewhat similar to that of *H. orimeles*, but it is absent from Shetland (112) and Orkney (111) and extends into the Isle of Man (71). On the Continent it occurs in south-east France and west Germany. These two species have so many characters in common and are so similar in distribution that they must surely have had the same ancestry. *H. chloranthum*, which differs from *H. orimeles* only in its greenish ligules, which never become radiant, replaces that species in central Scotland, where it is widespread. The range of these two species overlaps in a few places in west Scotland and near Moffat, Dumfries (72).

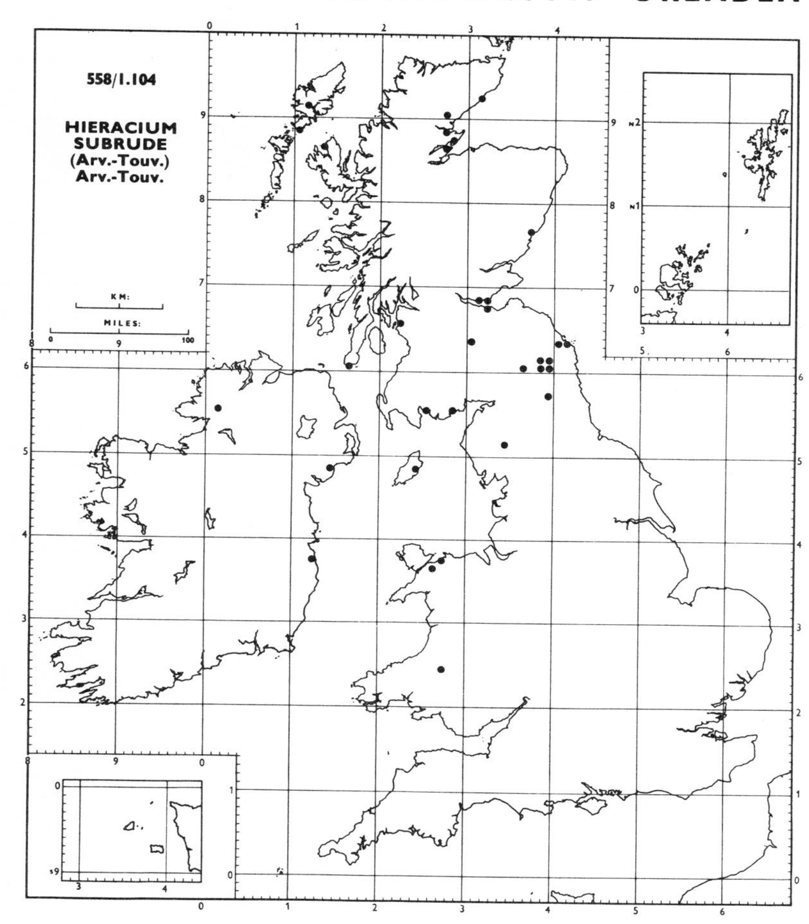

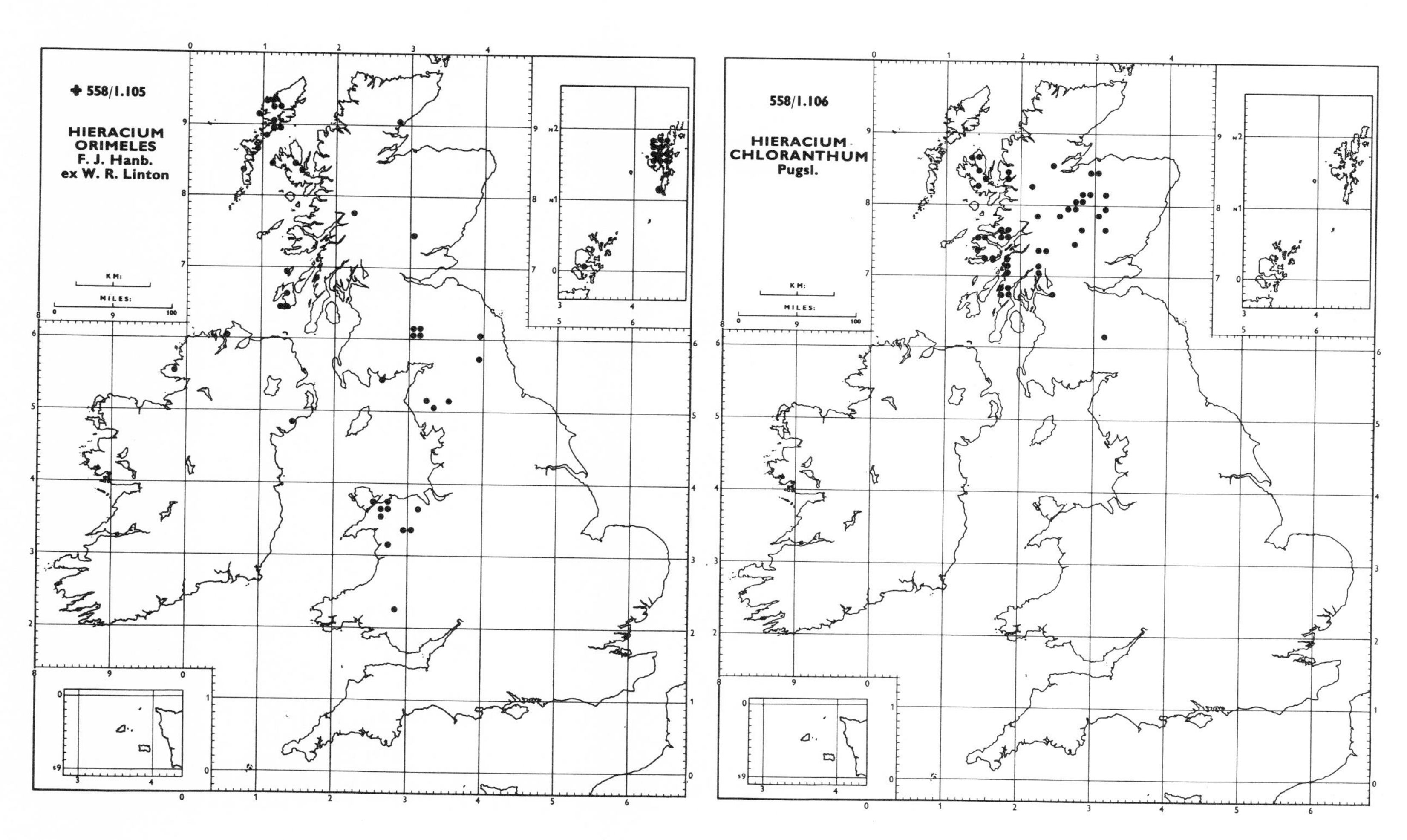

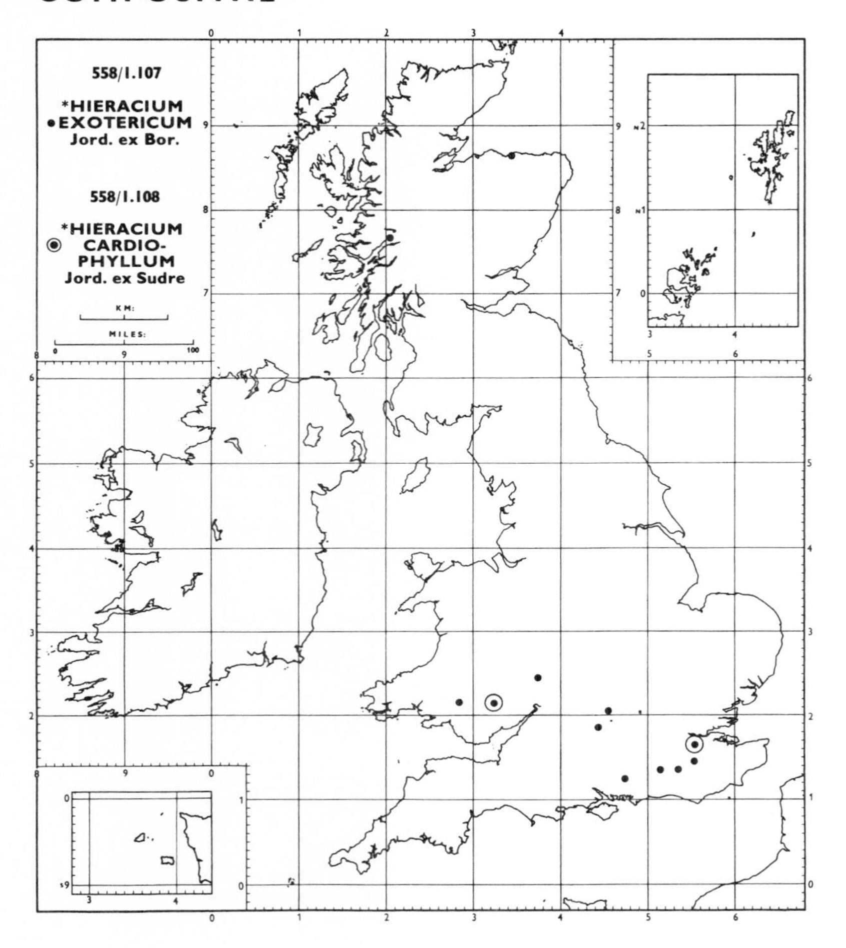

558/1.107–188 Section **VULGATA** F. N. Williams
This section, which is apparently a natural one, is the most complex in the Genus *Hieracium*. It contains the greatest number of species in the genus, 82 of these occurring in the British Isles. The great degree of tolerance as regards habitat shown by species of this section allows them readily to be introduced into new areas, thus making it difficult to be sure whether the plants are native or not. Species of this section are found throughout Europe, and are recorded in northern Asia and North America. They occur in almost every part of the British Isles. Owing to the doubtful status (alien or native) of many of the species in this section we have been hesitant in suggesting their possible origin. It is possible that some species have in fact originated in what do not appear to be "native" habitats.

H. exotericum Jord. ex Bor. *sensu stricto* occurs only in a few scattered localities where it is probably always introduced. It may, however, be native on river-side rocks at Abercrave, Brecon (42). It is difficult to give its native distribution, but it was originally described from woodland south of Paris, and has been recorded from much of continental Europe. **H. cardiophyllum** Jord. ex Sudre, native in the southern half of Europe, is well naturalized near Sevenoaks, Kent (16), and near Llangattock, Brecon (42). **H. grandidens** Dahlst. is an introduced species, which is well-naturalized and still spreading in scattered localities throughout Great Britain, with two records from north-eastern Ireland. It is probably of more frequent occurrence than the map suggests, as we have only recently been collecting records of it. It is native in the mountains of central Europe and is widely naturalized elsewhere on that Continent. *H. grandidens* flowers from May to October, but only the

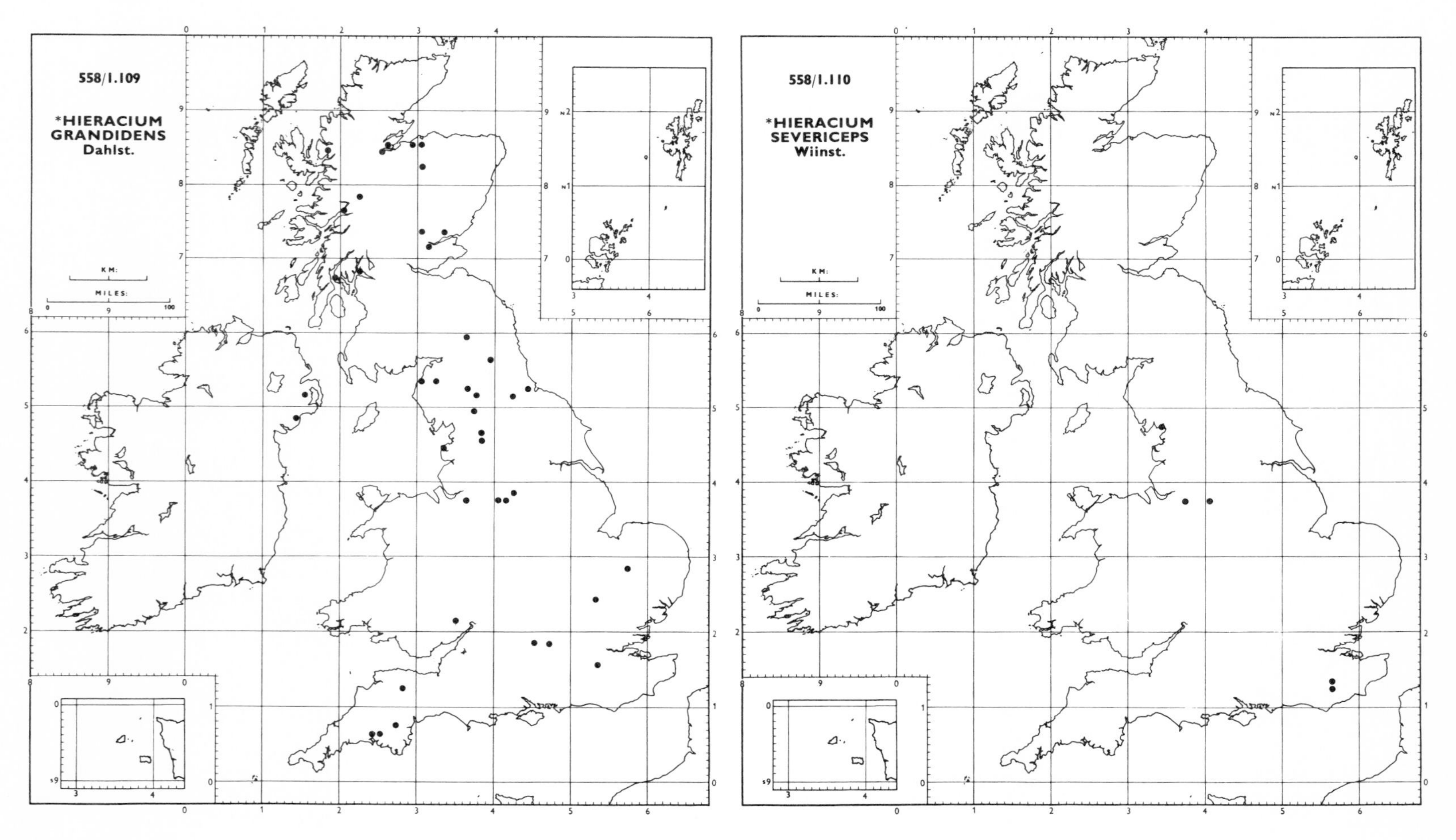

early-flowering (May and June) plants show the characteristic features. Specimens from localities in Kent (16), Derbyshire (57), Cheshire (58) and Lancashire (69) are identical with plants described from Sweden and Denmark as **H. severiceps** Wiinst. In all its known localities it is apparently an introduction, and its origin is unknown.

H. sublepistoides (Zahn) Druce is a frequent species on roadsides and railway banks, and is found occasionally in open woodland in central and southern England and Wales. There are isolated records from Scotland and Ireland. It is an introduction into many, if not all, of its localities in the British Isles. It is native in the Alps and Pyrenees.

H. cinderella (A. Ley) A. Ley is locally frequent on road and railway banks and in adjacent woods in the English–south Wales border country, with an isolated locality in Denbigh (50). Specimens from rocks near Saintfield, Down (H. 38), also belong to this species. Its habitats in the British Isles do not make it possible to say whether it is a native plant, but it is not known elsewhere.

H. patale Norrl. occurs in considerable abundance on roadsides, railway banks and woods on the southern border of Dartmoor (3), and on walls at Onich, Inverness (97). It is presumably an introduction in both localities. It is a native of Fennoscandia.

H. subcrassum (Almq. ex Dahlst.) Johans. occurs on roadside banks and woods at Titley, Herefordshire (36), Marshbrook, Shropshire (40), and Grisedale, Lake District (69); in all these areas it may be native. It also occurs in Austria, Hungary and Sweden.

H. hjeltii Norrl. is known in the British Isles only from specimens collected near the railway at Silverdale, Lancs (60) in 1912, where it was presumably introduced. *H. hjeltii* is native in Fennoscandia.

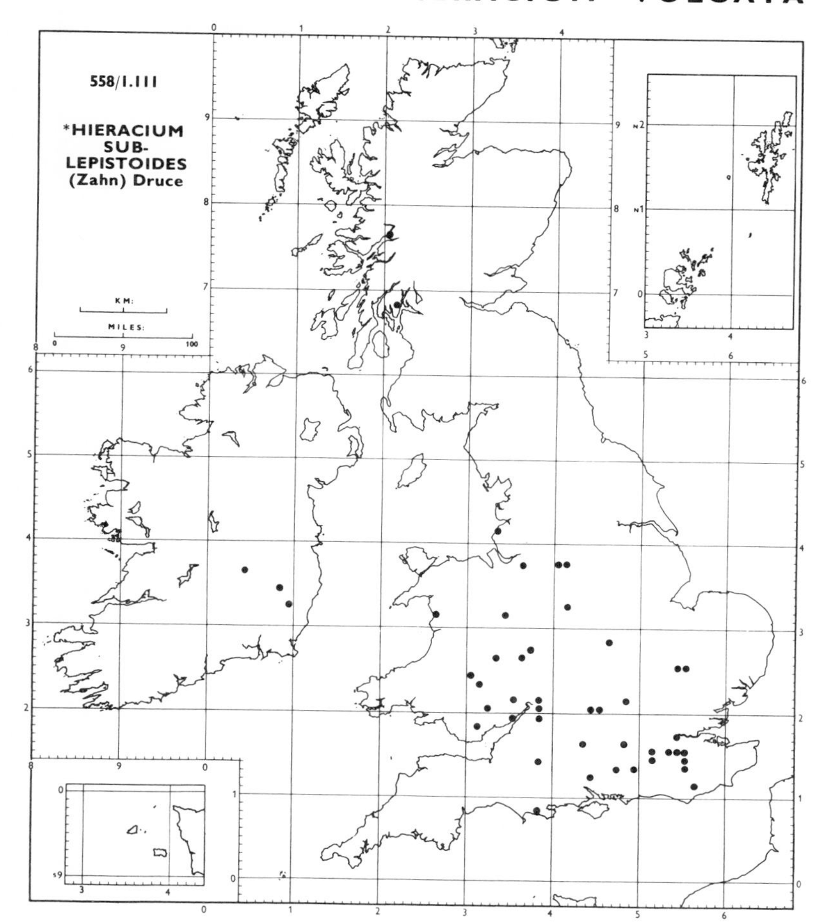

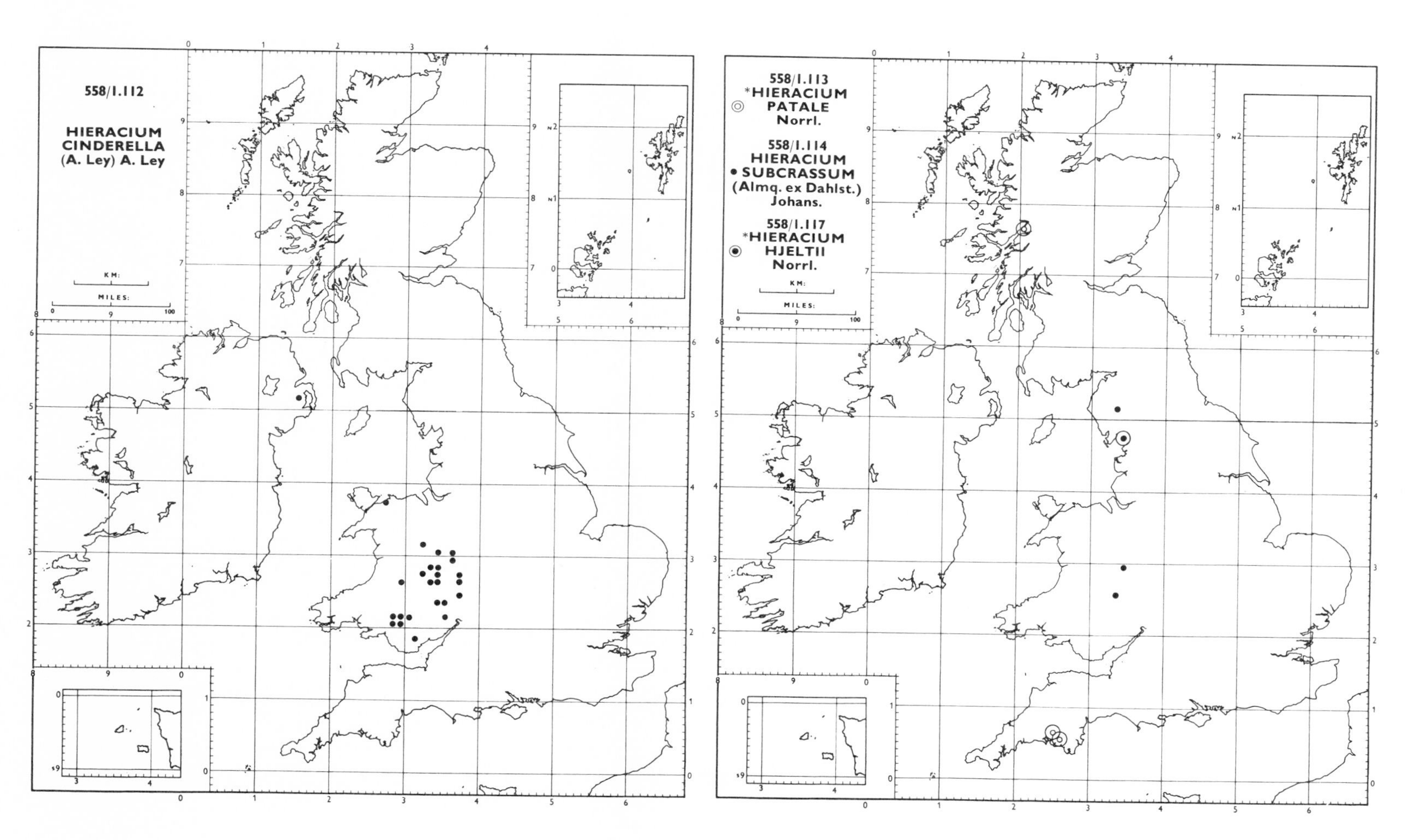

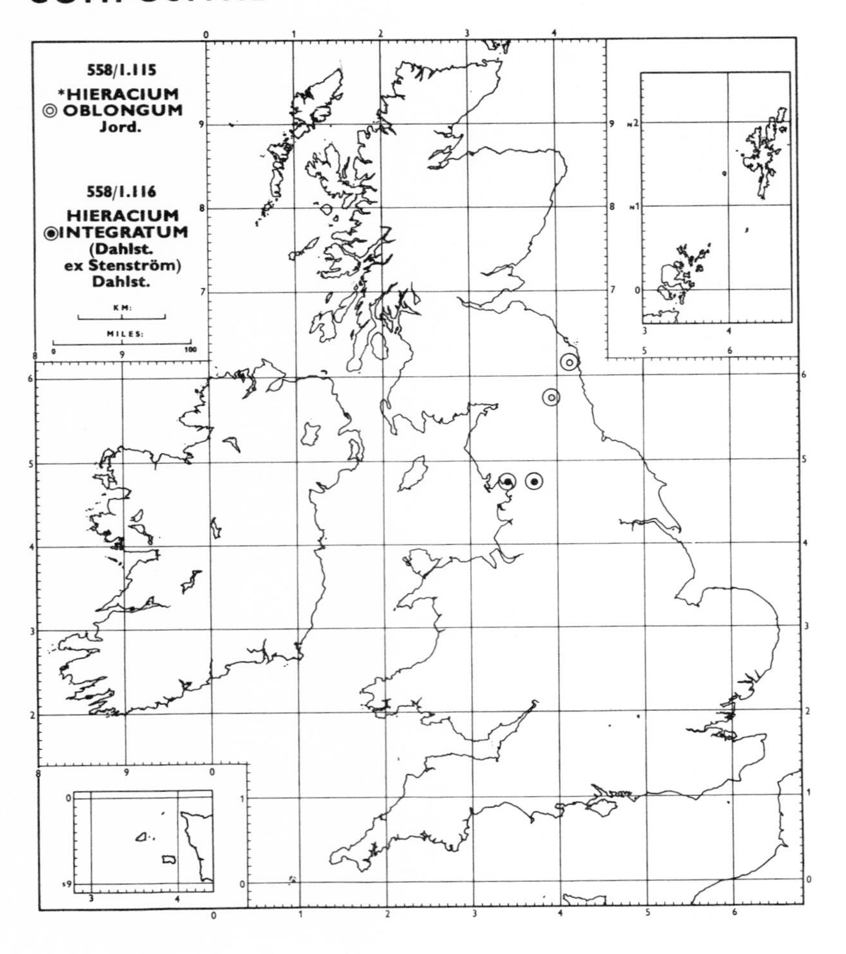

H. oblongum Jord. is naturalized on grassy banks of the reservoir at Hallington and on the walls near Alnwick, Northumberland (67, 68). It occurs throughout the greater part of continental Europe.

H. integratum (Dahlst. ex Stenström) Dahlst. appears to be a native plant of the limestone at Ingleborough (64) and possibly also at Arnside Knott (69). It is distributed throughout northern, central and eastern Europe.

H. pellucidum Laest. is a common native species of the limestone areas of south Wales, the Derbyshire dales (57) and northern England. At Wymington, Bedfordshire (30), and at Oxford (23) it is an introduction. It also occurs throughout northern Europe. Both in Great Britain and on the Continent plants with different basic leaf-shapes occur in the same colony.

H. camptopetalum (F. J. Hanb.) P. D. Sell & C. West is endemic to a few localities in northern Scotland.

H. snowdoniense P. D. Sell & C. West is endemic to rocky faces of the mountains of Caernarvonshire (49).

H. pruinale (Zahn) P. D. Sell & C. West is a native species known only from specimens gathered at Glen Dole, Clova (90), in 1890.

H. uistense (Pugsl.) P. D. Sell & C. West is a native endemic to western Scotland and Shetland (112).

H. candelabrae W. R. Linton is a native endemic to the limestone of Ingleborough (64).

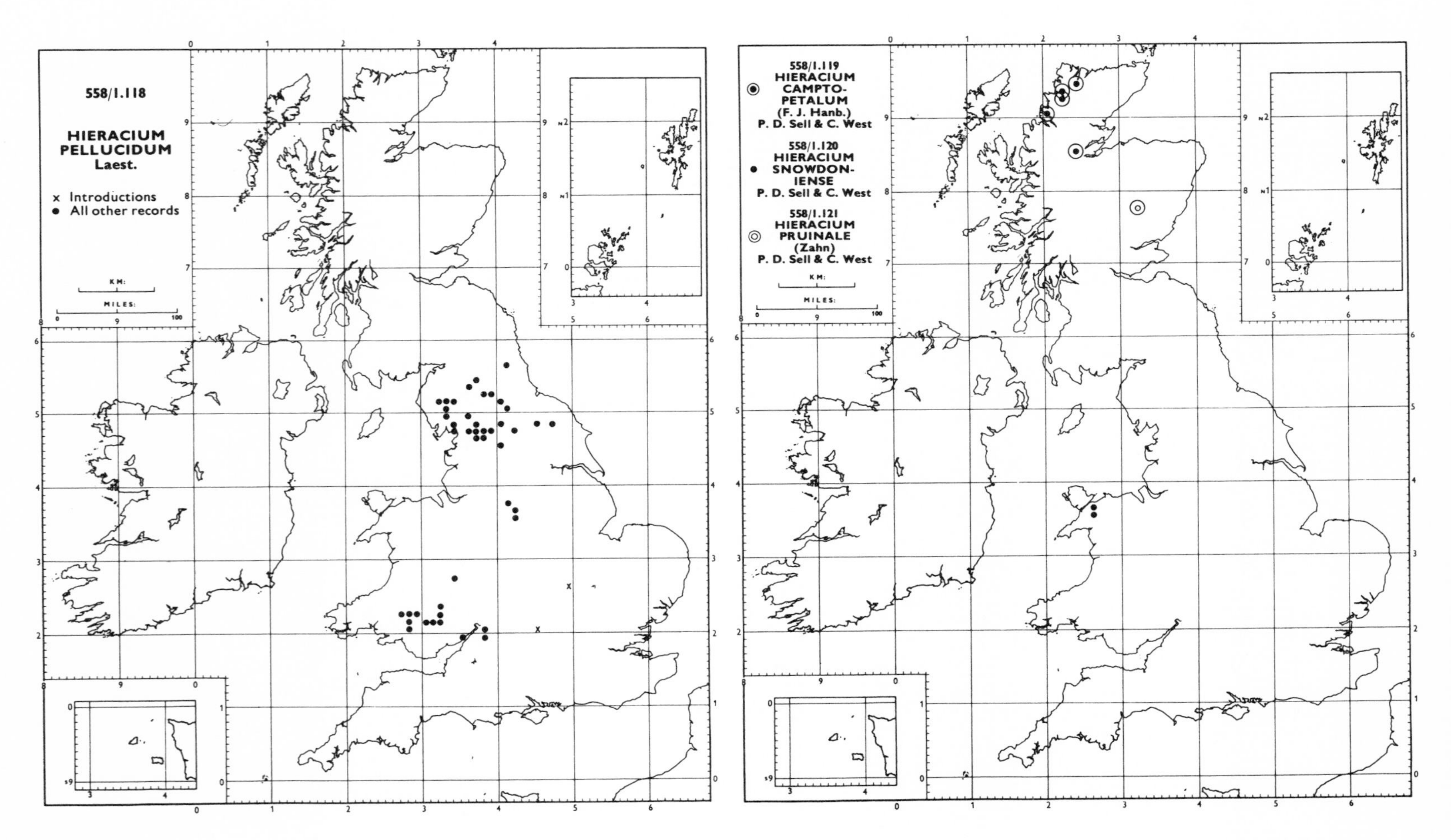

H. stenstroemii (Dahlst.) Johans. (*H. glevense* (Pugsl.) P. D. Sell & C. West) is an abundant native species of rocky areas in south-east Wales and adjacent parts of England. It also occurs in Scandinavia.

H. pollinarium F. J. Hanb. is a native endemic to a few places in Sutherland (108).

H. itunense Pugsl. was described from specimens collected at Barras, Westmorland (69), in 1903, where it was last seen in 1929. It is doubtful whether it was a native plant.

In the British Isles **H. prolixum** Norrl. is a native species known only from sand hills at Tongue Bay and from cliffs of Ben Hope, Sutherland (108). It also occurs in Scandinavia and east Europe.

H. subprasinifolium Pugsl. is a native endemic species of grassy limestone slopes in the Staffordshire and Derbyshire dales (39, 57).

H. discophyllum P. D. Sell & C. West is a native species on limestone cliffs at Pwll Byfre, Brecon (42), and at Fan Fechan, Carmarthen (44). Plants from cliffs in a few other localities in that area have slightly larger, more glandular capitula, but are probably conspecific. Plants from Sweden labelled *H. orbicans* Almq. (a later homonym) may also be conspecific.

H. asteridiophyllum P. D. Sell & C. West is a native endemic species on limestone cliffs near Crickhowell, Brecon (42).

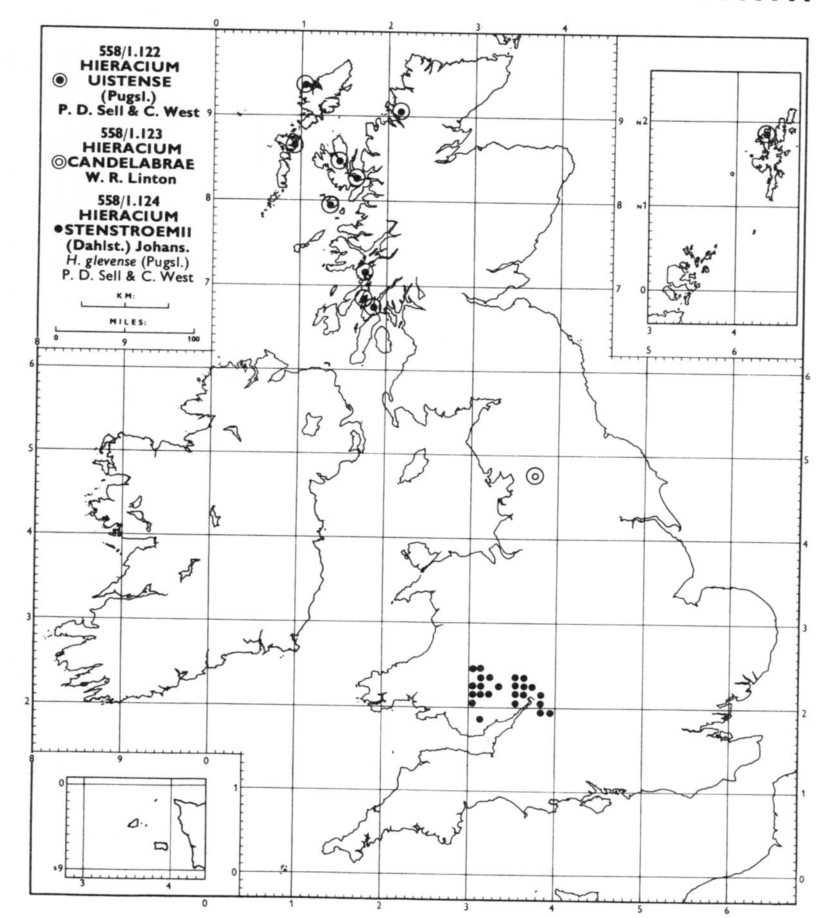

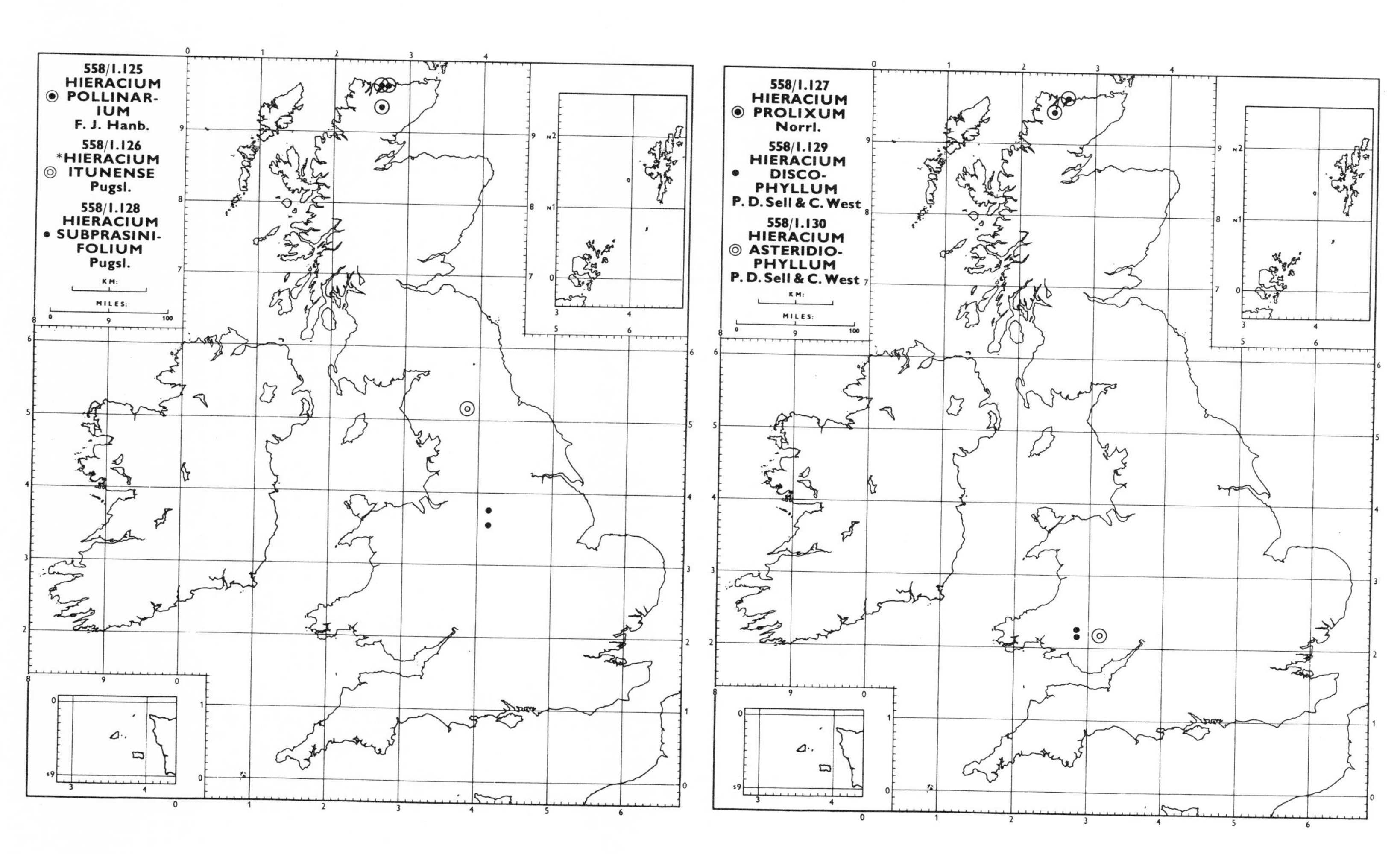

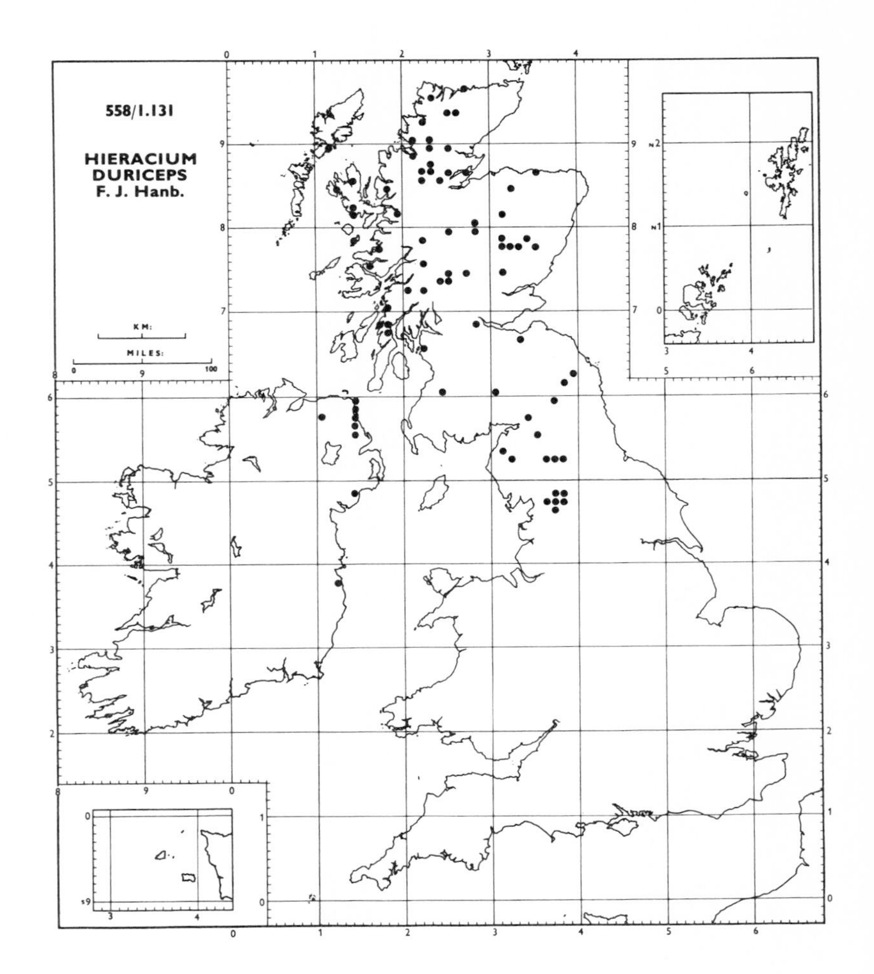

H. duriceps F. J. Hanb. is a widespread native species on cliffs and rocky stream-sides in north England and Scotland, and also occurs in eastern Ireland. Elsewhere it is known from Sweden and central Europe.

H. scotostictum Hyland. is an introduced species, well naturalized on railway banks and other grassy places around London, and in a few other scattered localities. It was first recorded in the London area in 1920, and since that date has steadily extended its range. It is probably native in Spain and central Europe, but it is widespread over the remainder of the Continent. The areas to which it is indigenous are difficult to define.

H. gougetanum Gren. & Godr. is the plant naturalized on the banks of the River Liffey, near Dublin (H. 21), long known as *H. maculosum*. It is a native plant of Spain, Pyrenees and central Europe.

H. pictorum E. F. Linton (*H. semicrassiceps* Pugsl.) is a common, native variable species of rocky sub-alpine streams and cliffs of the Scottish Highlands. It also occurs in Sweden.

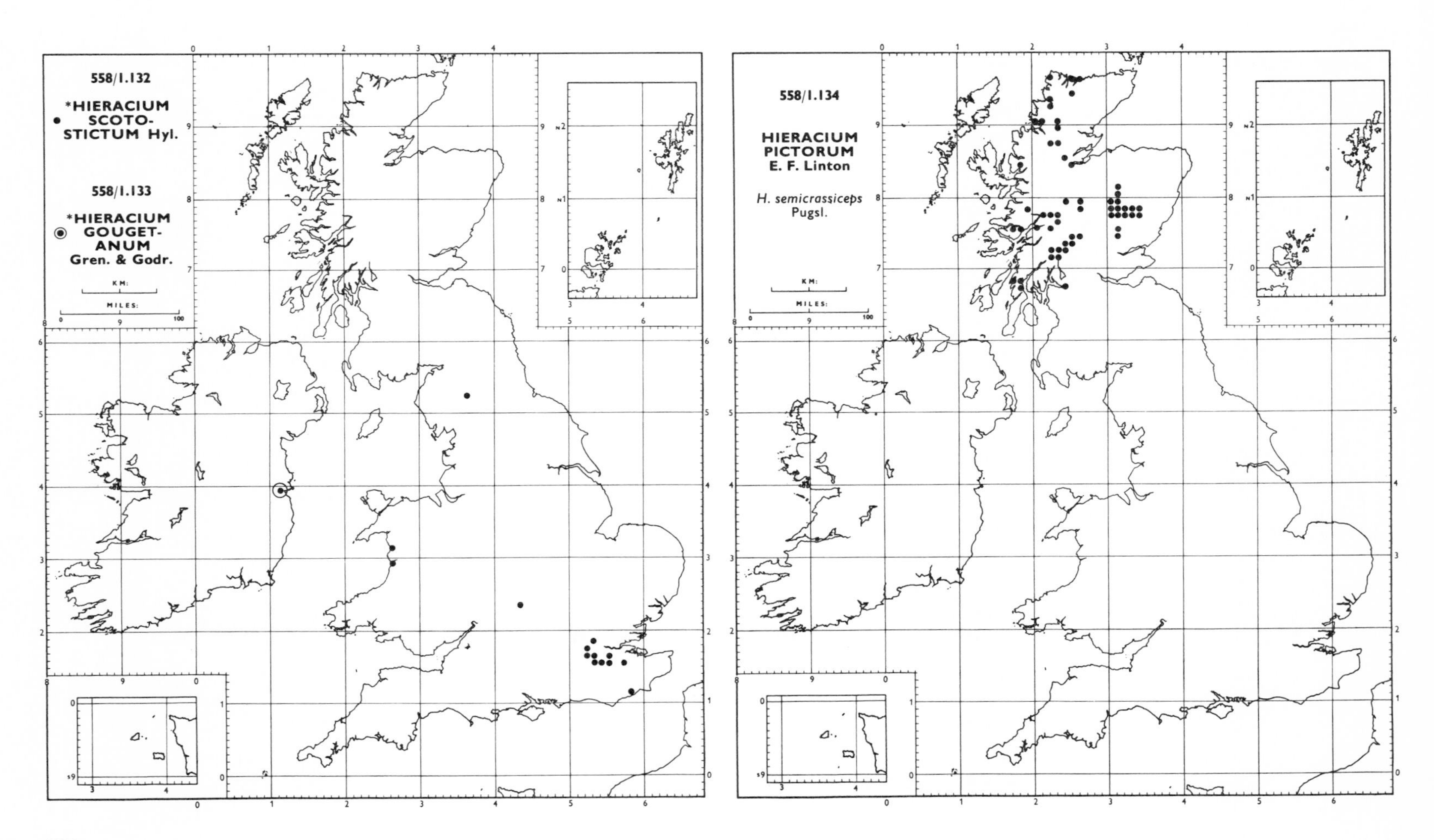

H. pollinarioides Pugsl. is a local endemic species on (? basic) coastal cliffs of the north and west of Sutherland (108) and Ross (105).

H. piligerum (Pugsl.) P. D. Sell & C. West (*H. variicolor* var. *piligerum* Pugsl.) is a locally frequent native endemic of cliff ledges and rocky stream-sides of the Highlands of central Scotland, with a few outlying localities further north. It is closely allied to the Scandinavian *H. variicolor* (Stenström) Omang.

H. cymbifolium Purchas is a common endemic species of limestone cliffs and slopes in the Staffordshire and Derbyshire dales (39, 57) and in mid- and north-west Yorkshire (64, 65). There are single records from adjacent areas in Cheshire (58) and Westmorland (69).

H. pseudostenstroemii Pugsl. is a frequent, native endemic species on limestone cliffs and stream-banks in mid- and north-west Yorkshire (64, 65), with a single locality in Westmorland (69).

H. pseudosarcophyllum Pugsl. is a native endemic species restricted to rocky stream-sides near Moffat, Dumfries (72).

H. sanguineum (A. Ley) W. R. Linton is a local, native, endemic species of limestone cliffs and rocks of Brecon (42), Yorkshire (64), Burren, Clare (H. 9) and Sligo (H. 28).

H. pachyphylloides (Zahn) Roffey is a local, native endemic species of cliffs and grassy railway banks at Symonds Yat, Gloucester (34), and at the Great Doward, Hereford (36).

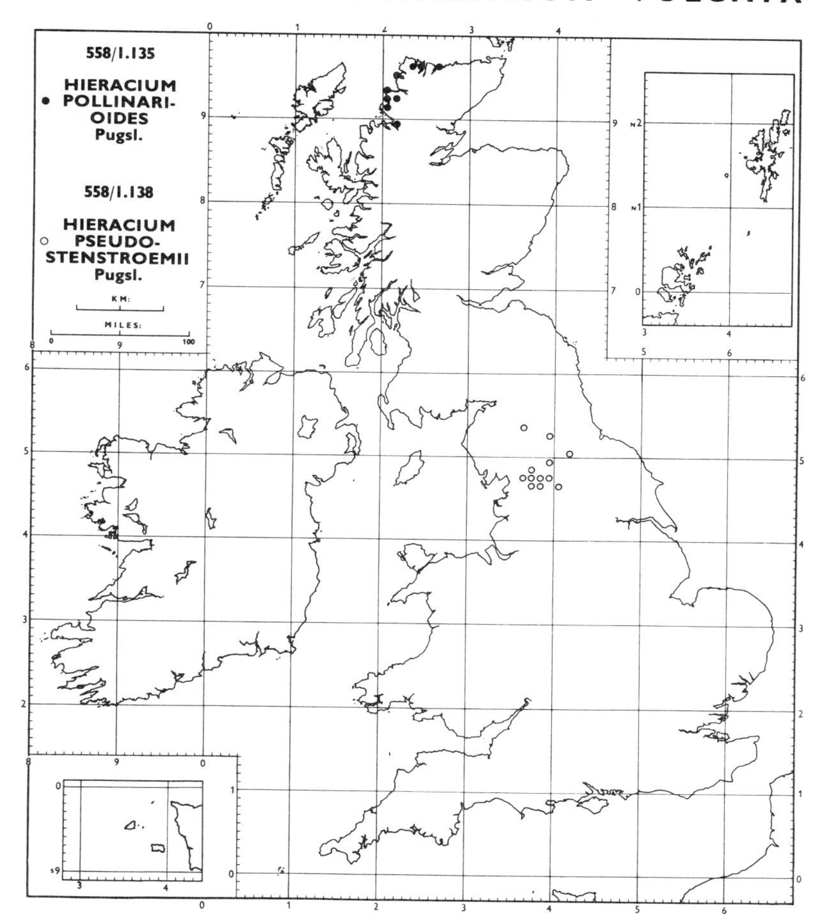

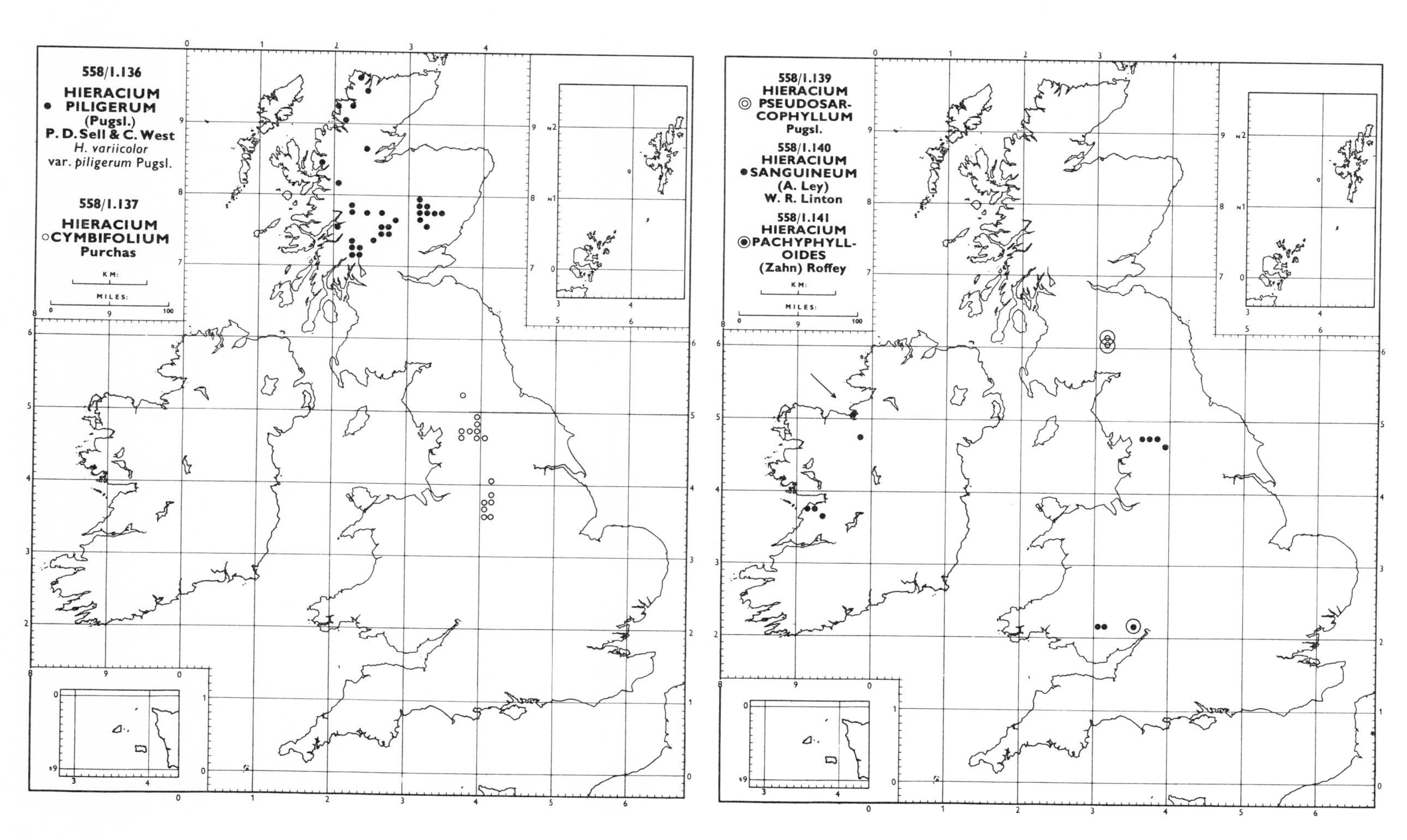

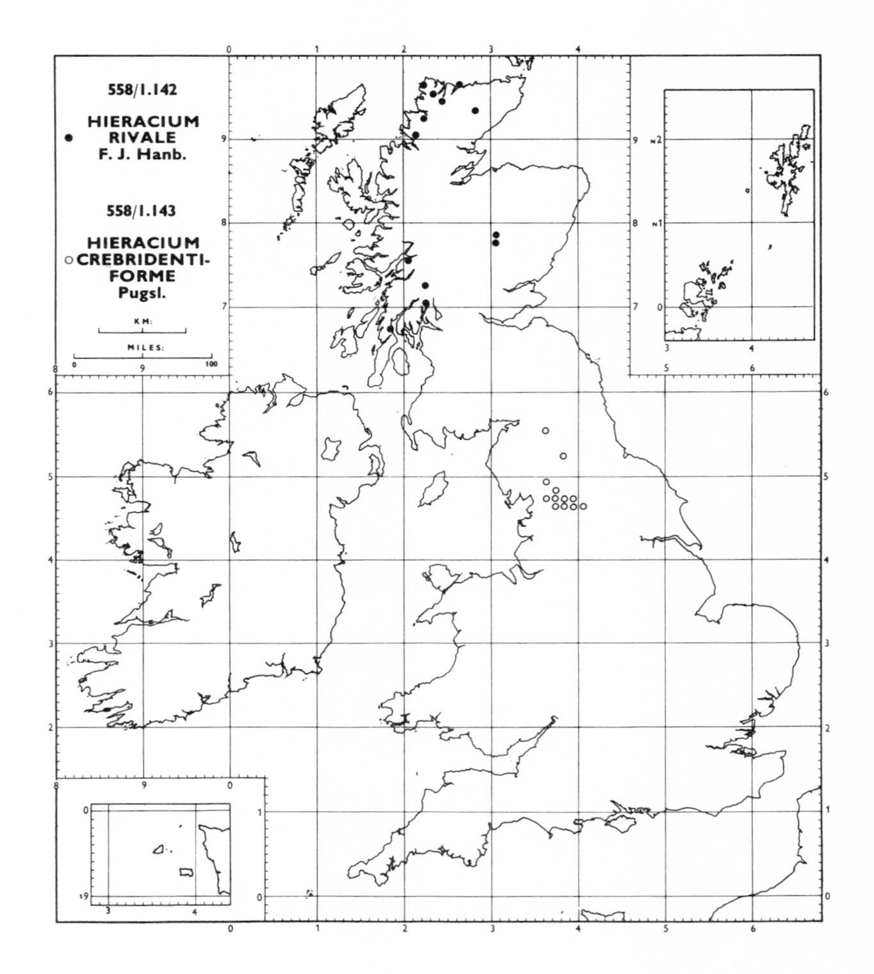

H. rivale F. J. Hanb. has been said by various authors to be a widespread species in Scotland, but, as here interpreted, it is a local, native endemic plant on cliffs and by rocky streams in that country.

H. crebridentiforme Pugsl. is a rather common, native endemic species on limestone cliffs and other rocky places in Yorkshire (64, 65), with isolated records from Teesdale (66) and from Slaggyford, Northumberland (67). The closely allied endemic **H. auratiflorum** Pugsl. is common in mid- and north-west Yorkshire (64, 65) and extends into Westmorland (69), Cumberland (70) and Northumberland (67). These two species are almost certainly derived from the same ancestor.

H. maculosum (Stenström) Omang is a very local native species of the limestone near Ingleborough (64) and Whinsill Fell, Westmorland (69). It also occurs in Fennoscandia and Austria.

H. mucronellum P. D. Sell & C. West is a native endemic species of the limestone of Bettyhill and Durness, Sutherland (108).

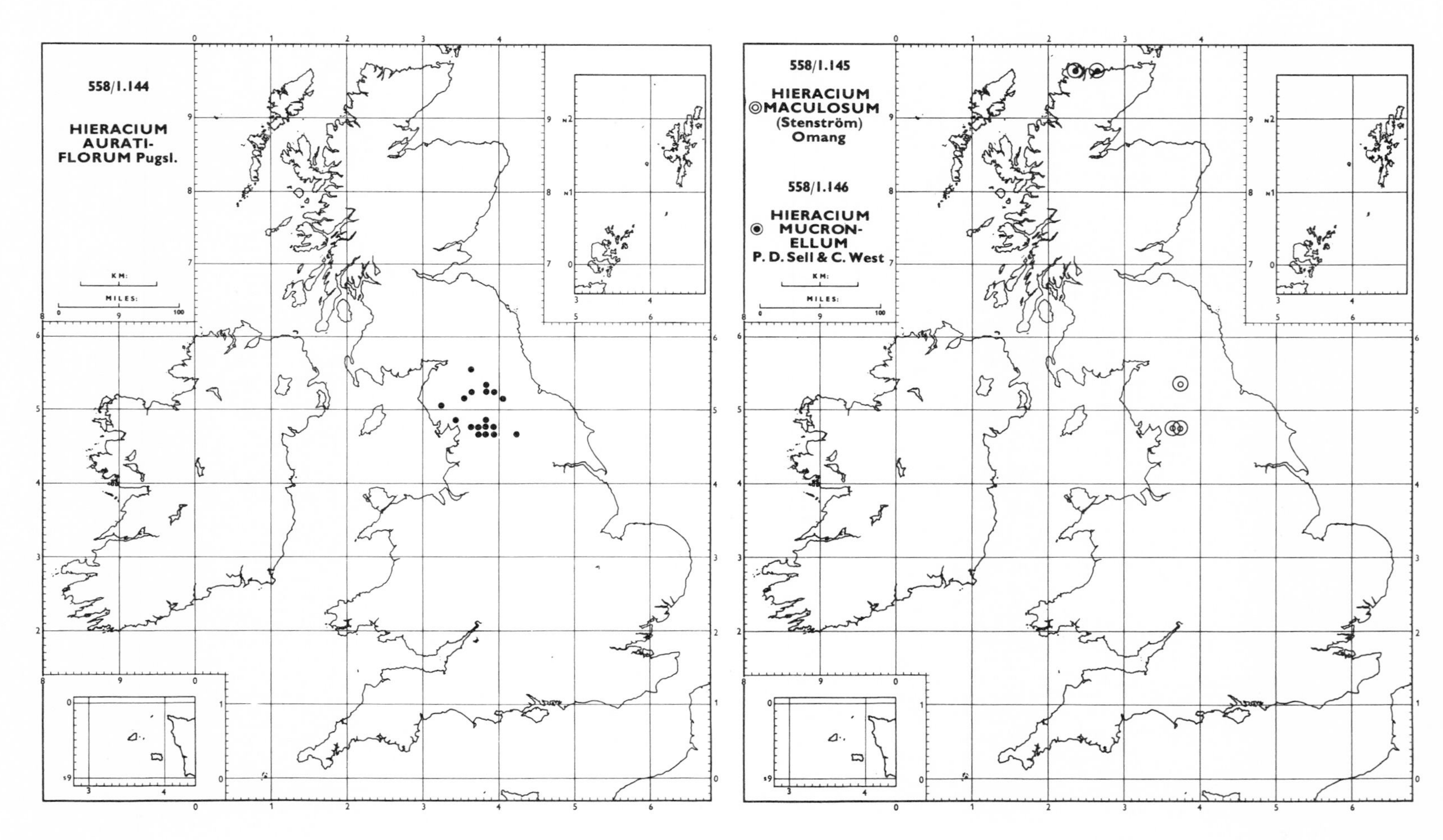

H. pauculidens P. D. Sell & C. West is a native, endemic, coastal species at Coalbackie and Skerray, Sutherland (108).

H. variifolium P. D. Sell & C. West is a native, endemic species on cliffs and rocky places in Sutherland (107, 108).

H. anguinum (W. R. Linton) Roffey is a native, endemic species restricted to rocky stream-sides near Moffat, Dumfries (72).

H. aggregatum Bakh. is a very local, native, endemic, alpine species of rocky ledges and stream-sides in the mountains of south Aberdeen (92), Clova (90) and the Cairngorms, with an isolated locality in Argyll (98).

H. subtenue (W. R. Linton) Roffey is a widespread, but local, native endemic of cliff ledges and rocky stream-sides in central and north-west Scotland.

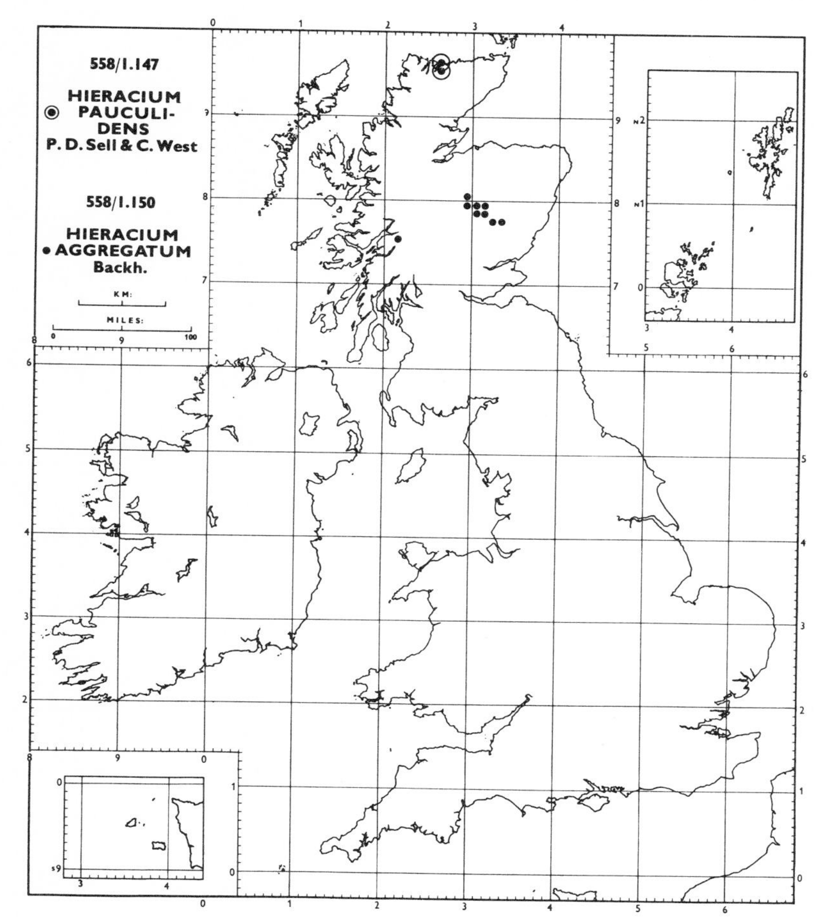

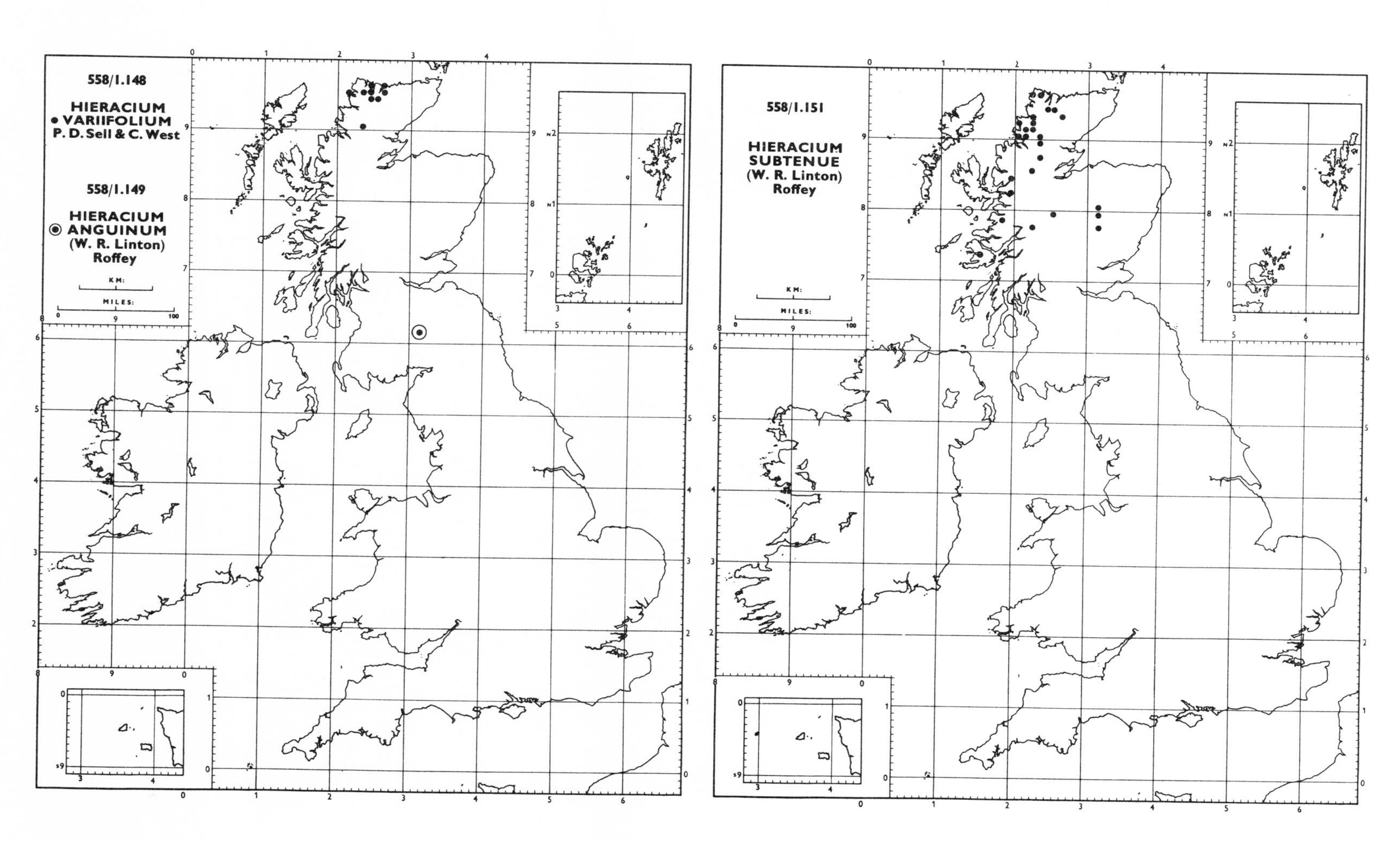

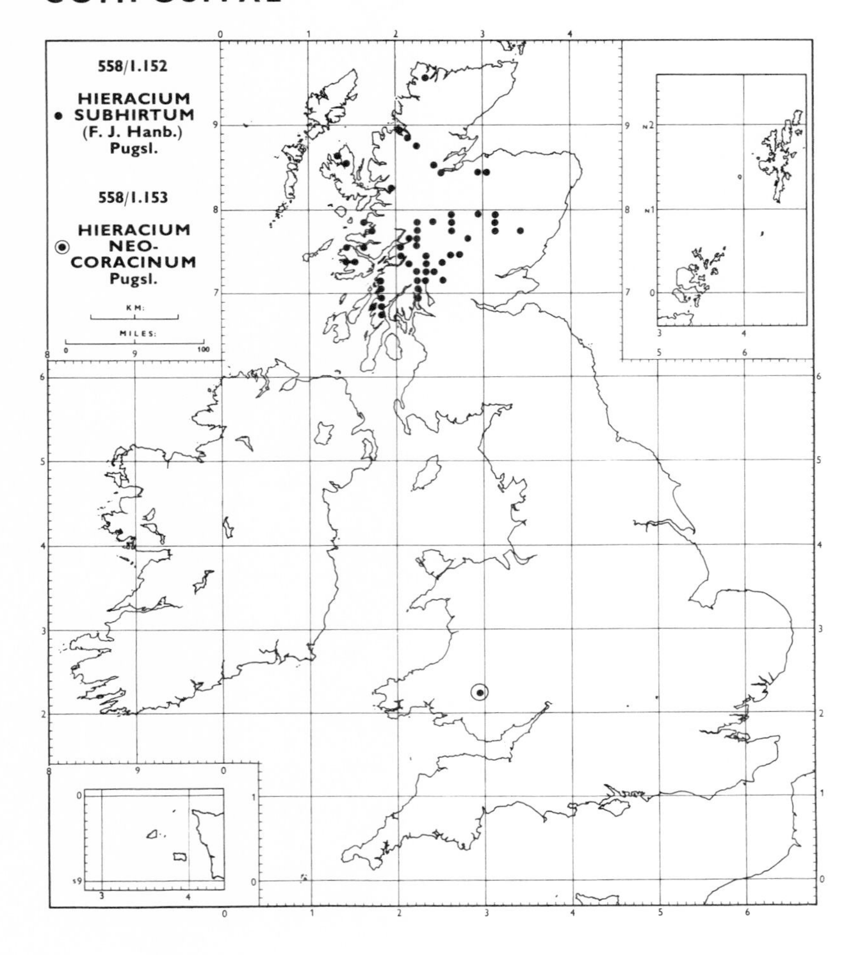

H. subhirtum (F. J. Hanb.) Pugsl. is a rather common, native endemic species of cliffs and rocky stream-sides in central Scotland, with several scattered localities further north.

H. neocoracinum Pugsl. is a scarce, native endemic species restricted to Craig Cerrig-gleisiad, Brecon (42).

H. oistophyllum Pugsl. is a local, native species of cliffs and grassy slopes, often partially shaded, in northern England and central Scotland, with isolated localities in Dumfries (72) and Ross (106). In Yorkshire this plant seems to be confined to the limestone, and in its other localities it appears to prefer basic soils. It also occurs in Scandinavia and Russia. **H. lintonianum** Druce is a closely allied, native, sub-alpine endemic species of the mountains of east-central Scotland.

H. uisticola Pugsl. is a very local, native endemic of west and central Scotland and of Uist in the Outer Hebrides (110).

H. breadalbanense F. J. Hanb. is a very local, native, endemic, alpine species of central Scotland.

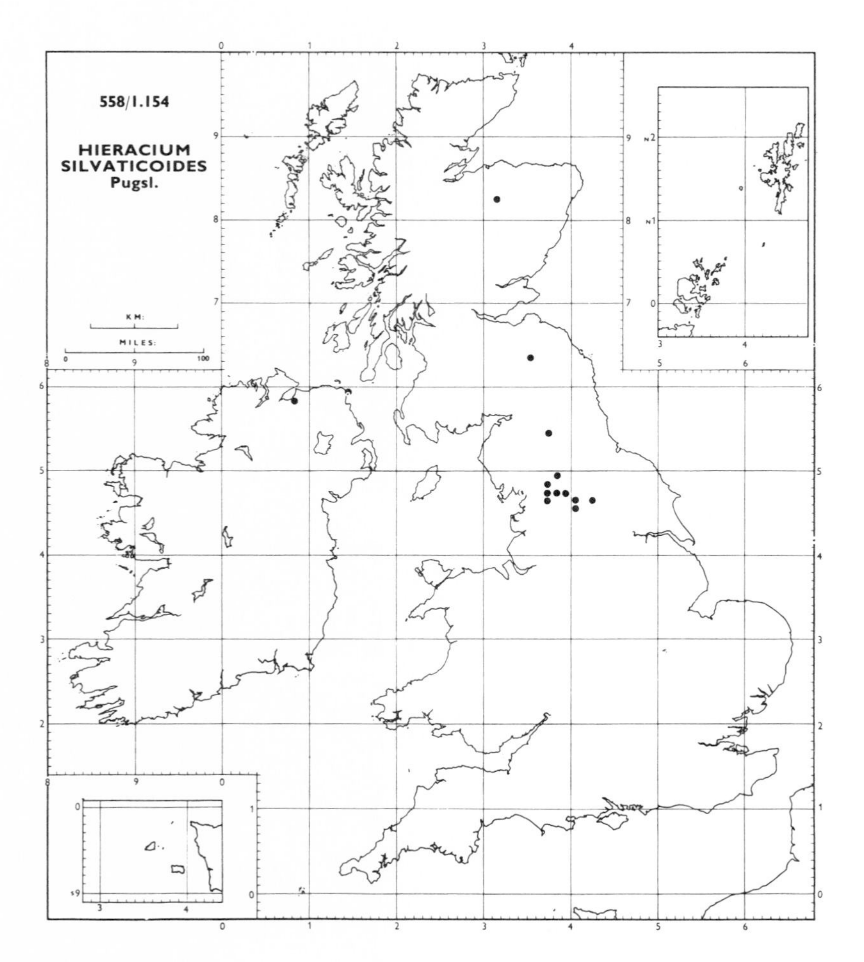

H. silvaticoides Pugsl. is a local, native endemic of the Yorkshire limestone, with two adjacent localities in Westmorland (69) and one in Cumberland (70). There are also single records from Melrose, Roxburgh (80), Bridge of Avon, Banff (94), Torr Head, Antrim (H. 39), and Benevenagh, Derry (H. 40).

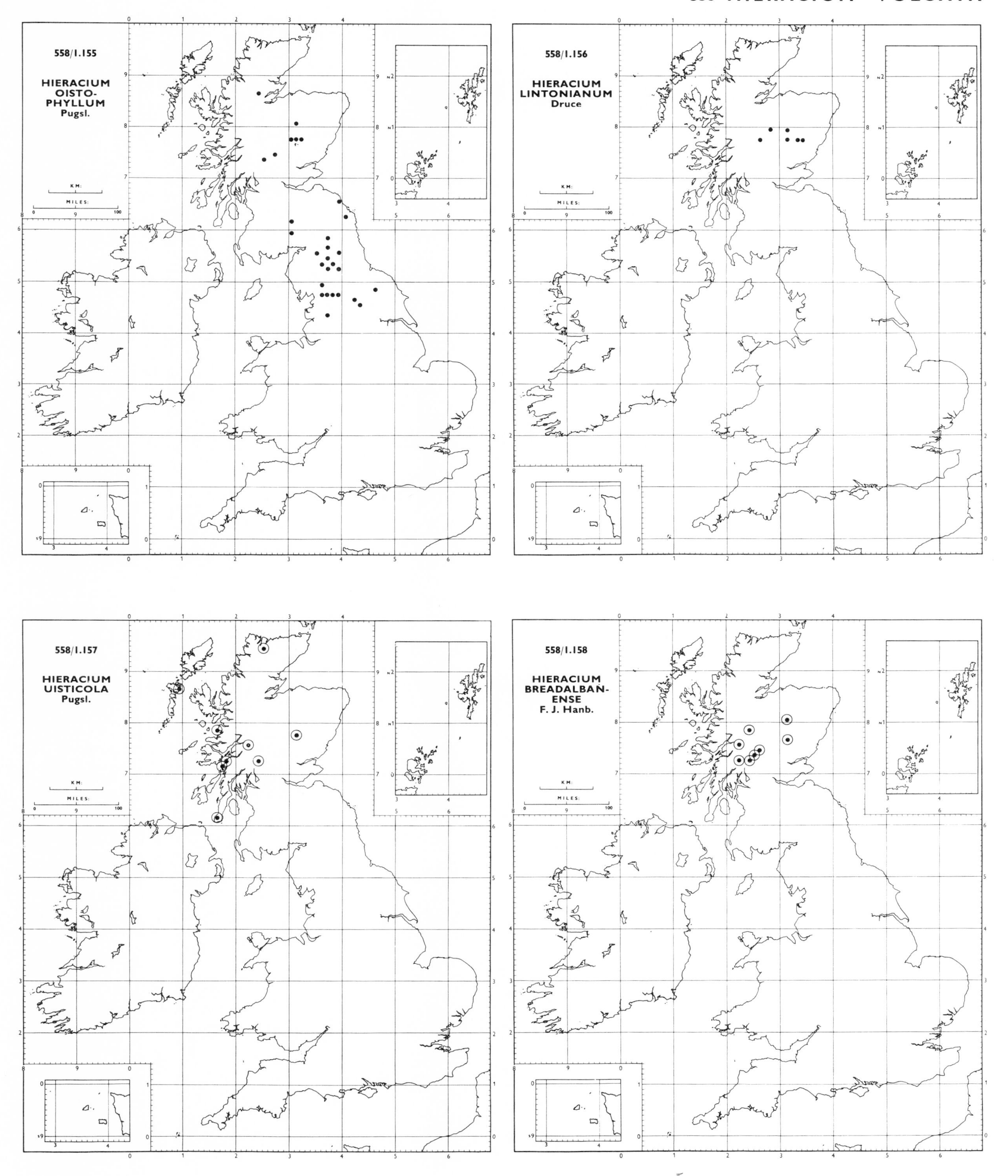
558/1.155
HIERACIUM OISTO-PHYLLUM Pugsl.
558/1.156
HIERACIUM LINTONIANUM Druce
558/1.157
HIERACIUM UISTICOLA Pugsl.
558/1.158
HIERACIUM BREADALBAN-ENSE F. J. Hanb.

H. cuneifrons (W. R. Linton) Pugsl. is a local, native endemic species of grassy rocky places in Brecon (42), with adjacent localities in Glamorgan (41) and Carmarthen (44).

H. radyrense (Pugsl.) P. D. Sell & C. West is known only from two localities in Glamorgan (41), where it is presumably native.

H. dipteroides Dahlst. is a native endemic species, which occurs in damp woods and by stream-sides in several scattered localities in central Scotland and on Beinn Gharbh, Sutherland (108).

H. rhomboides (Stenström) Johans. is a scarce native species, recorded only from Ribblehead, Yorkshire (64), Teesdale (65, 66), near Roddam, Northumberland (68), Swindale Beck, Westmorland (69), and the Unich Water, Angus (90). It also occurs in Sweden.

H. caesiomurorum Lindeb. is a common, widespread native species of grassy and rocky places in central and north Scotland. It is also widespread in Scandinavia.

H. stenophyes W. R. Linton is a very local, native endemic species restricted to rocky stream-sides near Moffat, Dumfries (72).

H. fulvocaesium Pugsl. is a very distinct, native endemic species, restricted to grassy slopes at Bettyhill, Sutherland (108).

H. maculoides P. D. Sell & C. West is a local, native endemic species occurring in a few localities on the Yorkshire limestone, and on limestone rocks at Silverdale, Lancashire (60), and Westmorland (69).

H. oxyodus W. R. Linton is a rare, native endemic species in a few localities in west-central Scotland, with an isolated record from Rhiconich, Sutherland (108).

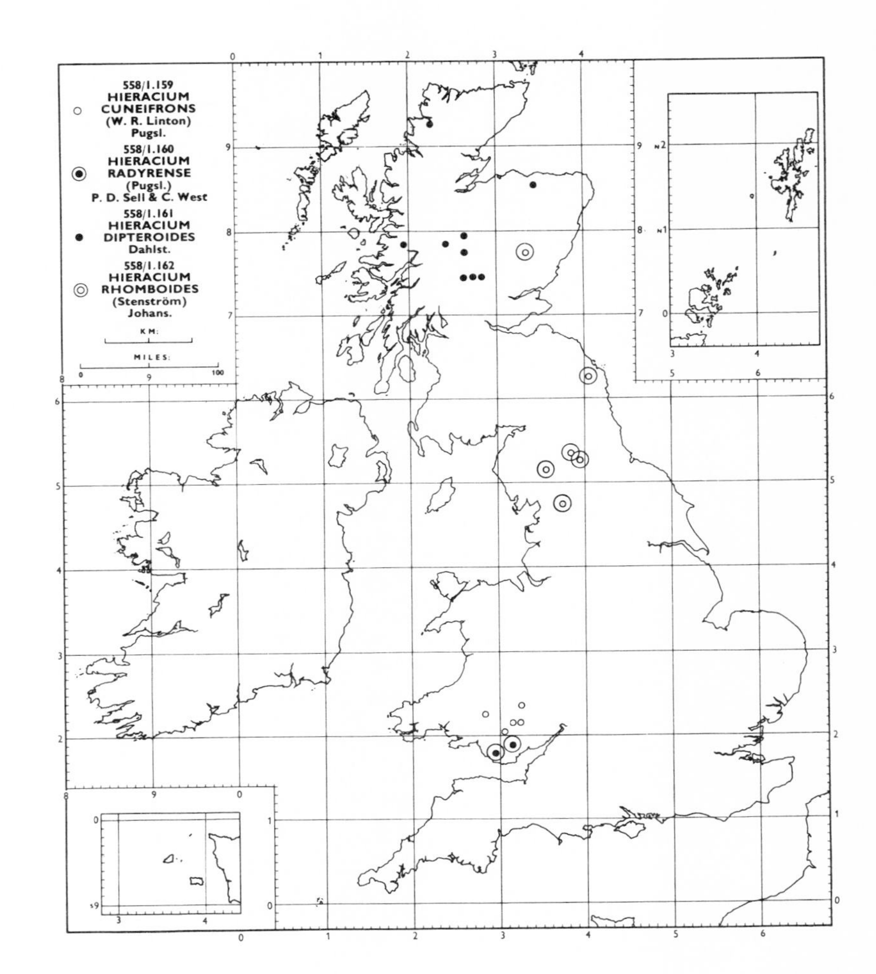

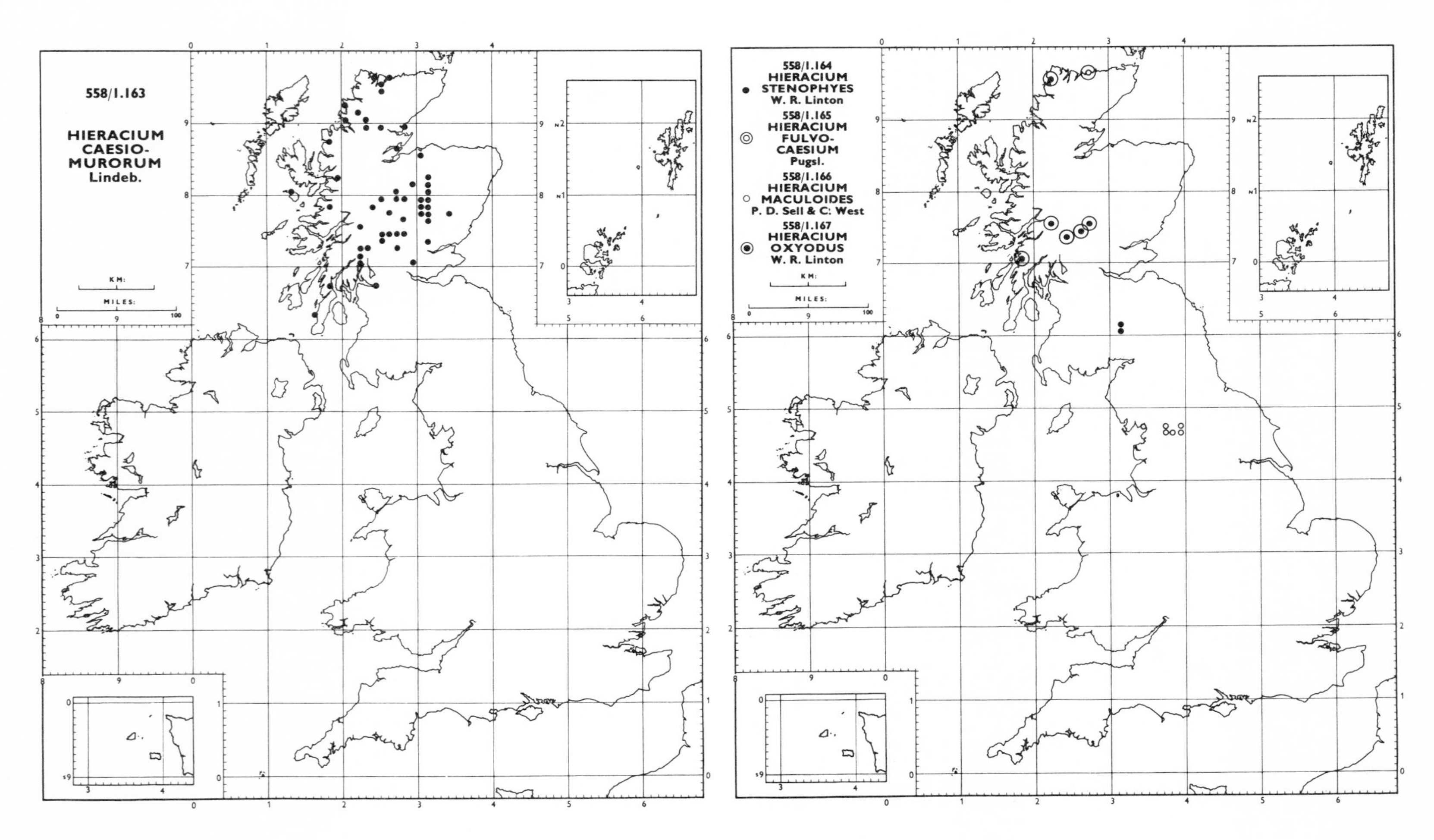

H. euprepes F. J. Hanb. is a very variable native species scattered over western and northern Britain from south Wales to Orkney (111), and in a number of cliff localities in north-eastern Ireland. There is a single specimen from Glencar, Kerry (H. 1). It occurs at altitudes varying from sea-level to about 3,000 ft (900 m.) in rocky and grassy places, and has a preference for basic soils. Outside the British Isles it is known only from the Faeroes.

H. submutabile (Zahn) Pugsl. is a locally common plant of south Wales, where it is presumably native. It occurs also in a few other places in Wales, central England (probably introduced) and near the Bristol Channel. The very closely allied **H. rectulum** A. Ley is a native endemic restricted to a limestone cliff near Llangadock, Carmarthen (44).

H. pulchrius (A. Ley) W. R. Linton is a very local, endemic species of cliff ledges in Brecon (42) and on Fan Fechan, Carmarthen (44).

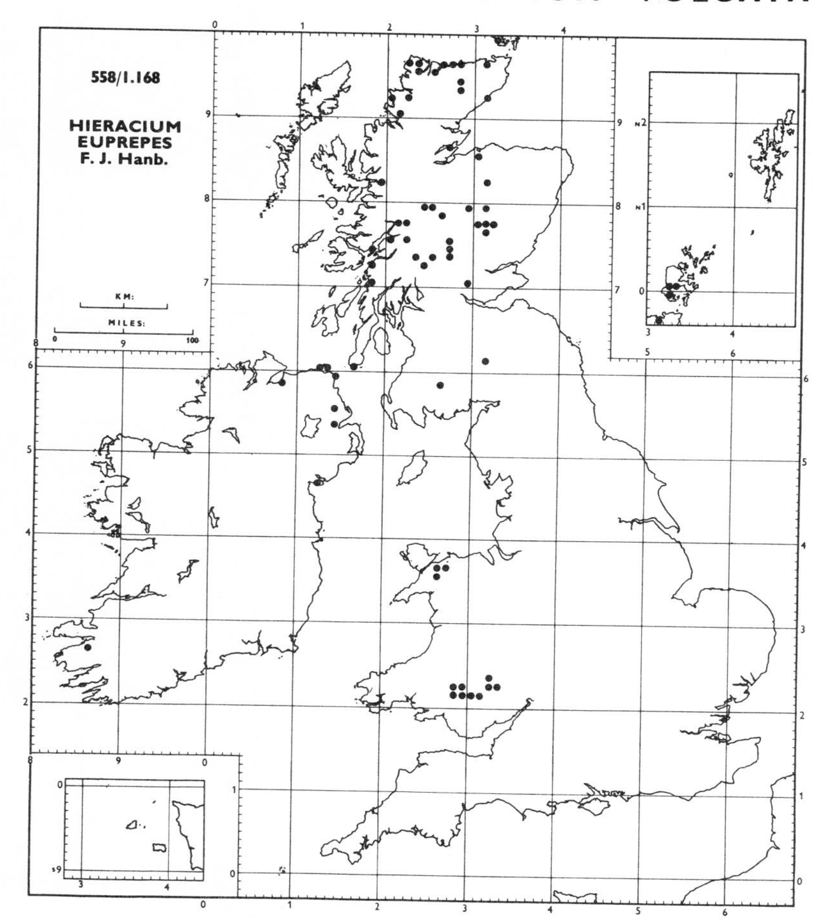

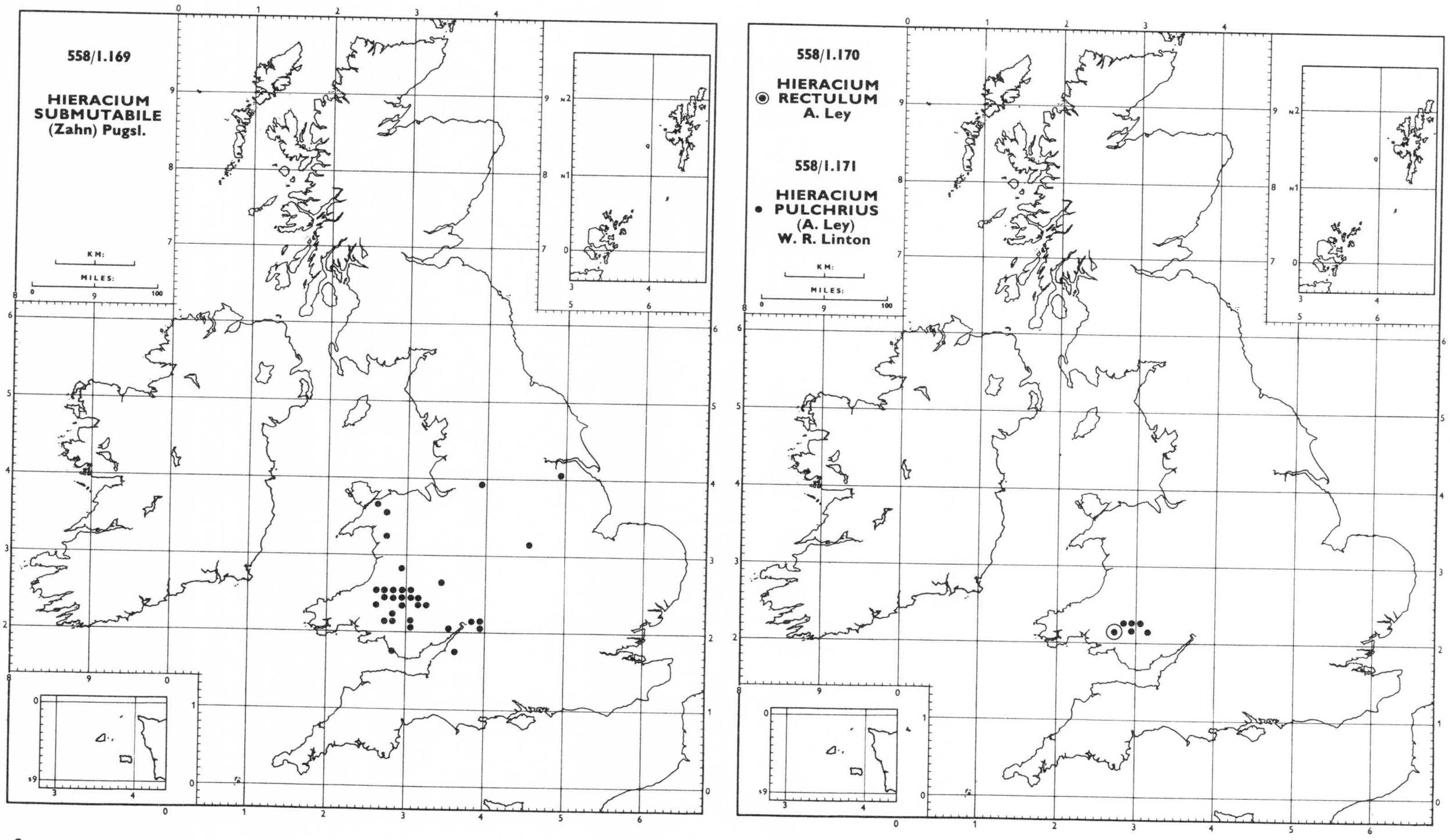

H. maculatum Sm. is a variable, introduced species, scattered throughout England, with a few records from Wales, Scotland and Ireland. It occurs on walls, quarries, grassy slopes, waste ground, slag heaps and occasionally in woodland, usually not far from a road or railway. It is a native of central Europe, and is introduced in many other parts of the Continent and in North America. **H. glanduliceps** P. D. Sell & C. West, which, in the past has been called *H. maculatum*, is a closely allied, native endemic species of the Yorkshire limestone.

H. diaphanum Fries is an extremely variable species of common occurrence throughout Wales and central England. There are scattered records from the remainder of England, six from Scotland and one from Ireland. Many of the English plants, which occur on roadsides, railway and canal banks and other waste ground, have flaccid green leaves, and an inflorescence of many capitula, which are clothed with long, slender glandular hairs (*H. anglorum* (A. Ley) Pugsl.). Such plants have probably been introduced. A few scattered colonies have rigid caesious-green leaves, with an inflorescence of few capitula, which are clothed with rather short glandular hairs (*H. diaphanum* Fries *sensu stricto*). Some colonies, such as those in Scotland, are almost certainly native, but others are questionably so. The great majority of the plants in Wales have the rigid, caesious leaves and inflorescence with few capitula of the last-mentioned form, but have glandular hairs intermediate in length (*H. daedalolepioides* (Zahn) Roffey). Plants of this form are found predominantly on cliff ledges and in other natural situations, and are almost certainly native. *H. diaphanum* Fries *sensu stricto* also occurs in Sweden and in east-central Europe. **H. subminutidens** (Zahn) Pugsl. is a very local, native endemic of rocky stream-sides in Brecon (42), and is almost certainly derived from one of the forms of *H. diaphanum*.

H. diaphanoides Lindeb. is a widely distributed native plant in northern England and in south-west and central Scotland. It may also be native in Cwm Llebrith, Caernarvon (49). It is introduced in a few places in south England, and Wales. Where native it usually grows on cliffs or by rocky streams, but in south-east England it may have been introduced with grass seed. Elsewhere it occurs in Fennoscandia and over the greater part of central and eastern Europe.

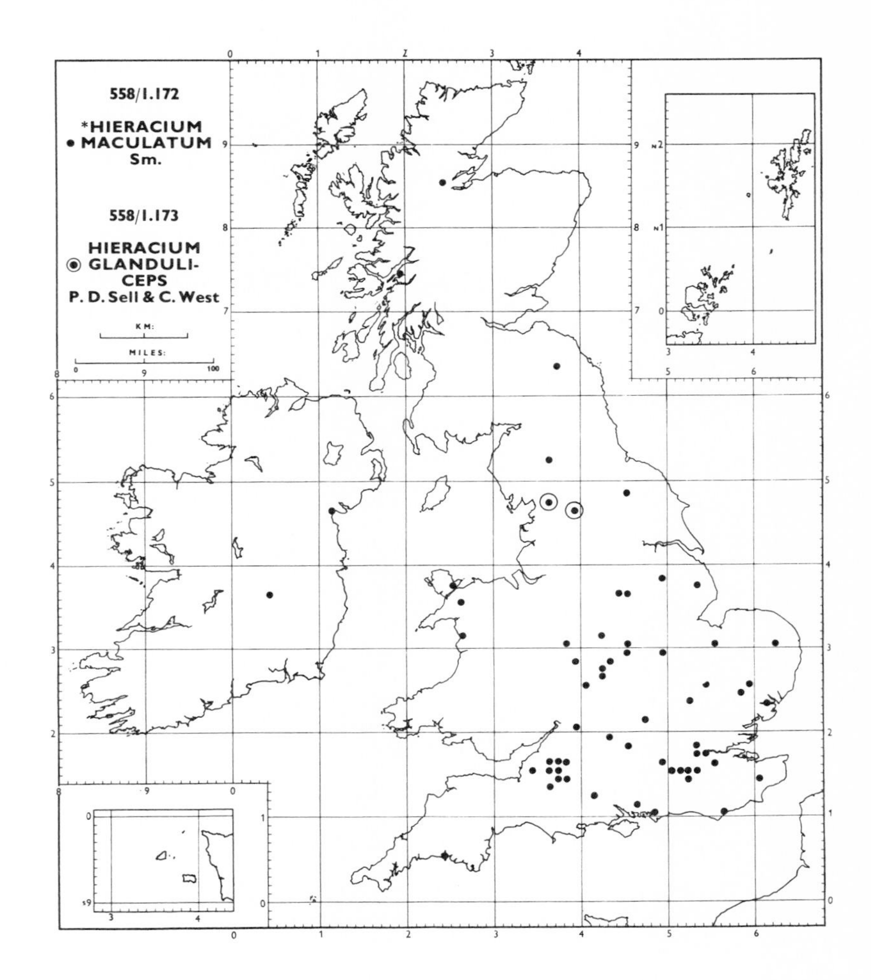

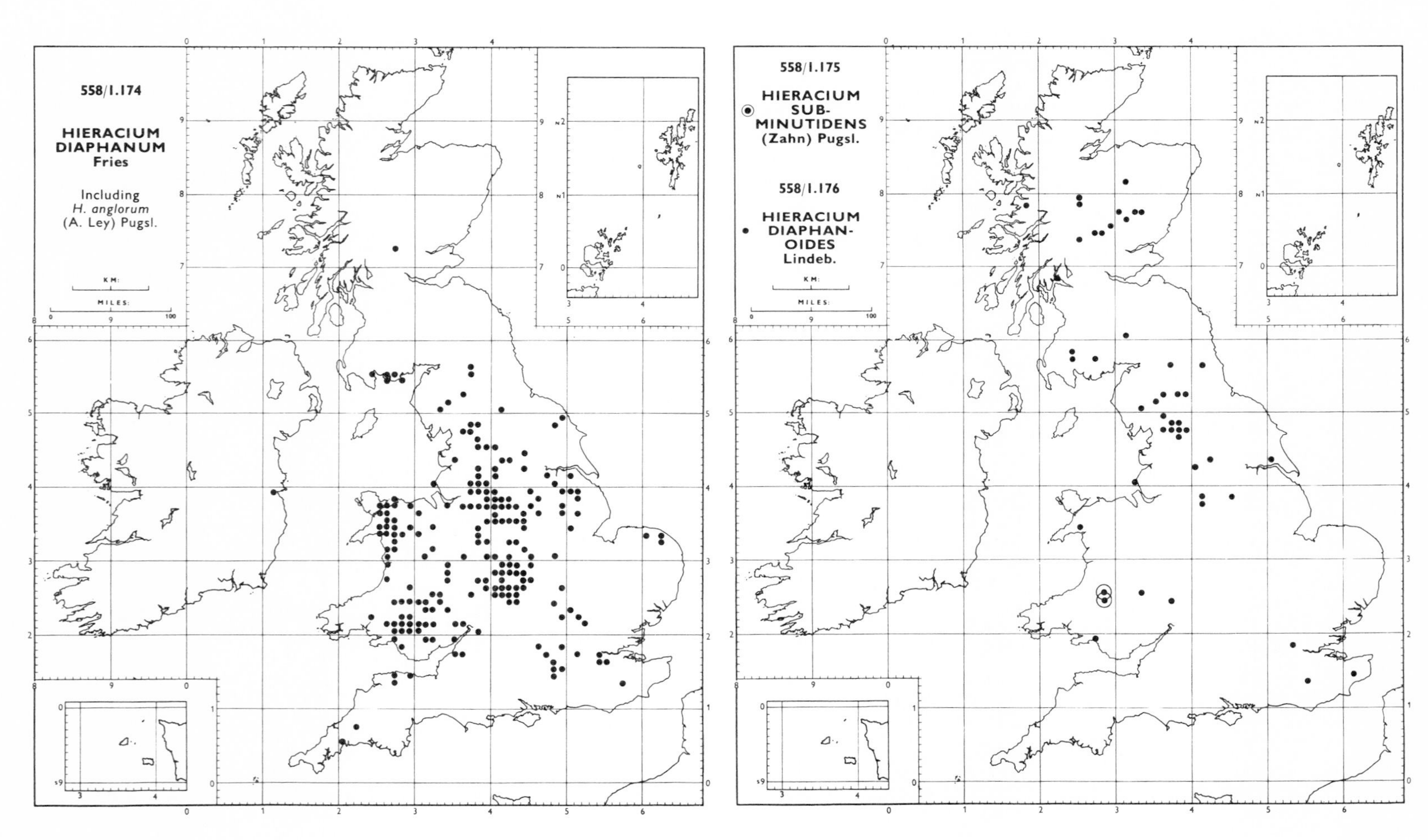

H. strumosum (W. R. Linton) A. Ley (*H. lachenalii* auct. mult.; *H. acuminatum* auct.) is the commonest, most variable, but nevertheless most characteristic *Hieracium* in the English lowlands in June and July. It also occurs throughout Wales and in the English uplands north to the Lake District, with isolated localities at Tralee, Kerry (H. 1), near Saintfield, Down (H. 38), and Blairgowrie, Perth (89). In Wales and the English uplands it occurs in rocky places and is perhaps native, but elsewhere it occurs on walls and grassy places, particularly roadsides and railway banks, where it must surely have been introduced. The plant of the Yorkshire limestone has a characteristic facies, but cannot be satisfactorily defined as a distinct species. *H. strumosum* occurs throughout much of continental Europe, where its indigenous distribution is difficult to determine. The closely allied **H. cheriense** Jord. ex Bor. (*H. tunbridgense* Pugsl.) is a common species in south-east England, where it partially replaces *H. strumosum*. It is locally common in south-west England, and it also occurs at Radyr, Glamorgan (41), Brecon (42), Saintfield, Down (H. 38), near Kinnan, Argyll (98), and as a weed in the Botanic Garden at Cambridge (29). Its habitats on roadsides, railway banks and wood margins suggest that it is nowhere indigenous. In south-east England it appears to be spreading. It is a widespread species of central Europe from France to Hungary.

H. subamplifolium (Zahn) Roffey is a locally abundant, native endemic species of shady rocky places in south Wales and adjacent areas of England, with isolated localities in Devon (4) and Somerset (6).

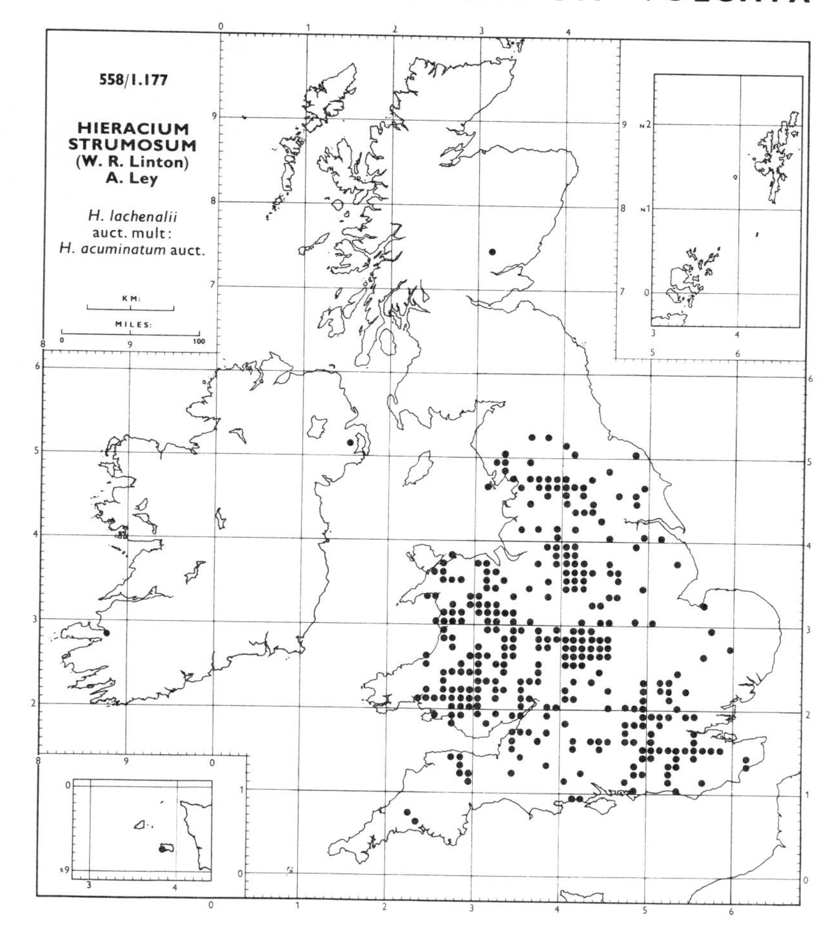

558/I.177
HIERACIUM STRUMOSUM (W. R. Linton) A. Ley
H. lachenalii auct. mult: *H. acuminatum* auct.

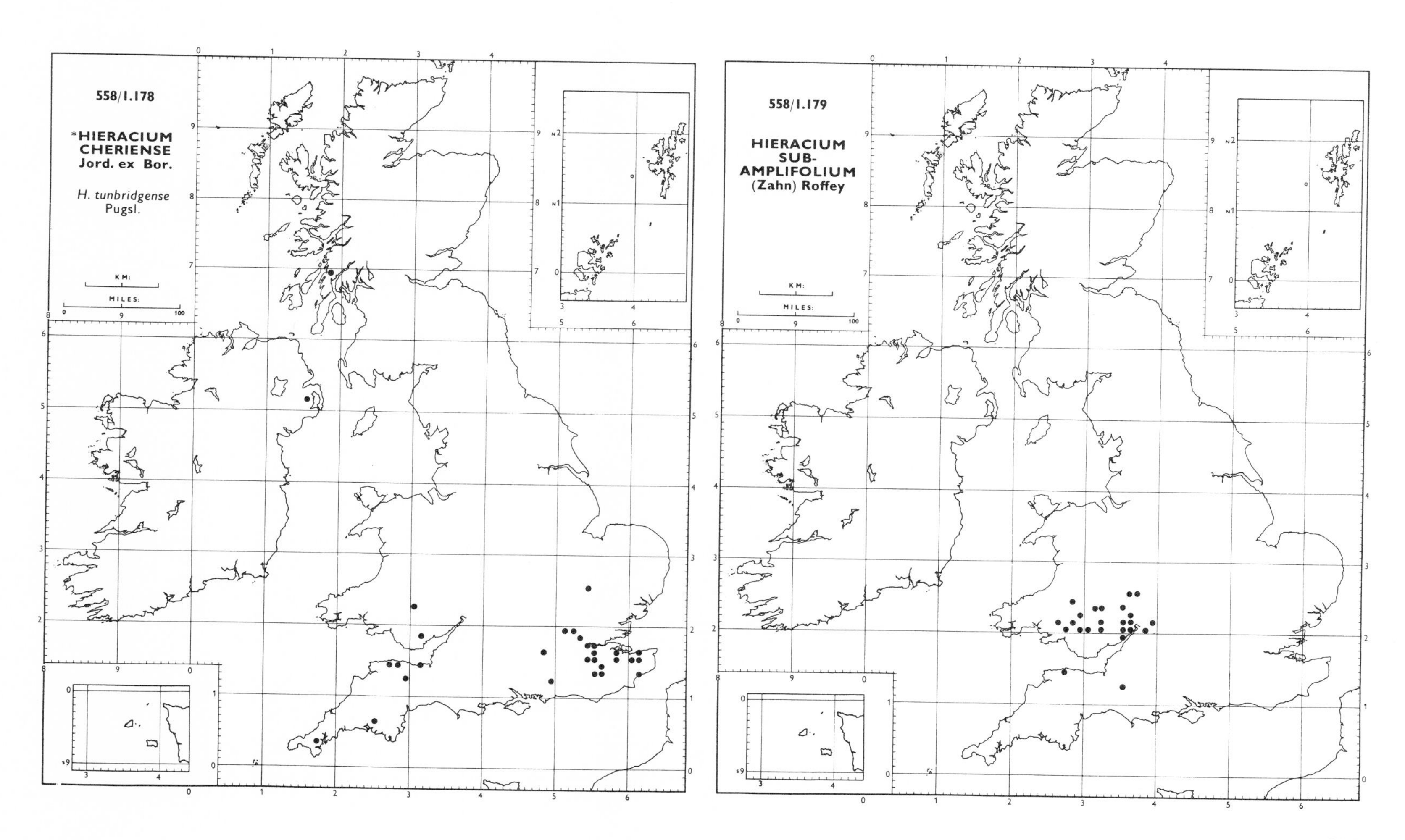

558/I.178
*HIERACIUM CHERIENSE Jord. ex Bor.
H. tunbridgense Pugsl.

558/I.179
HIERACIUM SUBAMPLIFOLIUM (Zahn) Roffey

H. surrejanum F. J. Hanb. is a very local species restricted to the Sussex–Surrey border (13, 17). In view of its strong similarity to the introduced species, **H. lepidulum** (Stenström) Omang, and its similar habitats, it must questionably be regarded as a native plant. *H. lepidulum* is a frequent species of roadsides, railway banks and grassland in south-east England, with a few widely scattered localities elsewhere in England and a solitary record from south Wales. On the Continent it has a wide distribution in Scandinavia and central Europe. It is interesting that the early records of this species in Britain were made in the area where the closely allied *H. surrejanum* grows. All other records were made within the last 20 years. In some colonies in Britain, as on the Continent, individuals with more densely glandular heads and usually red-tinted leaves (var. *haematophyllum* Dahlst.) occur with the typical plant.

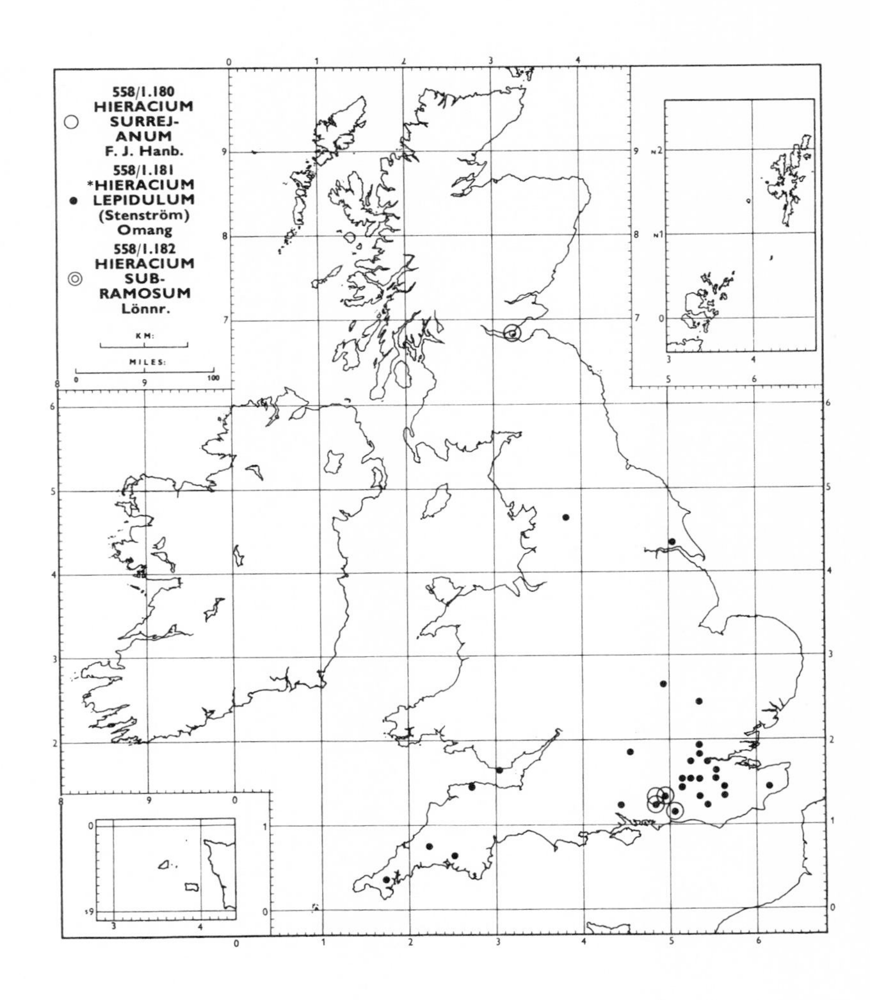

H. subramosum Lönnr. is known in the British Isles only from specimens collected at Pettycur, Fife (85), in 1876, where it was presumably native. Repeated searches for this species in recent years have produced negative results. Elsewhere it is known only in south Scandinavia.

H. rubiginosum F. J. Hanb. is a very variable, native species occurring in abundance on the limestone in north England, Derbyshire (57) and south Wales, with scattered localities in north Wales, Scotland and northern Ireland. It is mainly to be found in rocky places, but sometimes occurs on grassy slopes and on walls. Outside the British Isles it is recorded only from the Faeroes.

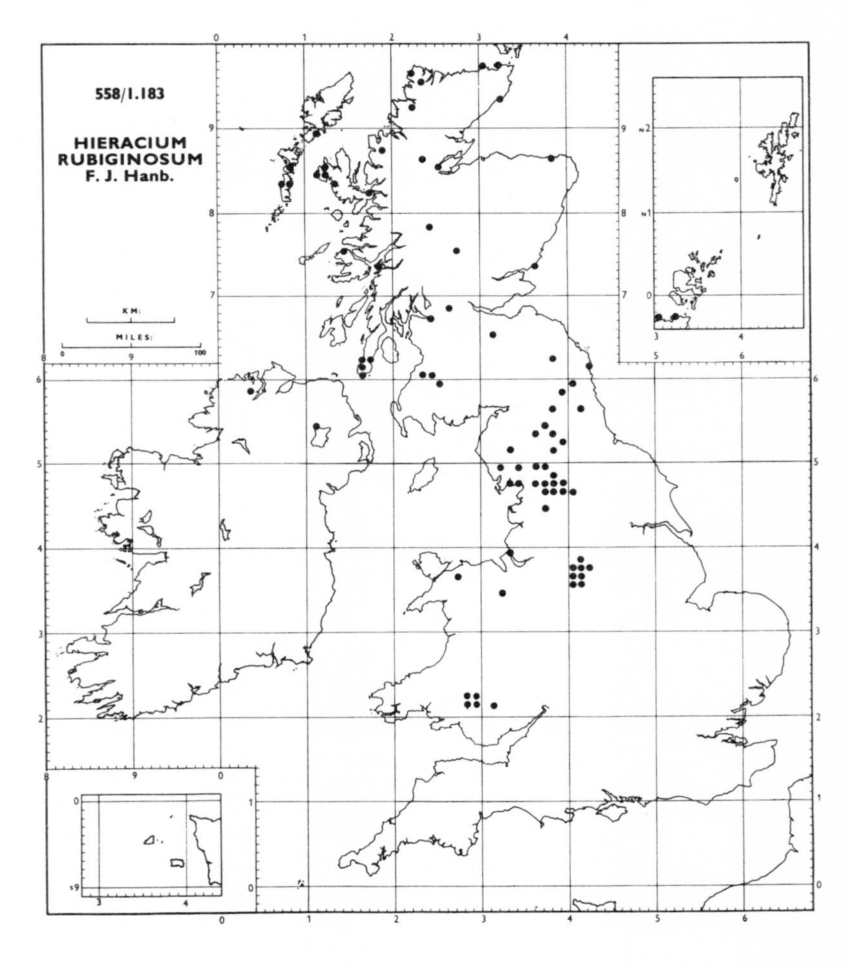

In late June, July and August **H. vulgatum** Fries is the commonest hawkweed throughout northern England and Scotland. It is scattered over north Wales, central England and north-east Ireland, with a few records further south in England, Wales and Ireland. It is a native plant of rocky and grassy places at all altitudes. In southern England it occurs in a few places on roadsides and railway banks, where it is presumably introduced. It occurs in abundance over large areas of Scandinavia and central Europe. It is interesting that neither this nor any allied species occurs in Orkney (111), Shetland (112), Faeroes or Iceland. This is a most variable species and has been divided into numerous varieties. Although this variation is in part geographical, the characters by which the variants can be recognized are neither constant nor easily defined. **H. triviale** Norrl. is a closely allied, native species, which occurs in a few localities in Wales and Scotland. It is also to be found in Scandinavia. **H. cravoniense** (F. J. Hanb.) Roffey is a common, native endemic in rocky and grassy places (often in shade) in northern England and Scotland. It has many characters in common with *H. vulgatum*, but can readily be distinguished from it in the field, not only by the caesious colour of its leaves, but also by the long, shaggy hairs of its phyllaries.

H. caesionigrescens Fries is a rare, native species of the limestone near Grassington, Yorkshire (64). It also occurs in south Scandinavia.

H. pollichiae C. H. Schultz (*H. roffeyanum* Pugsl.) is a well-established, introduced species of south-east England where it is found on walls, roadsides and in grassy places. It is a widespread native species of central Europe, and is introduced elsewhere on the Continent.

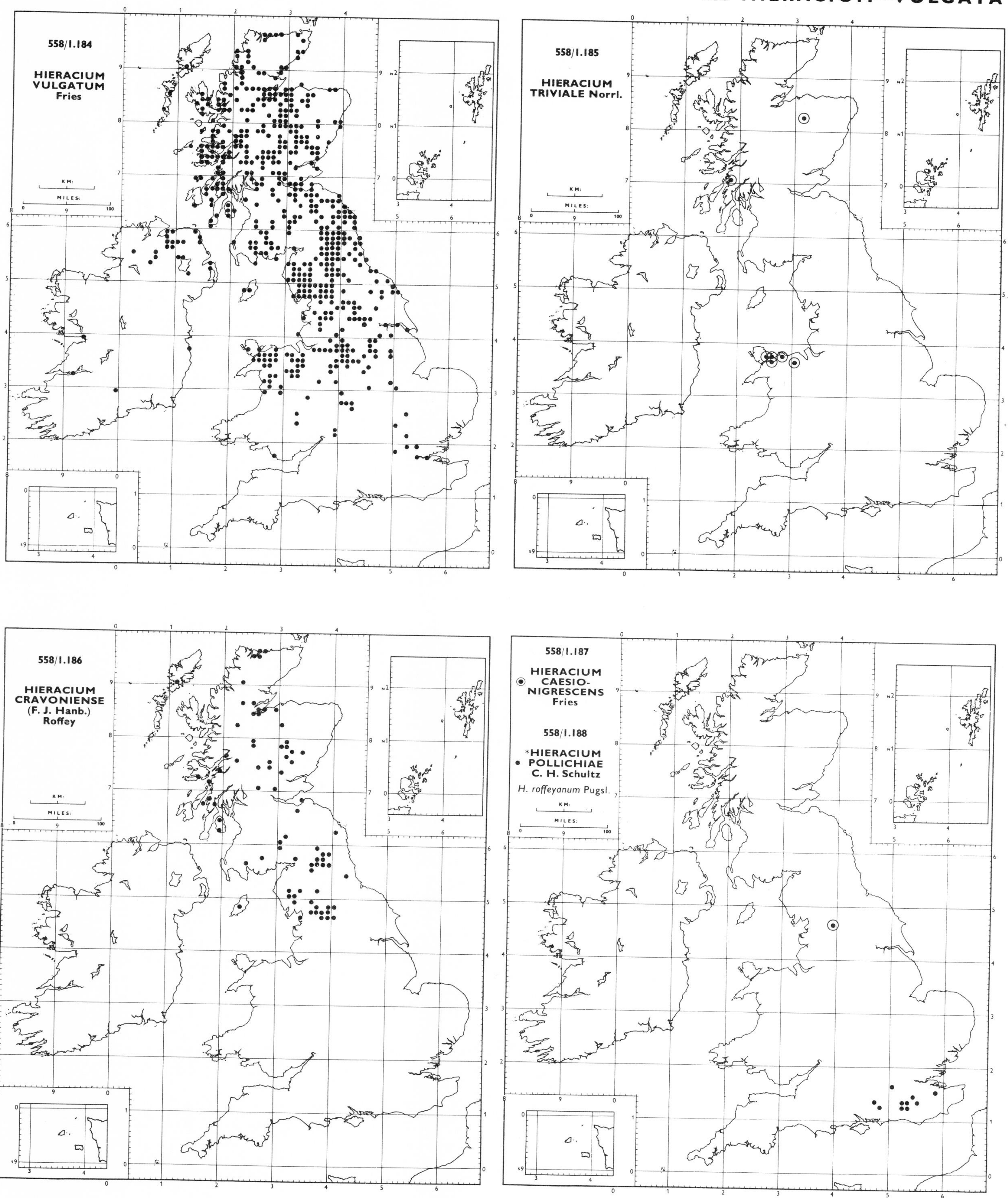

558/1.184
HIERACIUM VULGATUM Fries
558/1.185
HIERACIUM TRIVIALE Norrl.
558/1.186
HIERACIUM CRAVONIENSE (F. J. Hanb.) Roffey
558/1.187
HIERACIUM CAESIO-NIGRESCENS Fries
558/1.188
*HIERACIUM POLLICHIAE C. H. Schultz
H. roffeyanum Pugsl.
KM:
MILES:

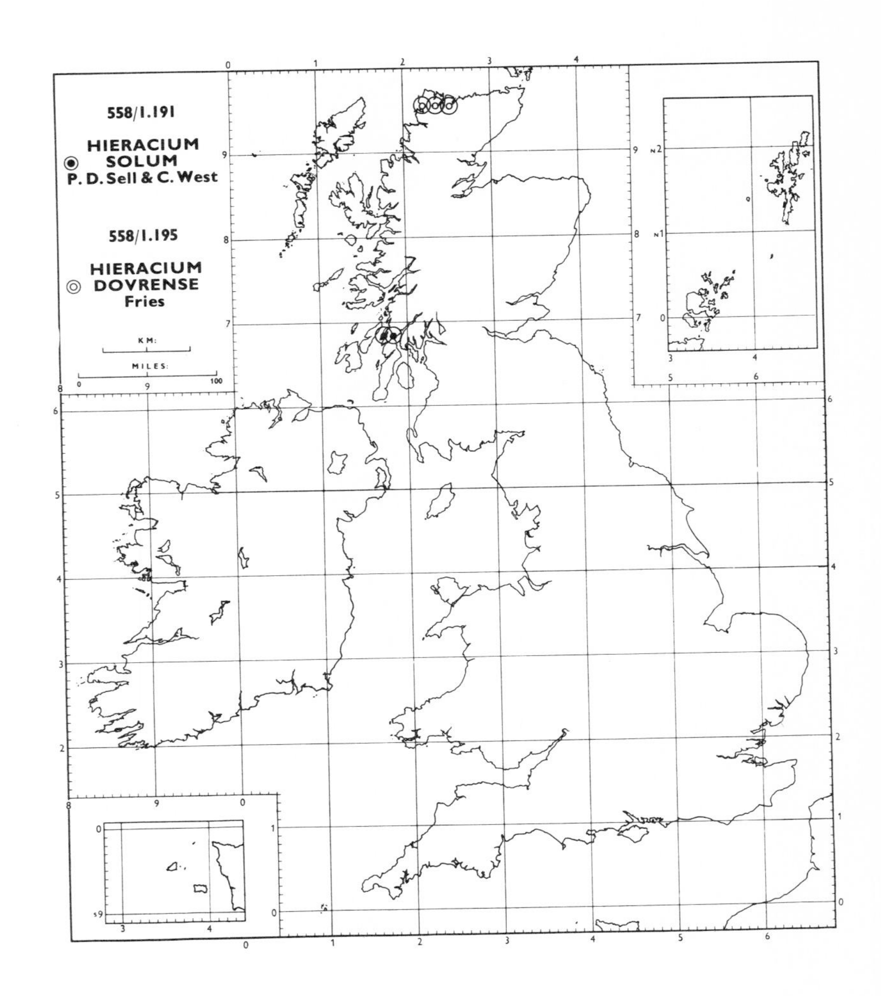

558/1.189–206 Section **ALPESTRIA** F. N. Williams

This section is an artificial group of 18 native species, 16 of which are endemic. Outside the British Isles species of this section occur in Greenland, Iceland, Faeroes, Scandinavia and in the mountains of central Europe. In central Europe the distinction between species of this section and those of the Section *Prenanthoidea* is not clear. The species of *Alpestria* are mainly of low stature, with an inflorescence of few capitula, probably brought about by the exposed situations which they inhabit. When cultivated some of them grow into large plants with an ample inflorescence, and they can then be separated only with difficulty from species of the Sections *Foliosa* and *Tridentata*. They flower in July and August, and show little variation. A detailed account of the species of *Alpestria* found in the British Isles has been published by P. D. Sell and C. West (1965).

H. zetlandicum Beeby, **H. breve** Beeby, **H. gratum** P. D. Sell & C. West, **H. difficile** P. D. Sell & C. West, **H. australius** (Beeby) Pugsl., **H. praethulense** Pugsl., **H. hethlandiae** (F. J. Hanb.) Pugsl., **H. attenuatifolium** P. D. Sell & C. West, **H. pugsleyi** P. D. Sell & C. West, **H. dilectum** P. D. Sell & C. West, **H. subtruncatum** Beeby, **H. northroense** Pugsl. and **H. vinicaule** P. D. Sell & C. West are endemic to Shetland (112) where, with one exception (*H. vinicaule*), they are restricted to steep rocky cliffs and grassy slopes and stream-sides near the coast where they are inaccessible to sheep. *H. vinicaule*, the most common of these local endemics, occurs also in places easily accessible to sheep, and appears to be untouched by them. This group of species has its closest affinity with species from the Faeroes and Iceland, *H. zetlandicum* being very closely allied to *H. ostenfeldii* Dahlst. and *H. hartzianum* Dahlst. of the Faeroes and *H. arrostocephalum* Omang of Iceland; *H. gratum* to *H. elegantiforme* Dahlst. from Iceland; and *H. praethulense* to *H. phrixoclonum* Omang and *H. halfdanii* Oskarss., from Iceland. **H. solum** P. D. Sell & C. West, which also resembles the species of this group, is endemic to two cliffs in Kintyre (101).

H. dovrense Fries, which in the British Isles is known only from near Rhiconich, Foinaven, and Ben Loyal, Sutherland (108), where it is very scarce, is a frequent plant in the Dovre region of Norway.

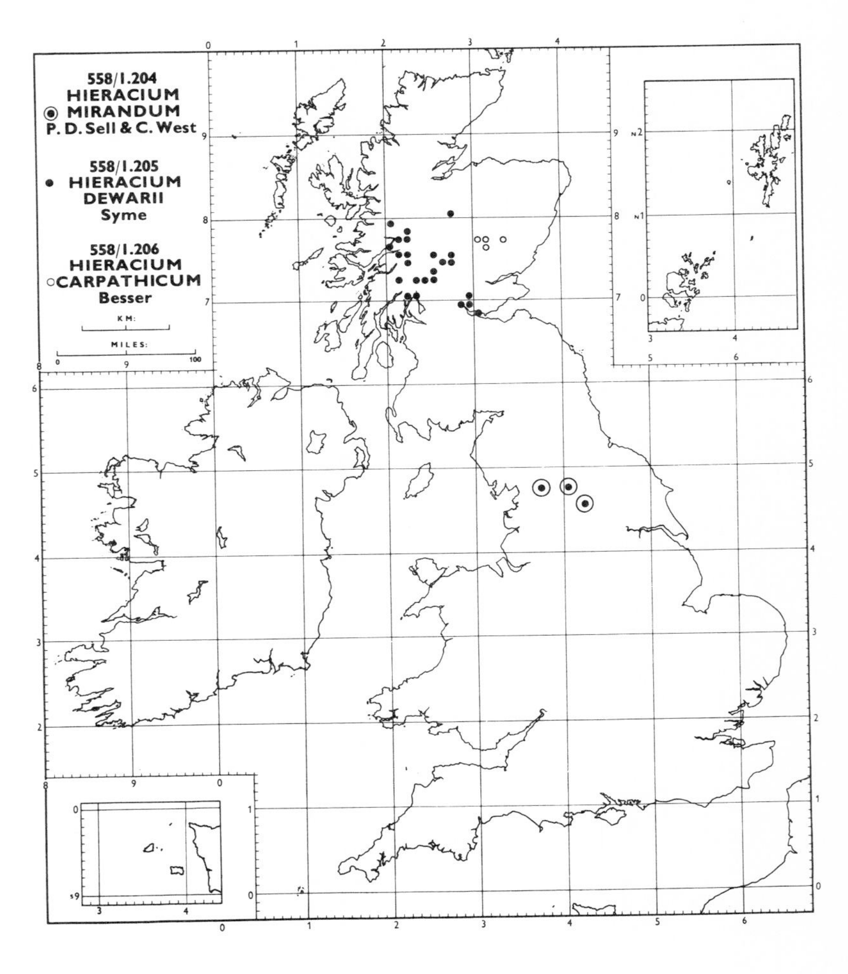

H. mirandum P. D. Sell & C. West, **H. dewarii** Syme and **H. carpathicum** Besser (*H. perthense* F. N. Williams) form a closely allied group of species, which approach those of the Section *Foliosa*. *H. mirandum* is known only from two localities in Yorkshire (64) and one in Derbyshire (57). *H. dewarii* is a widespread, but local endemic species of streambanks and rocky ledges in west-central Scotland. *H. carpathicum* is a very scarce plant in the British Isles known only from Clova, Angus (90), and Glen Shee, Perth (89). It also occurs in the Polish Tatra Mountains. Such a distribution defies explanation, but adequate series of specimens from both localities have been compared, and no distinguishing characters found.

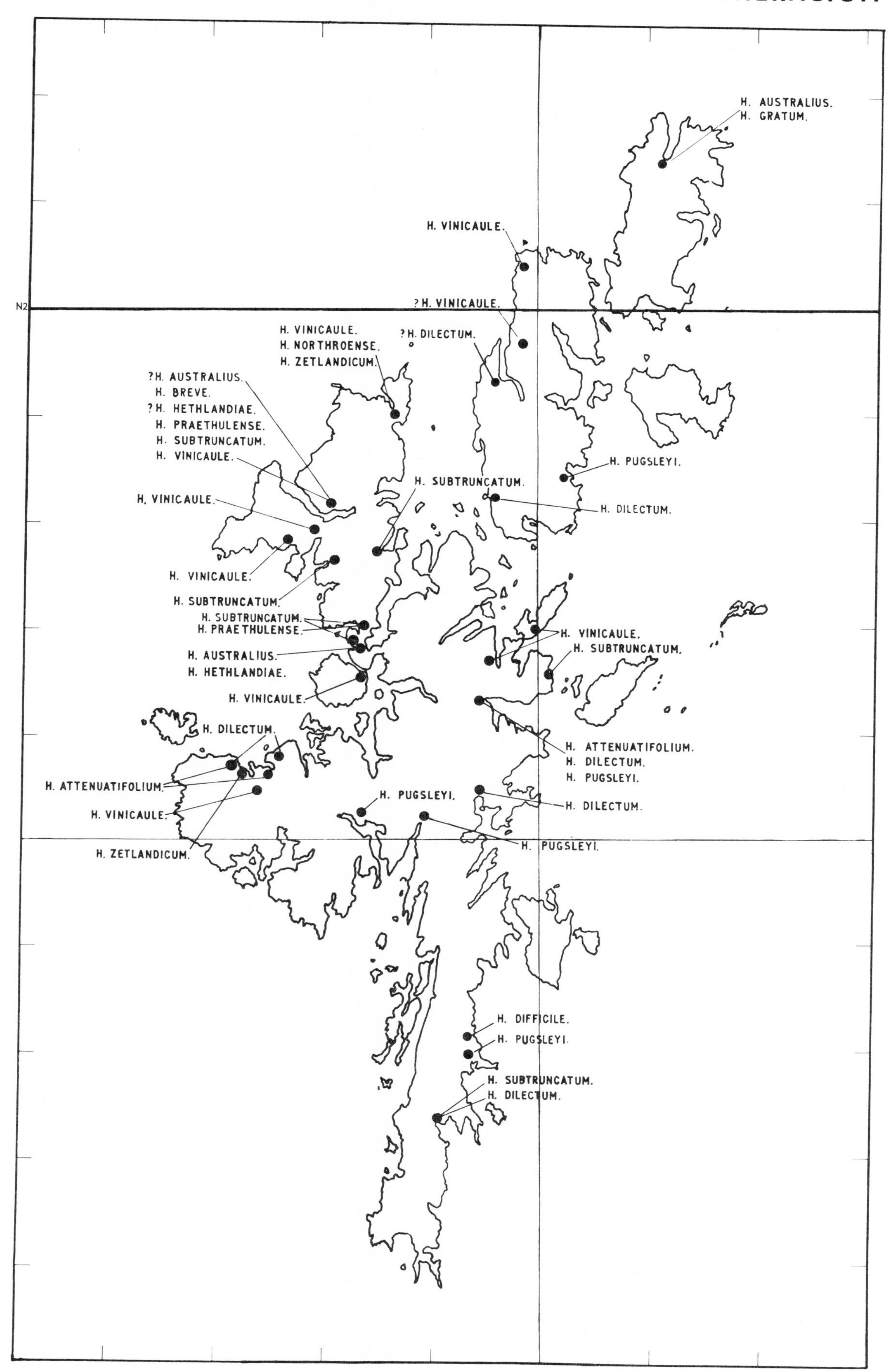

SPECIES OF THE SECTION ALPESTRIA IN SHETLAND

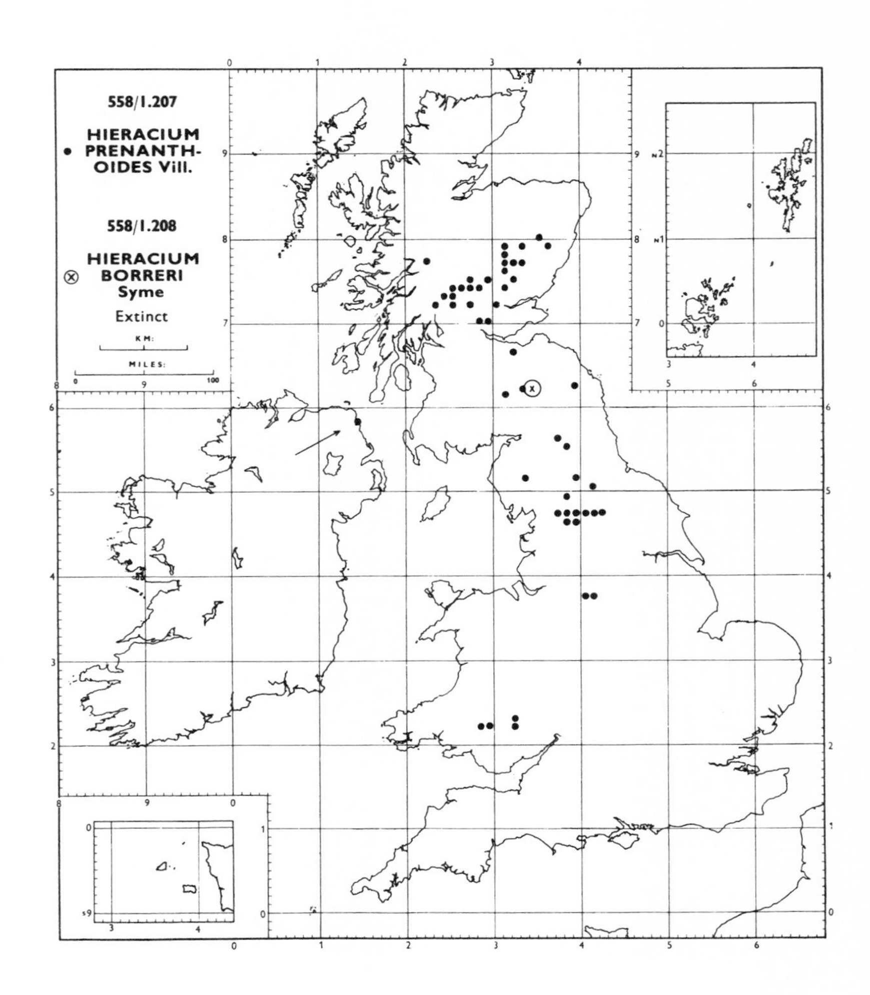

558/1.207–208 Section **PRENANTHOIDEA** Koch

This section is represented in the British Isles by only two species. Other species of the section occur throughout much of Europe and extend into western Asia. They flower in July and August. **H. prenanthoides** Vill. is a very distinct, locally common plant of grassy banks, rocky places and stream-sides in south Wales, Derbyshire (57), northern England and south-central and south-east Scotland, with an isolated record from boulder clay at Red Bay, Antrim (H. 39). In England and Wales it usually occurs on limestone. Some Scottish specimens have more flaccid leaves of a different shade of green, but they cannot be satisfactorily distinguished from the remainder. Elsewhere *H. prenanthoides* occurs in the French and Swiss Alps, and in Scandinavia. **H. borreri** Syme, a very distinct species, is known only from specimens obtained from Harehead Wood, Selkirk (79), during the last century. This wood has been thoroughly searched in recent years, but no trace of the plant has been found. *H. borreri* is questionably identical with *H. juranum* subsp. *pseudelatum* Zahn from the western Alps.

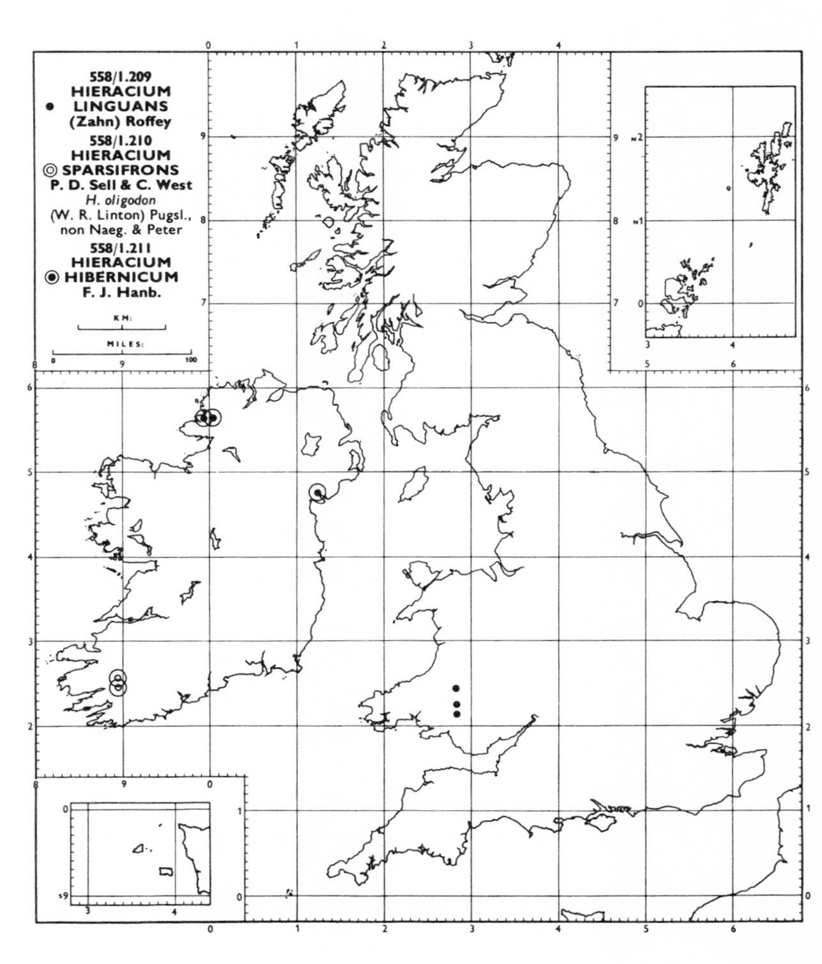

558/1.209–229 Section **TRIDENTATA** F. N. Williams

In the British Isles this section forms a natural group of 21 species. The species occur in a great variety of habitats, but not usually at a very high altitude, throughout Great Britain. In Ireland they are local and confined to the perimeter. Elsewhere they range over much of Europe as well as occurring in northern Asia and North America. Some species begin flowering at the end of June, but most of them do not do so until July and August.

H. linguans (Zahn) Roffey is a native endemic species restricted to rocky ledges in three localities in Brecon (42).

H. sparsifrons P. D. Sell & C. West (*H. oligodon* (W. R. Linton) Pugsl., non Naeg. & Peter) is a native endemic restricted to rocky riversides in a few places in south Kerry (H. 1).

H. hibernicum F. J. Hanb. is a rare, native endemic of rocky places in Down (H. 38) and west Donegal (H. 35).

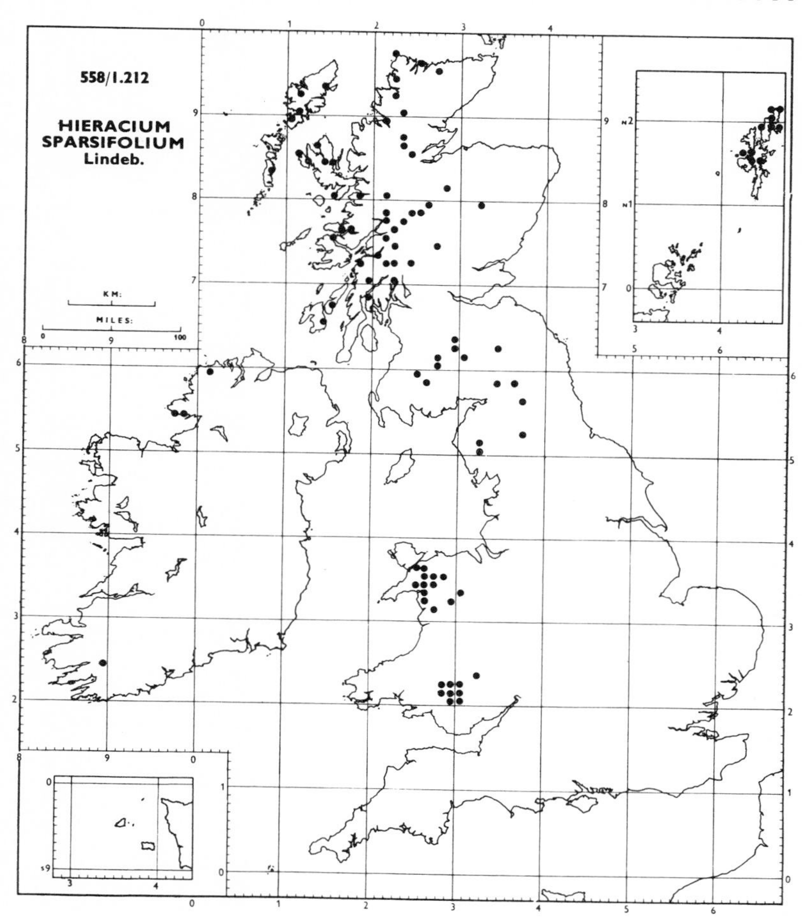

H. sparsifolium Lindeb. is a variable, widespread, native species of rocky and grassy places, especially by streams in upland regions, throughout much of Wales, north England and Scotland, with a few localities in western Ireland. The plants from south Wales are on the whole taller, and larger in their parts than those from elsewhere (var. *serpentinum* (W. R. Linton)). The plant most frequently found has leaves richly blotched with purplish-brown, but plants occur with unblotched leaves and they may completely dominate the population. *H. sparsifolium* is frequent in Shetland (112), Faeroes, Iceland and Norway, but is absent from Orkney (111) (cf. Sell and West, 1956).

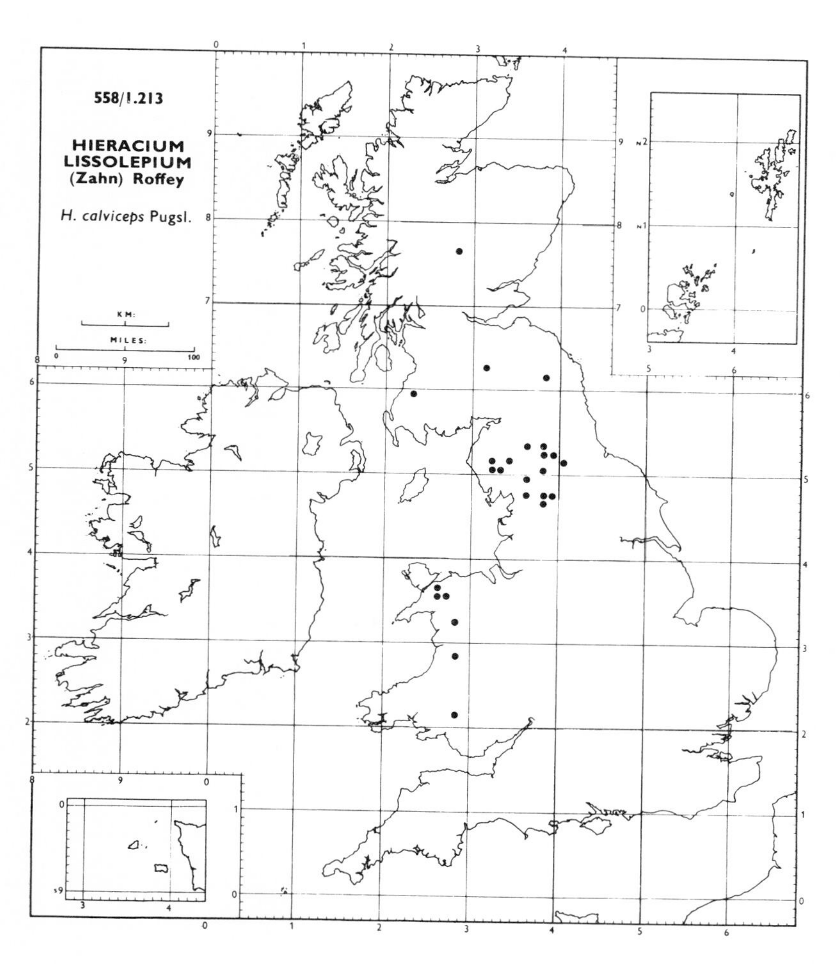

H. lissolepium (Zahn) Roffey (*H. calviceps* Pugsl.) is a frequent native species of rocky and grassy places, especially by streams on the limestone in northern England, with a few scattered localities in south and central Scotland and Wales. It is a widespread species in Scandinavia and central Europe.

H. uiginskyense Pugsl. is a local, native endemic species, usually found by rocky streams, in central and west Scotland, and Wales, with two isolated localities in Ireland. The closely allied **H. gothicoides** Pugsl. is a frequent species of grassy and rocky places in central Scotland, with isolated records from Westmorland (69), Wigtown (74) and Shetland (112).

H. substrigosum (Zahn) Roffey is a common, native endemic of rocky and grassy places, usually by streams, in Wales, particularly in the south. Plants collected from two localities near Braemar (92) seem to be conspecific.

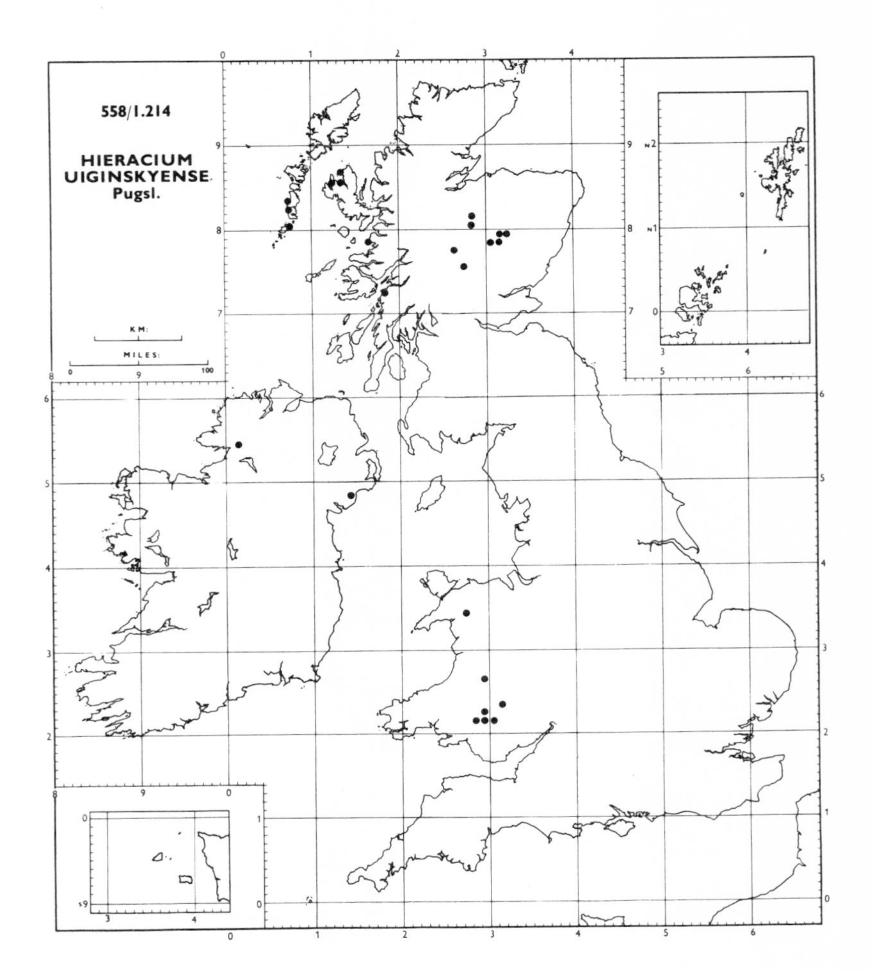

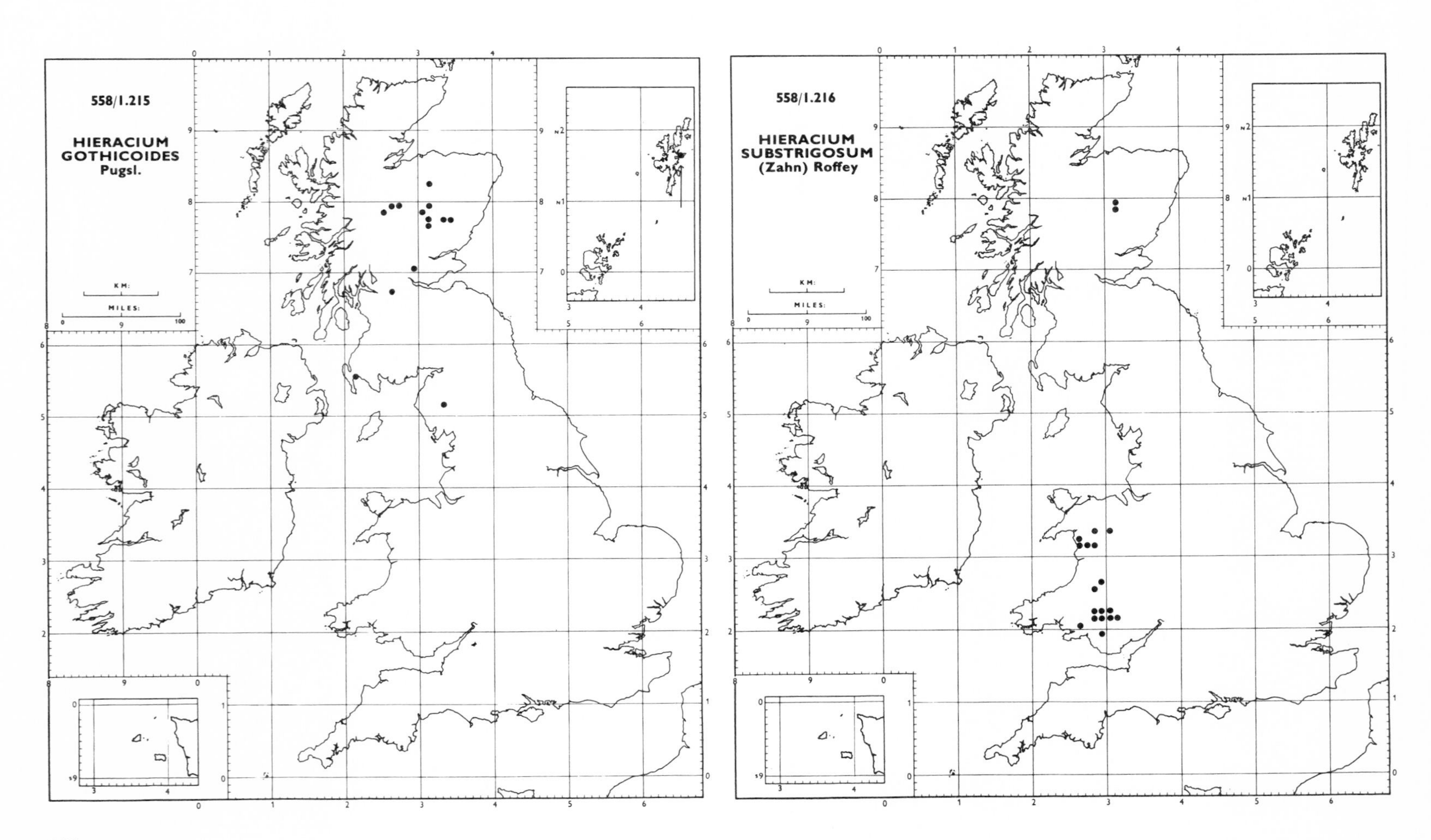

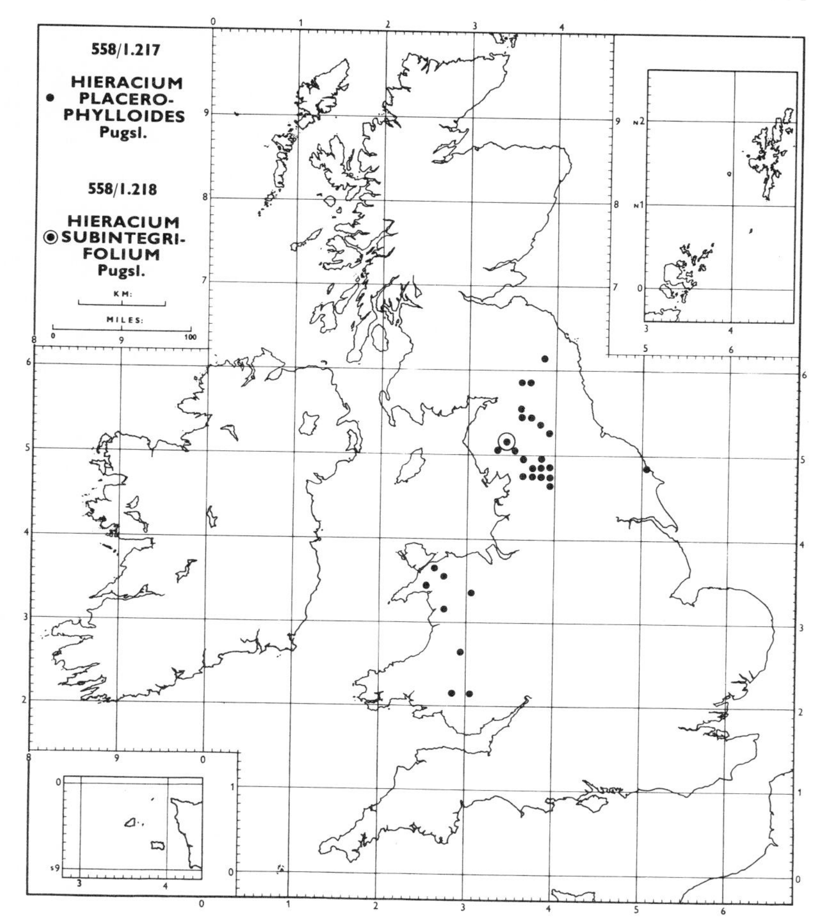

H. placerophylloides Pugsl. is a fairly common, native endemic species of rocky and grassy places, particularly by streams, on the limestone of northern England. In Wales it is of more local occurrence. The closely allied endemic, **H. subintegrifolium** Pugsl., is restricted to grassy banks at Glenridding, Westmorland (69).

H. nidense (F. J. Hanb.) Roffey is a rare, native endemic of rocky ledges and stream-sides in south Brecon (42) and adjacent parts of Carmarthen (44), with an isolated record from Blaenau Ffestiniog, Merioneth (48).

H. scullyi W. R. Linton is a rare, native endemic restricted to the rocky banks of the Roughty River, south Kerry (H. 1).

H. stewartii (F. J. Hanb.) Roffey (*H. longiciliatum* (F. J. Hanb.) Roffey) is a very variable, native endemic of grassy and rocky places, often by streams, in scattered localities in Wales, Scotland and Ireland.

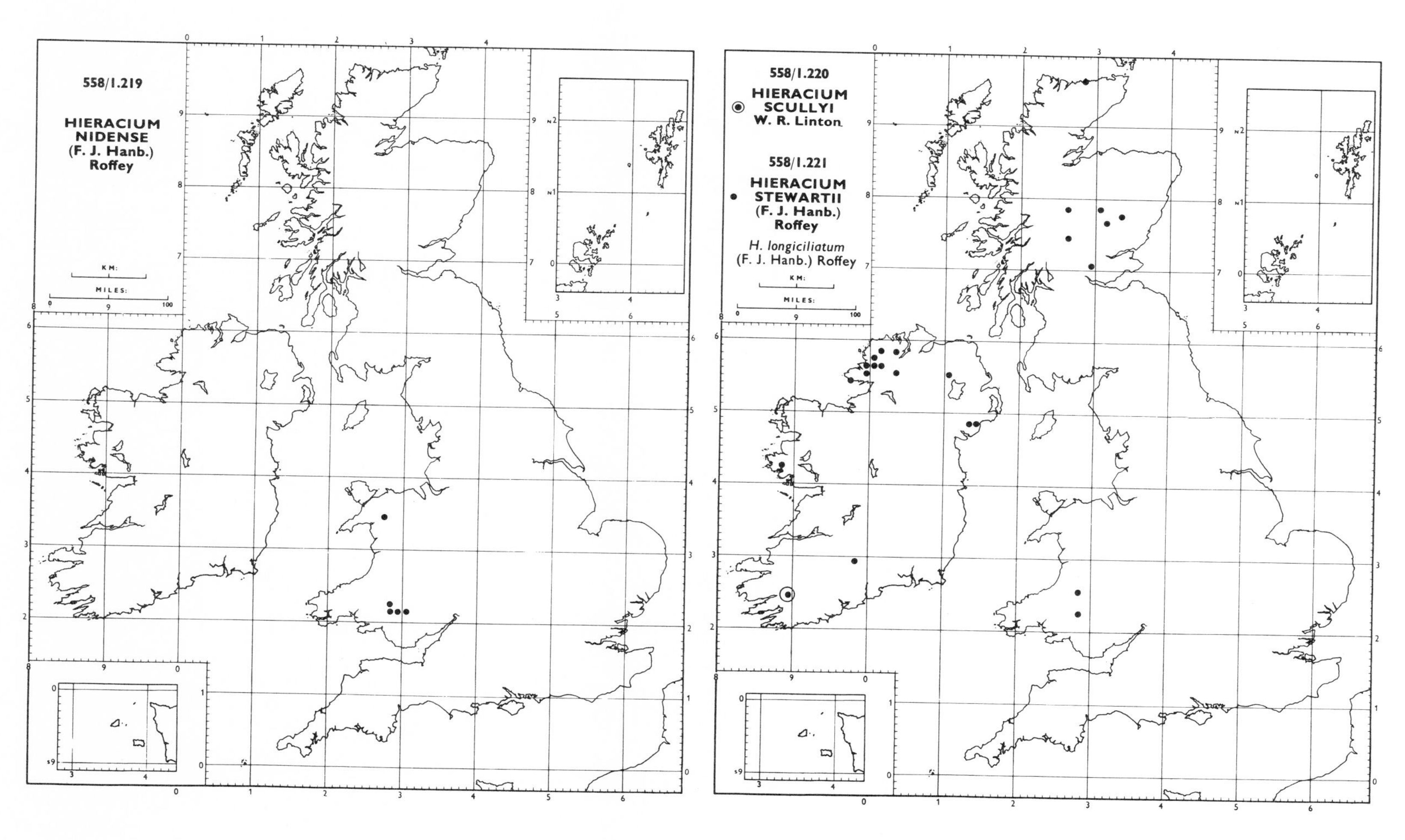

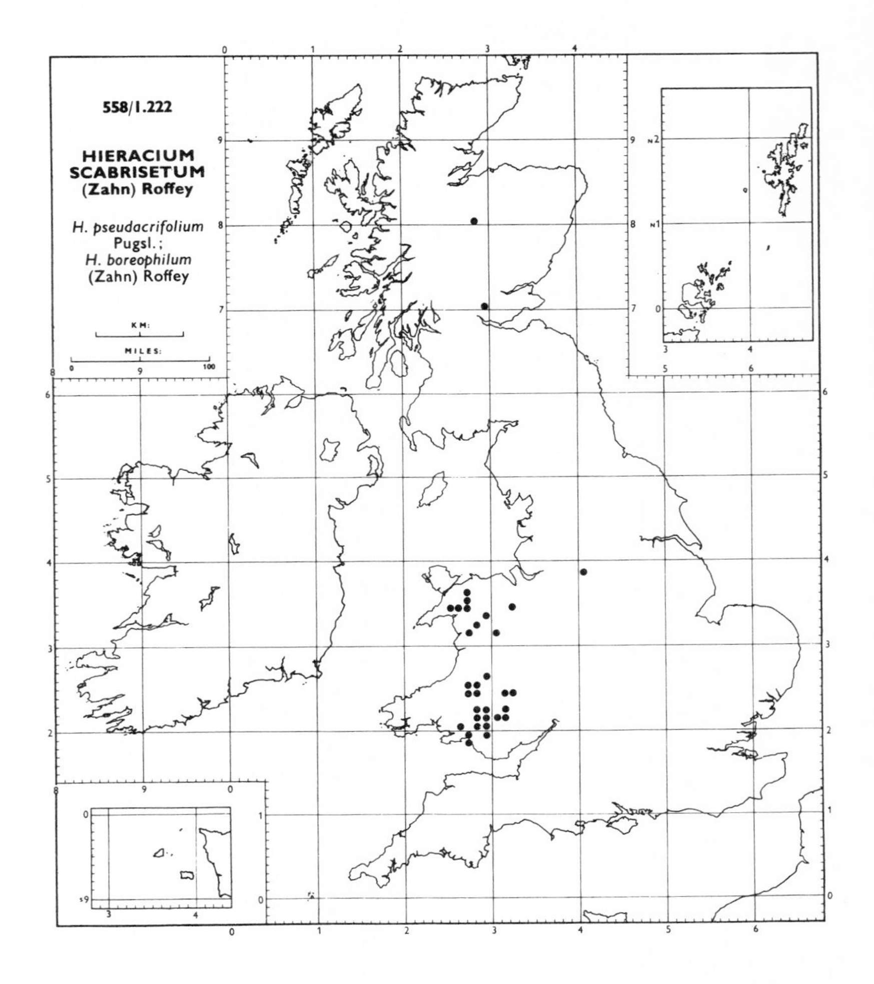

H. scabrisetum (Zahn) Roffey (*H. pseudacrifolium* Pugsl.; *H. boreophilum* (Zahn) Roffey) is a common, native species in grassy and rocky places, particularly by streams, in Wales, with an adjacent locality in Hereford (36) and outlying localities at Chapel-en-le-Frith, Derby (57), Glen Devon, West Perth (87) and between Kingussie and Kincraig, Inverness (96). It is also recorded from Norway.

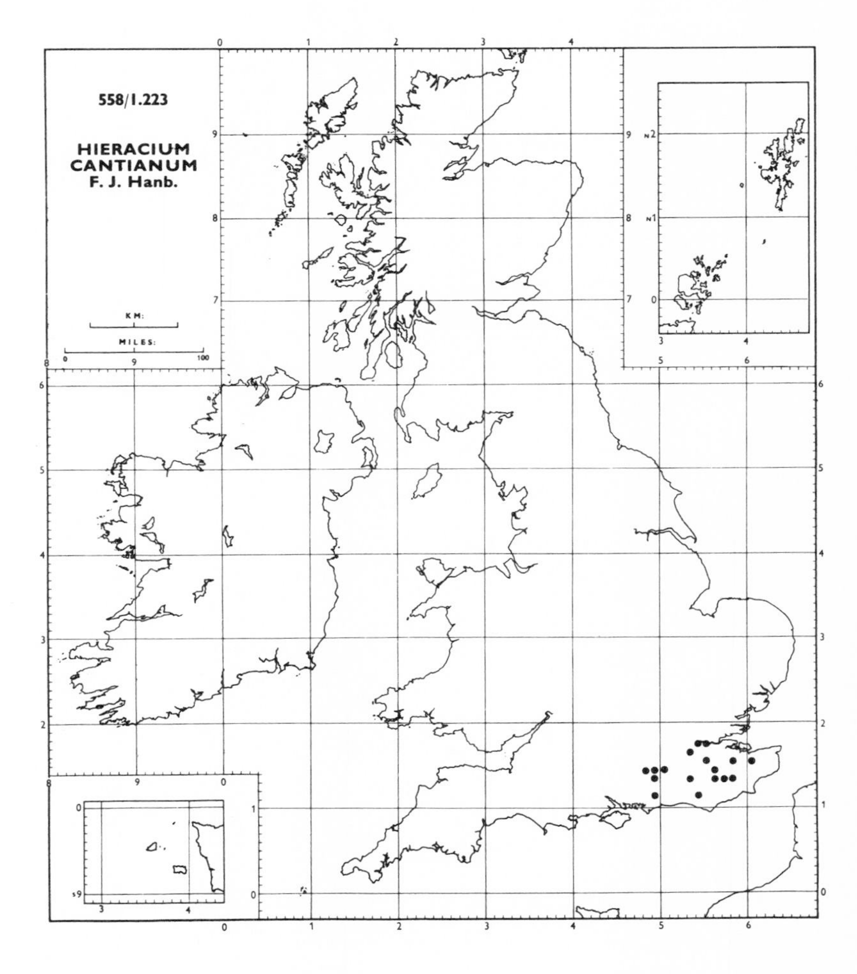

H. cantianum F. J. Hanb. is a locally frequent, native endemic species of open woodland of south-east England. **H. acamptum** P. D. Sell & C. West is a very rare, native endemic species from open woodland near Witley, Surrey (17). **H. trichocaulon** (Dahlst.) Johans. is a common native species of roadsides, open woodland and sandy heaths in southern England, with outlying localities in Lincolnshire (54), Brecon (42) and Merioneth (48). It also occurs in Scandinavia. **H. cambricogothicum** Pugsl. is known from railway banks and walls at Dunton Green and Beckenham in Kent (16), walls at Llanfairfechan, Caernarvon (49), and sandy banks of the River Findhorn near Forres, Moray (95). Only the last of these appears to be a natural habitat. This species is not recorded outside Britain. **H. eboracense** Pugsl. is a native endemic species of open woodland, grassy places and heaths. It is a fairly frequent plant of south Wales and south-eastern and northern England, with a few outlying localities in central England. **H. ornatilorum** P. D. Sell & C. West is a native endemic species restricted to Yorkshire. **H. calcaricola** (F. J. Hanb.) Roffey is a native endemic species, scattered over England and Wales in grassy places and heathland, with a tendency to have the same distribution pattern as that of *H. eboracense*. This group of 7 closely allied species is mainly confined to the English lowlands.

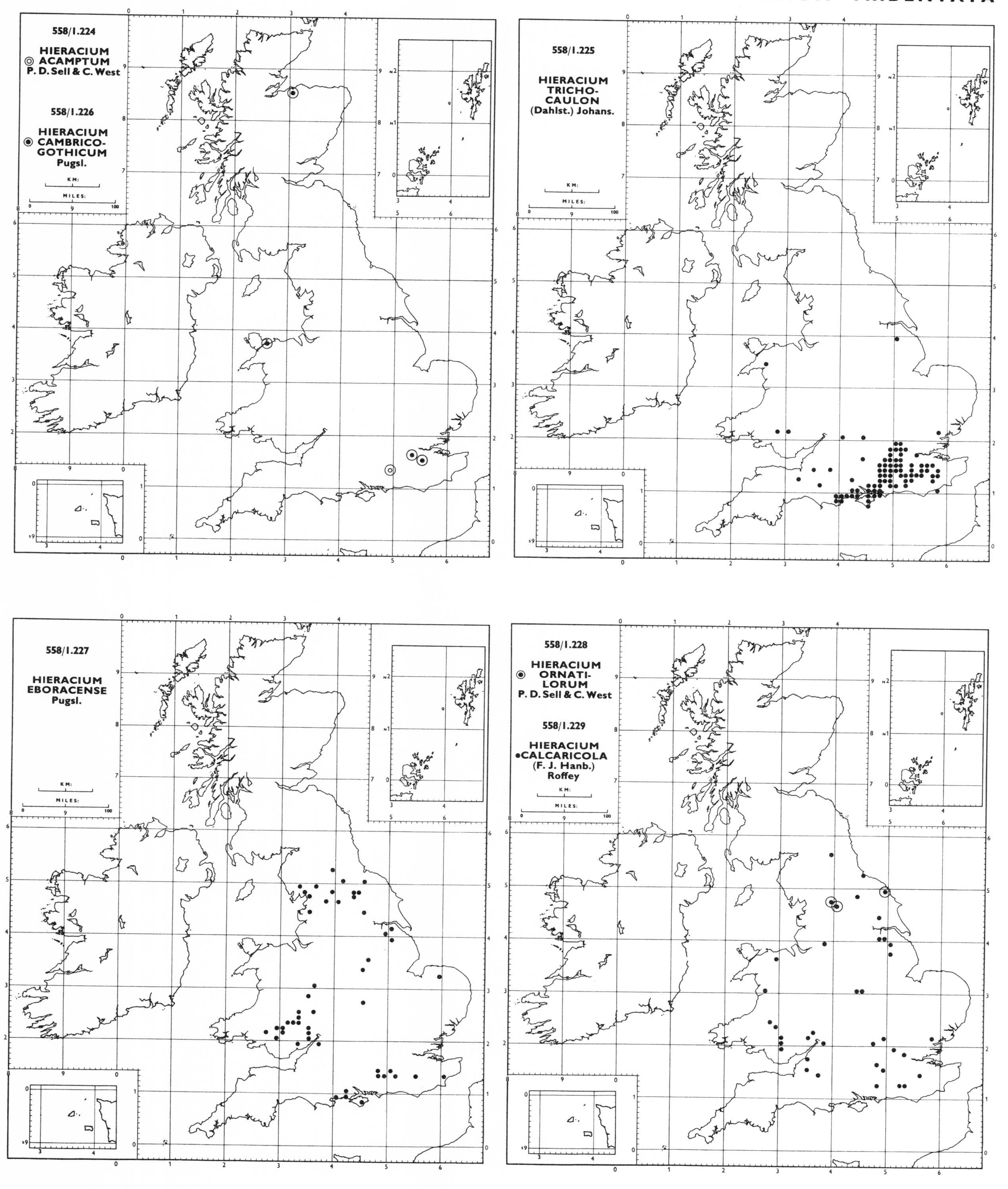
558/I.224
HIERACIUM
ACAMPTUM
P. D. Sell & C. West
558/I.226
HIERACIUM
CAMBRICO-
GOTHICUM
Pugsl.
558/I.225
HIERACIUM
TRICHO-
CAULON
(Dahlst.) Johans.
558/I.227
HIERACIUM
EBORACENSE
Pugsl.
558/I.228
HIERACIUM
ORNATI-
LORUM
P. D. Sell & C. West
558/I.229
HIERACIUM
CALCARICOLA
(F. J. Hanb.)
Roffey
KM:
MILES:

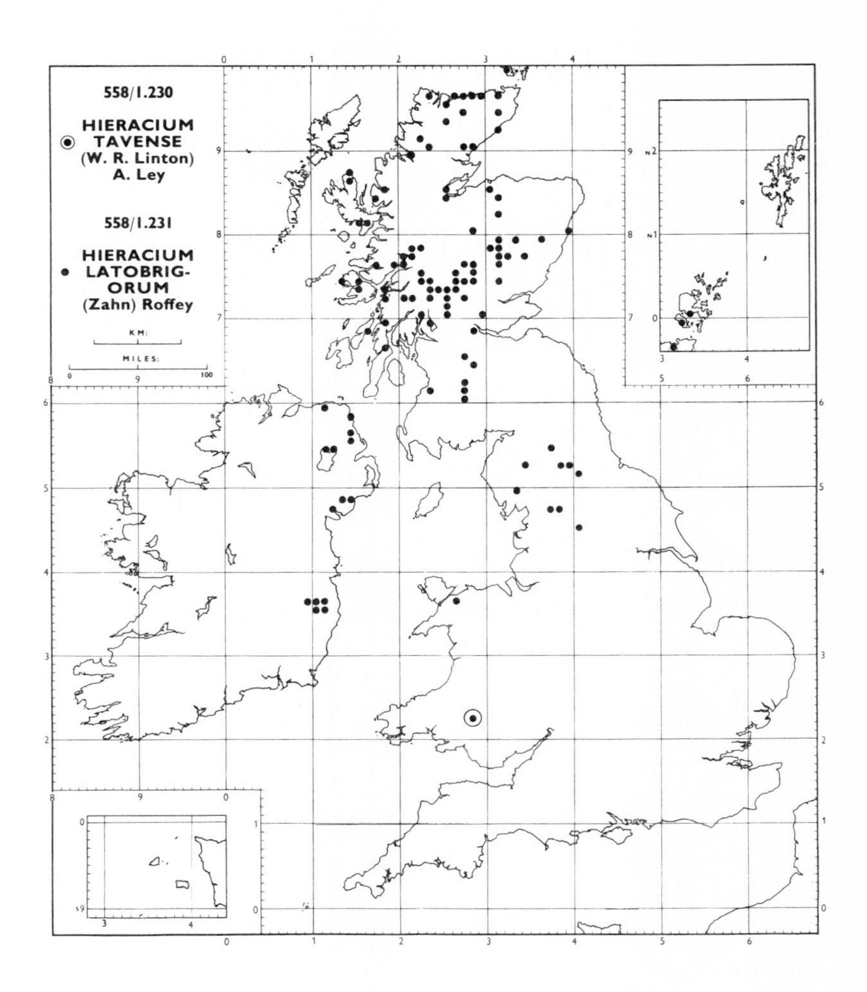

558/1.230–239 Section **FOLIOSA** Pugsl.

This section is a natural group of 10 closely allied native species. Plants belonging to the section are common in grassy and rocky places in Wales, northern England, Scotland and parts of Ireland. Although preferring hills and coastal areas they never ascend to a very great altitude. They also occur in Greenland, Iceland, Scandinavia, central Europe, and across Russia into Siberia. They flower in July and August, showing little seasonal variation.

H. tavense (W. R. Linton) A. Ley is a very rare endemic species restricted to the rocky banks of a stream in the Upper Tawe Glen, Brecon (42).

H. latobrigorum (Zahn) Roffey is a common species in Scotland, occurring also in Orkney (111), northern England, north Wales and eastern Ireland. It is a widespread species in central Europe.
H. subcrocatum (E. F. Linton) Roffey is a very common endemic species of Wales, northern England and Scotland, and is local in eastern Ireland. It is closely allied to *H. latobrigorum*, from which, however, it can always be distinguished by its black, not pure yellow, styles. It shows some variation in the indumentum of the phyllaries, Welsh plants tending to have more glandular hairs, and north English and south Scottish plants fewer glandular hairs, than the remainder.

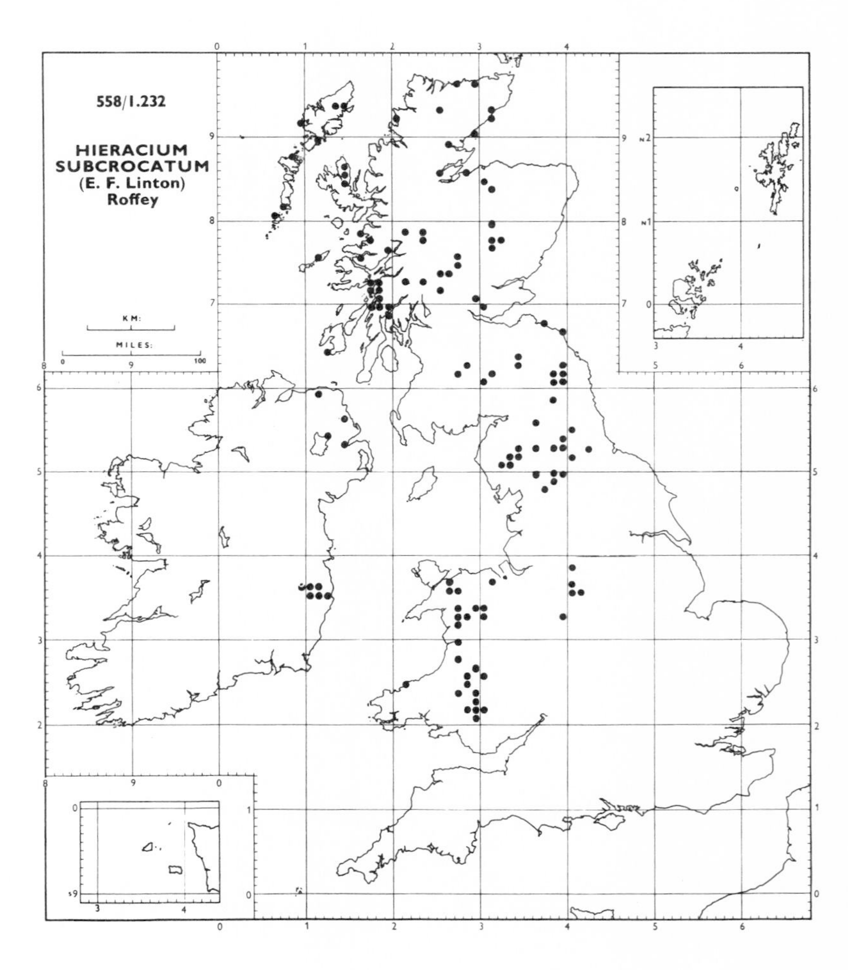

H. drummondii Pugsl. is a very local endemic confined to a few localities in south-central Scotland.

H. strictiforme (Zahn) Roffey is a very common species of central and northern Scotland and north-east Ireland, with scattered localities in southern Scotland, northern England, Wales and south-eastern Ireland. It also occurs in Norway.

H. reticulatum Lindeb. is a widely distributed species in central and northern Scotland, with isolated localities in southern Scotland, Northumberland (67) and south Wales. It also occurs in Scandinavia.

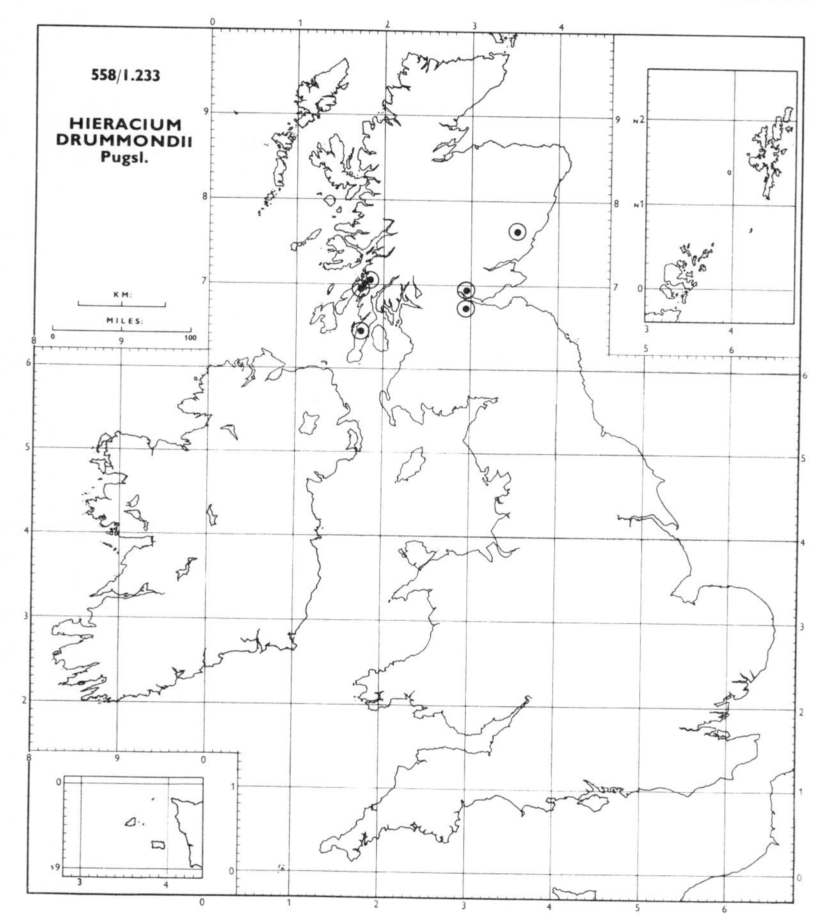

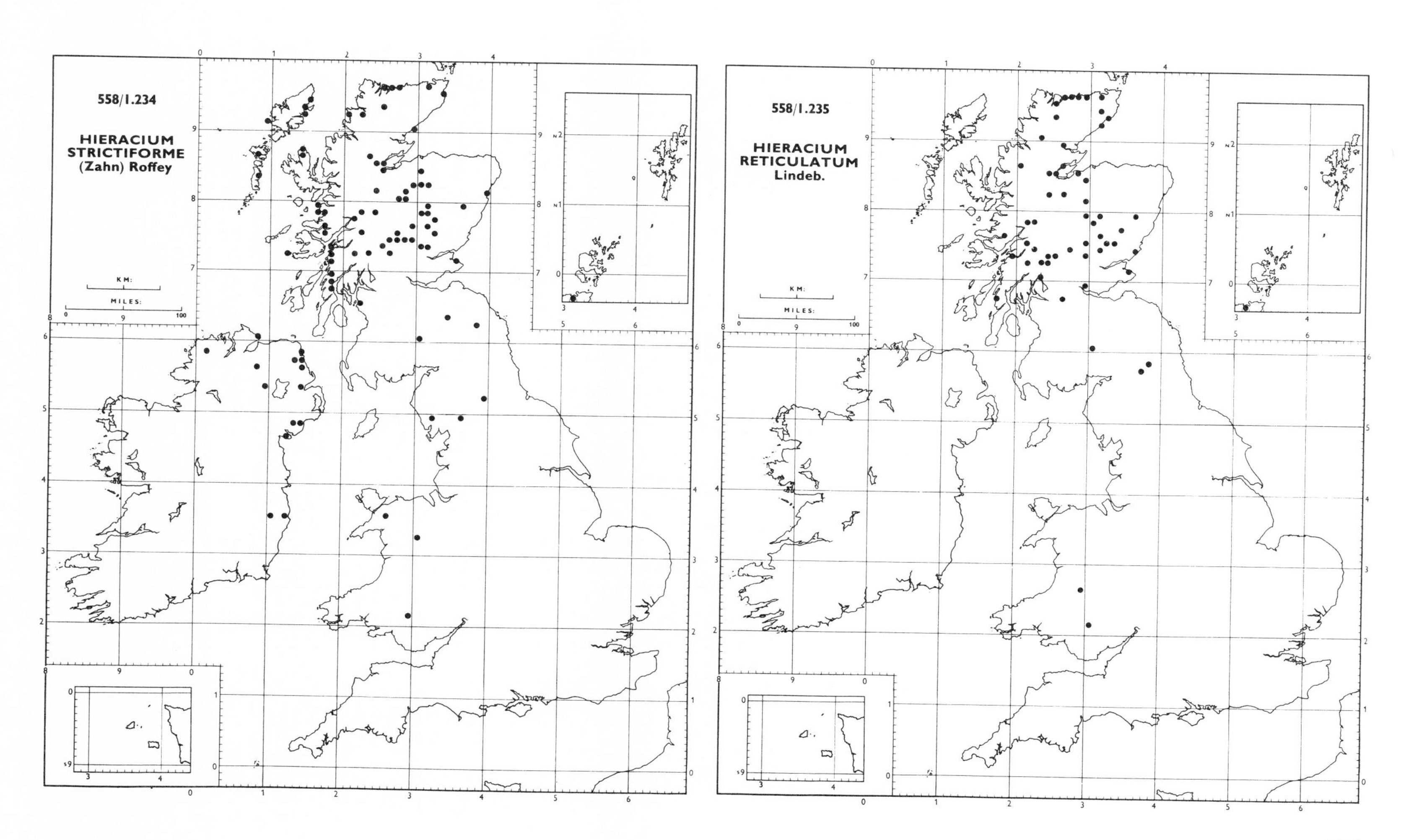

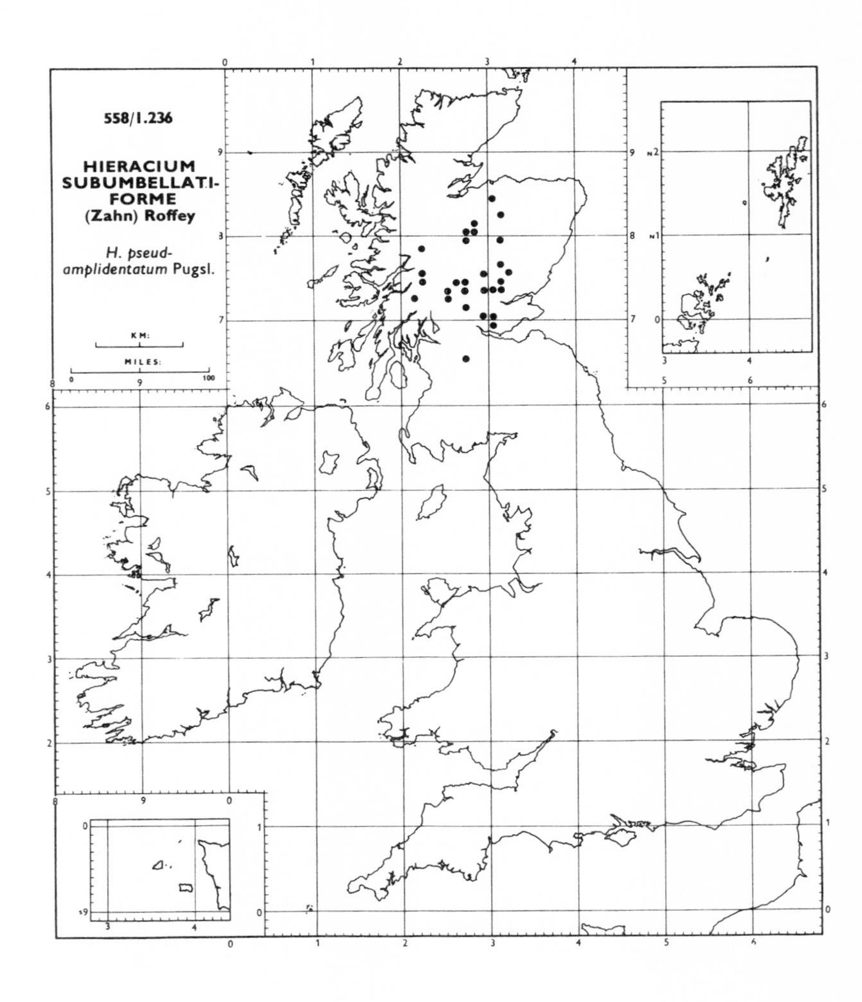

H. subumbellatiforme (Zahn) Roffey (*H. pseudamplidentatum* Pugsl.) is a widely distributed but local endemic of central Scotland.

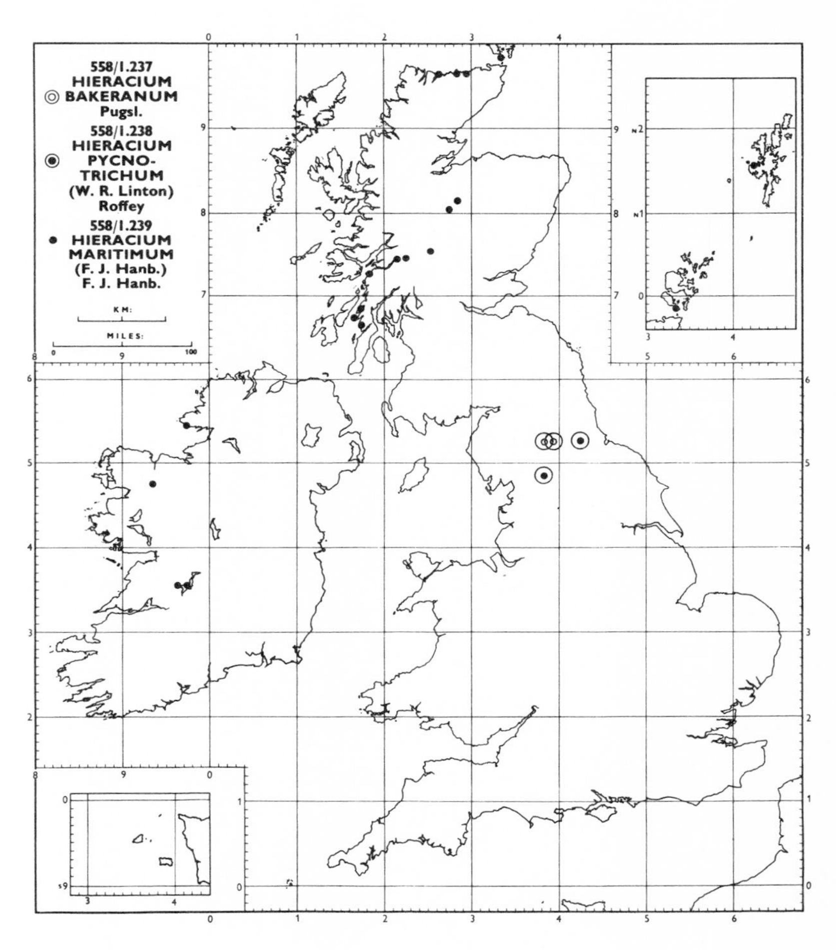

H. bakeranum Pugsl. is a very rare endemic confined to Upper Teesdale (65, 66).

H. pycnotrichum (W. R. Linton) Roffey is known only from specimens collected at Hawes, north-west Yorkshire (65), and near Bishop Auckland, Durham (66). It has not been seen in these localities for many years. It is also recorded from Scandinavia.

H. maritimum (F. J. Hanb.) F. J. Hanb. is a very local, native endemic of grassy banks and river-sides in northern and west-central Scotland, Orkney (111), Shetland (112) and western Ireland.

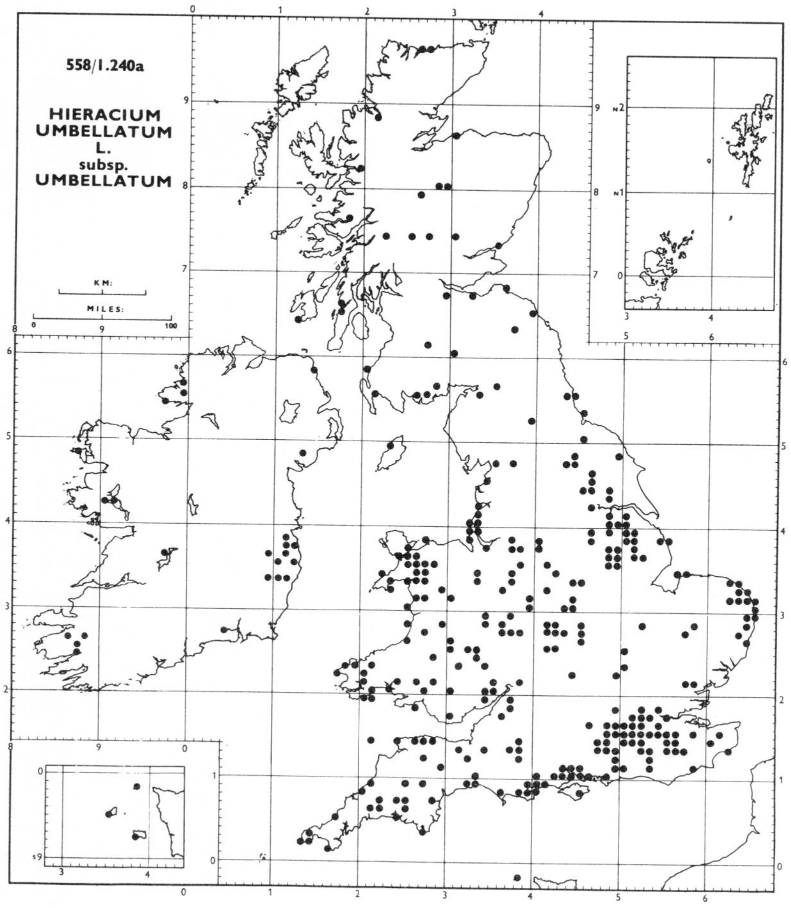

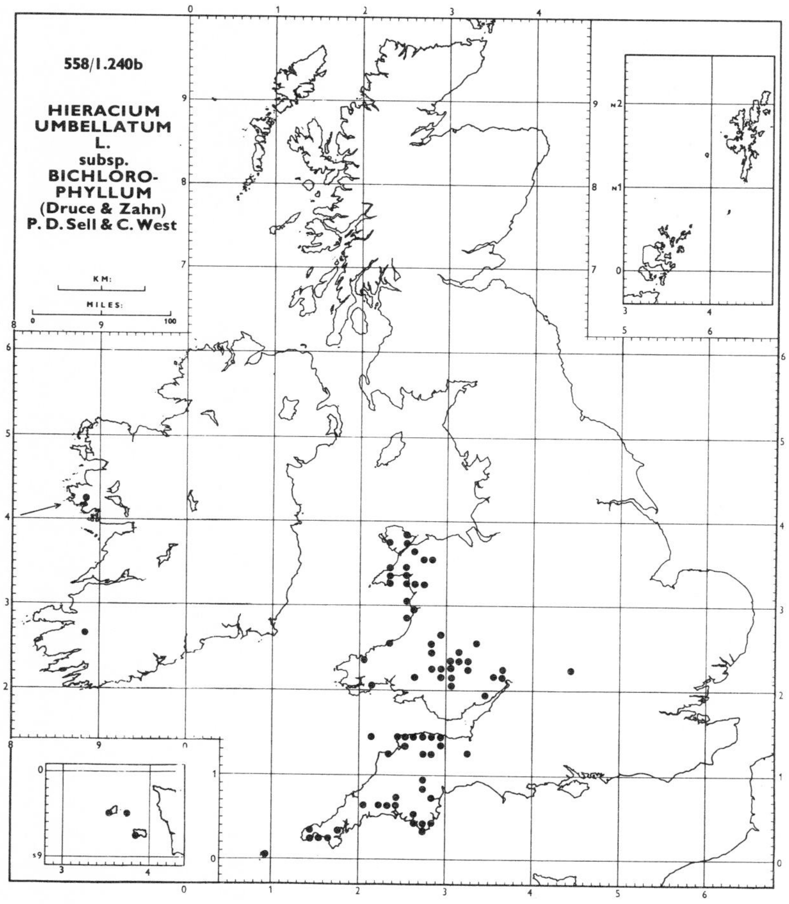

558/I.240a
HIERACIUM UMBELLATUM L. subsp. UMBELLATUM

558/1.240 Section **UMBELLATA** F. N. Williams

H. umbellatum L. is the only species of this section. It is also the only species occurring in the British Isles in which hybridization is known to take place. Some plants of the complex are, however, apomictic, and it is apparently not possible to distinguish these morphologically. It occurs on sandy heaths and dunes, and on grassy and rocky places throughout Great Britain (though it is rarer in the north), the Channel Islands and round the perimeter of Ireland. It varies considerably in stature, in the shape and dentation of the leaves and in the size of the inflorescence. In Wales and adjacent England and the south-west Peninsula, the Channel Islands and western Ireland, many of the populations have broad flaccid leaves with a lax inflorescence. Where such variants grow with typical *H. umbellatum* intermediates usually occur; we have therefore treated this broad-leaved variant as subsp. **bichlorophyllum** (Druce & Zahn) P. D. Sell & C. West. Other variants may have some ecological significance, but further experimental research on the group is needed before any taxonomic decisions can be made. *H. maritimum* of the Section *Foliosa* strongly approaches *H. umbellatum* in appearance and should certainly be considered when any experimental work on the group is carried out. *H. umbellatum* occurs throughout Europe, and in Asia and North America.

558/I.240b
HIERACIUM UMBELLATUM L. subsp. BICHLORO-PHYLLUM (Druce & Zahn) P. D. Sell & C. West

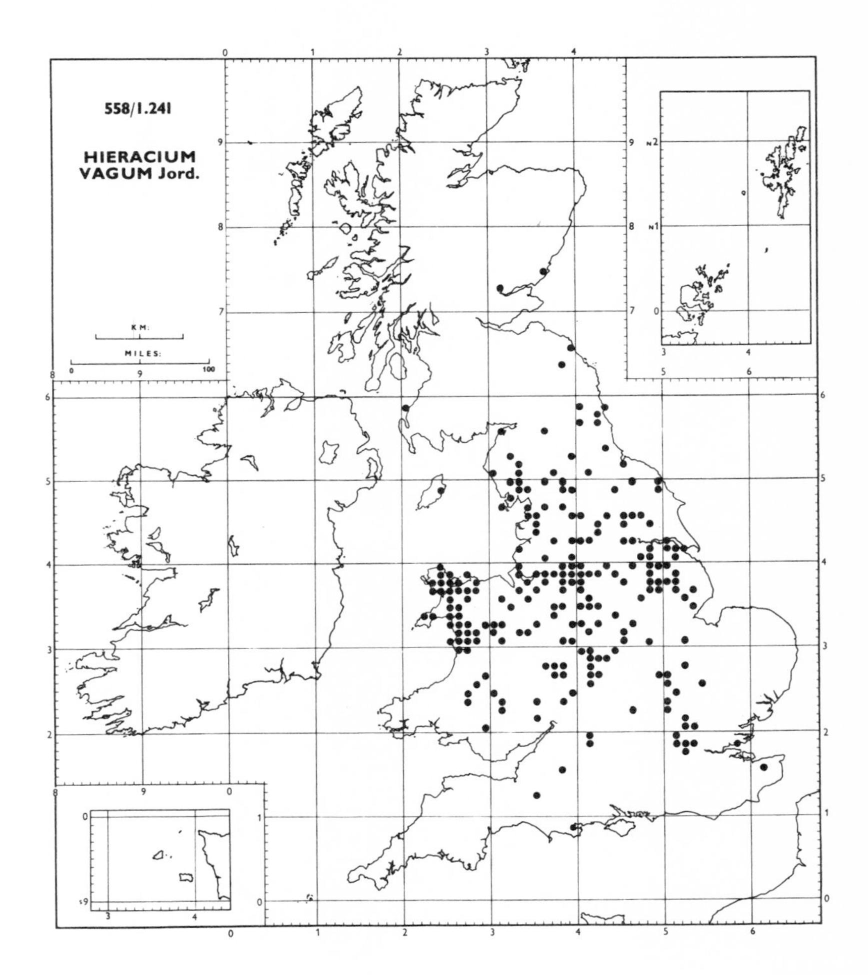

558/1.241–245 Section **SABAUDA** F. N. Williams

This section is a natural group of 5 closely allied species. Representatives of the section are common throughout England and Wales, and local in Scotland. Only one species occurs in Ireland where it is confined to the eastern part of the country. In many of their localities, especially railway and roadside banks, these plants are possibly introduced, but on sandy heaths, in old woodland and by rocky streams and on cliffs they are certainly native. They flower from late July to September, and can show considerable variation. Plants intermediate in character can sometimes be found in this group, and it is possible that some of the species are sexual. Species of the section occur throughout much of Europe and in Asia Minor.

H. vagum Jord. is a common species of central and north England and Wales, with a few outlying localities. It is native in rocky places and open woodland and has spread from these to roadsides, railway banks and other grassy places. It occurs throughout a large part of continental Europe.

H. salticola (Sudre) P. D. Sell & C. West is a locally abundant species of roadsides, railway banks and other grassy places in central England, particularly in the Birmingham area, south to Middlesex (21), Gloucestershire (34) and Hampshire (11), and north to Westmorland (69) and Angus (90). In these areas it is questionably native. Specimens collected at Aberdeen Links in 1887 and by the River Dee, near Kingcausie, Aberdeen (92), in 1878 would appear to be records of this species as a native plant. It is a widespread native species of France and Germany.

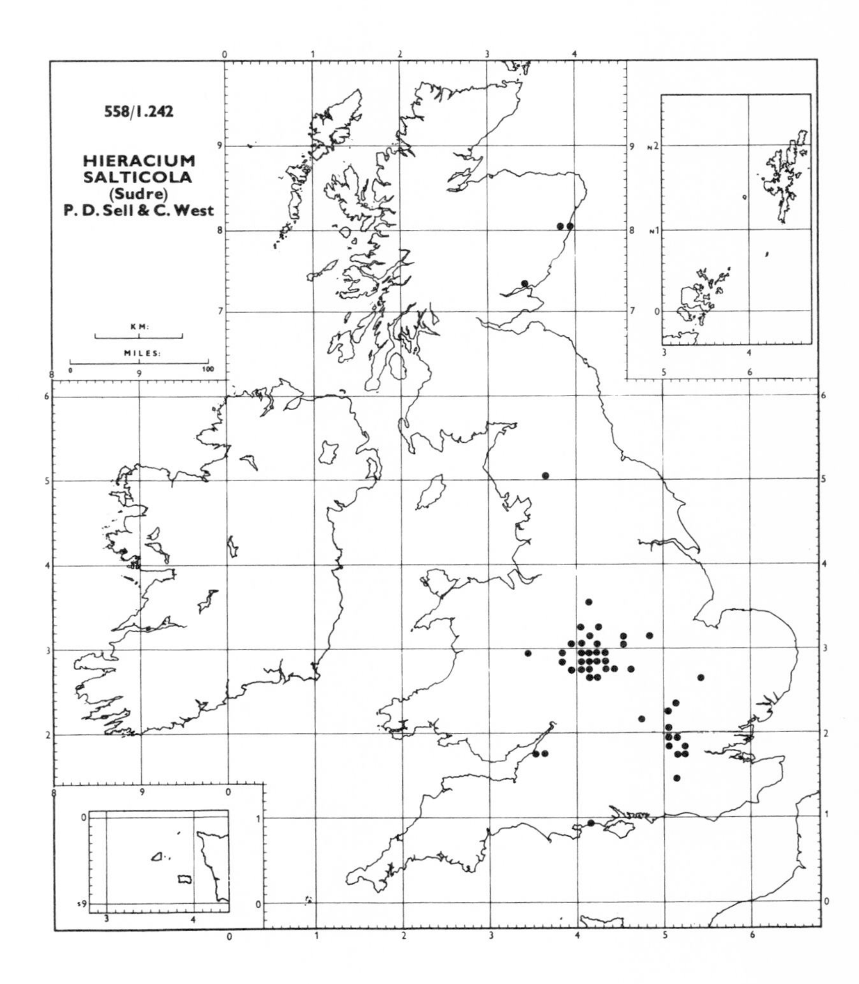

H. rigens Jord. is a locally common species of sandy heaths, open woodland, railway banks and grassy places in south-east England, with a few outlying localities in central England. There are single records from Cumberland (70), Angus (90) and Kincardine (91). It is probably native in at least some of its British localities. It also occurs in the Pyrenees and in central Europe.

H. virgultorum Jord. is a very local plant of open woodland and grassy banks in south-east England, where it is presumably native. It is widespread on the Continent, from Spain and France to Russia.

H. perpropinquum (Zahn) Druce is a common, very variable native species of open woodland and heathland, particularly on sandy soils, throughout the greater part of England and Wales. In Scotland it is a rather local species north to Ross (106). In Ireland it is restricted to a small area in the south-east, with an outlying record in the north-east. It occurs throughout much of western and central Europe.

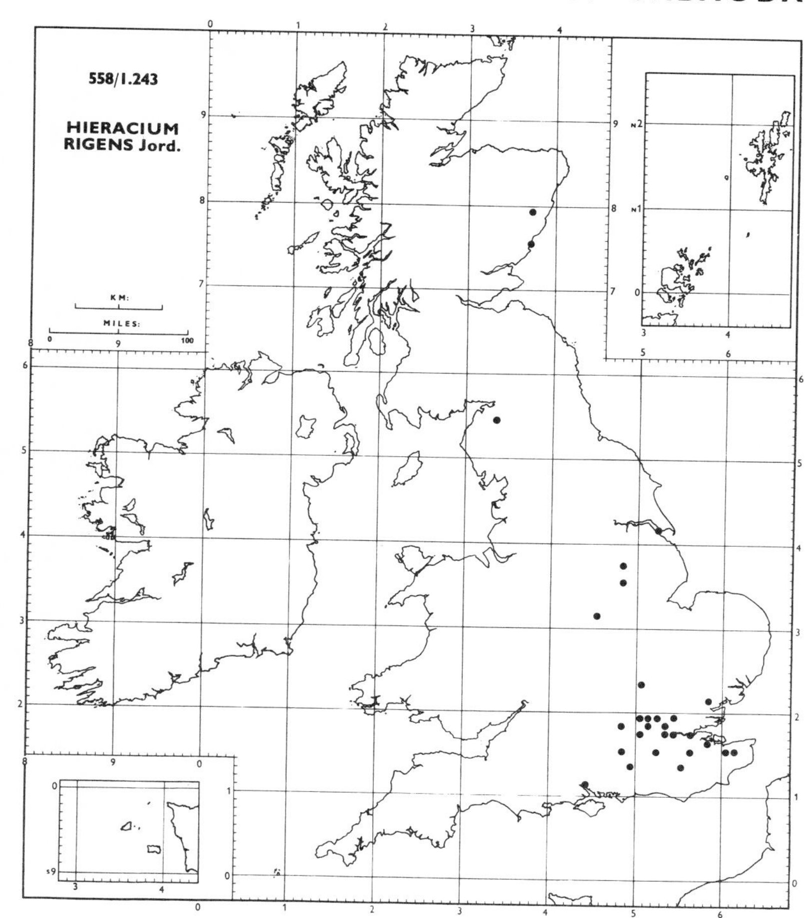

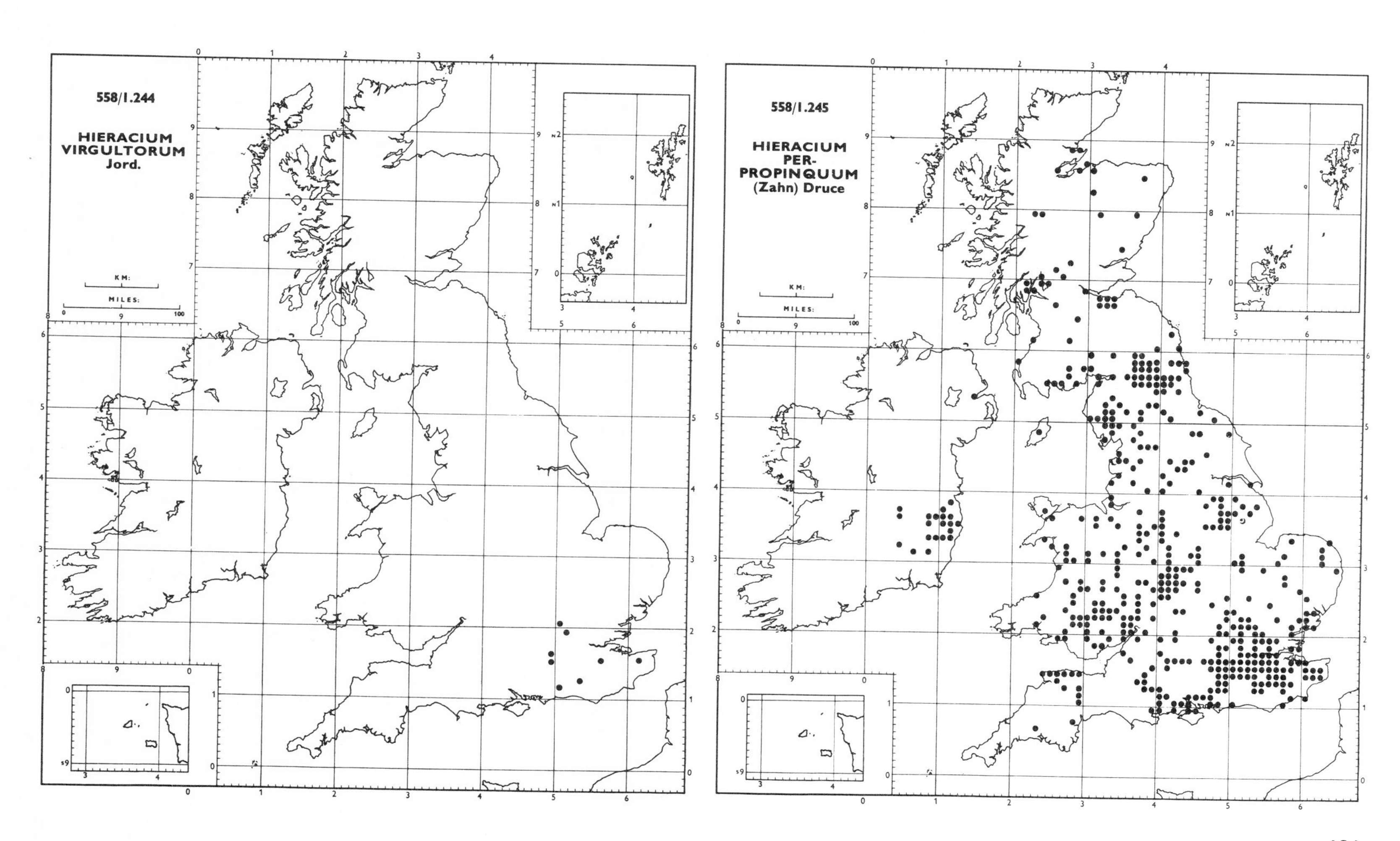

558/2 PILOSELLA Hill*

We have here followed C. H. and F. W. Schultz in separating the subgenus *Pilosella* (Hill) S. F. Gray from *Hieracium* as a distinct genus. Not only are its distinguishing characters without intermediates, but species belonging to the genus can be distinguished from *Hieracium* at a glance. Unlike species of *Hieracium*, which have little or no pollen, the species of *Pilosella* produce copious pollen. Hybrids are recorded between all known forms, many of them having been produced artificially. Many, if not all, can however, reproduce apomictically and spread vegetatively. The treatment of *Pilosella* must, therefore, for practical reasons, be different from that of *Hieracium*. We propose that a few of the more easily recognized groups should be treated as species, and where taxa within these groups are sufficiently widespread, or have enough ecological significance, they should be treated as subspecies.

P. peleterana (Mérat) C. H. & F. W. Schultz (*Hieracium peleteranum* Mérat) is a very local native species in a few coastal localities in southern England and the Channel Islands, and is recorded from near Borthwen, Merioneth (48). It also occurs inland at Craig Breidden, Montgomery (47), Wetton Mill, Staffordshire (39), Dove Dale, Derby (57), and Linton Falls, Yorkshire (64). The plants from south Devon (3), Stafford, Derby and Yorkshire are sufficiently distinct to be regarded as subsp. **tenuiscapa** (Pugsl.) P. D. Sell & C. West (*Hieracium peleteranum* var. *tenuiscapum* Pugsl.). *P. peleterana* is widespread in western Europe, from Scandinavia to Spain and eastward to Germany and Switzerland. Throughout its range *P. peleterana* hybridizes with *P. officinarum*.

P. officinarum C. H. & F. W. Schultz (*Hieracium pilosella* L.) is adequately mapped in the *Atlas* (p. 298) where a true absence of the species is shown in the Fenlands, parts of northern Scotland and Shetland (112). It shows much variation, some of the variants deserving the rank of subspecies. At the moment, however, we have insufficient data to map them adequately.

P. flagellaris (Willd.) P. D. Sell & C. West subsp. **flagellaris** (*Hieracium flagellare* Willd.) occurs on roadside and railway banks in a few scattered localities in southern and central England, and on railway banks in the Forth area, Scotland. In these localities it is an introduced plant. It has recently been found by W. Scott in an isolated locality in Shetland (112), where it is apparently native. The Shetland plant is a very distinct variant, which we have named subsp. **bicapitata** P. D. Sell & C. West. *P. flagellaris* is a widespread native species of central and eastern Europe from Germany to Russia, and is naturalized elsewhere.

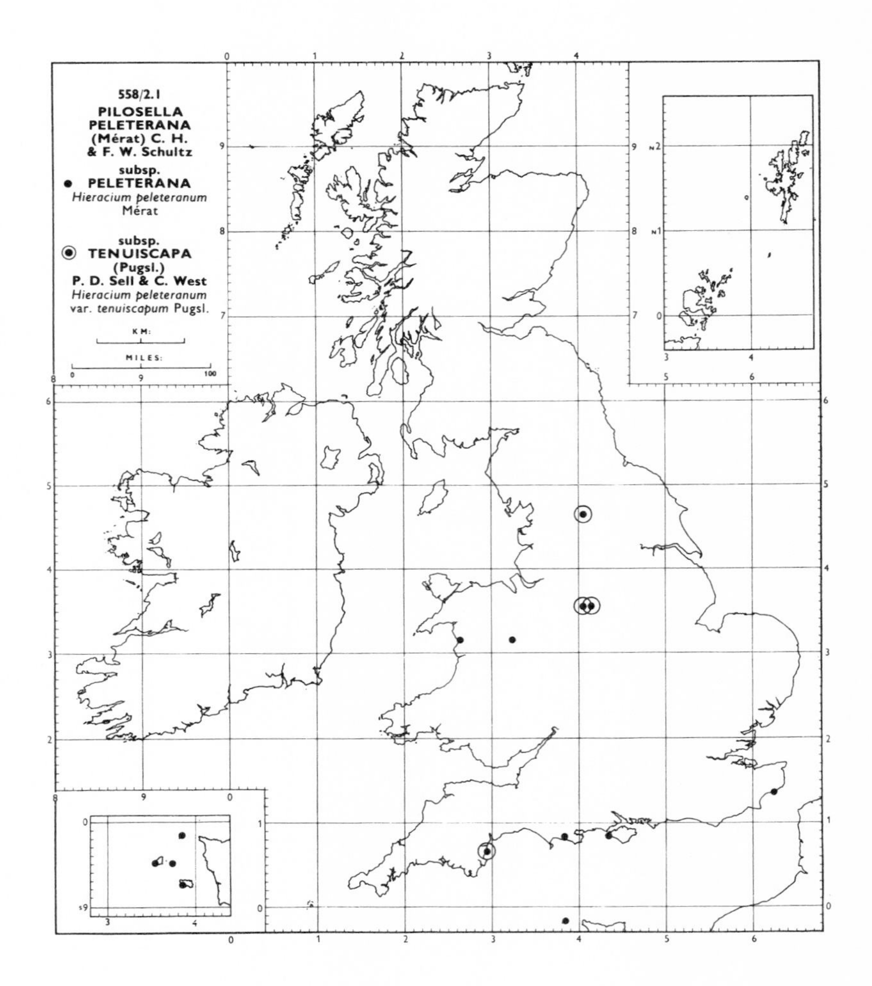

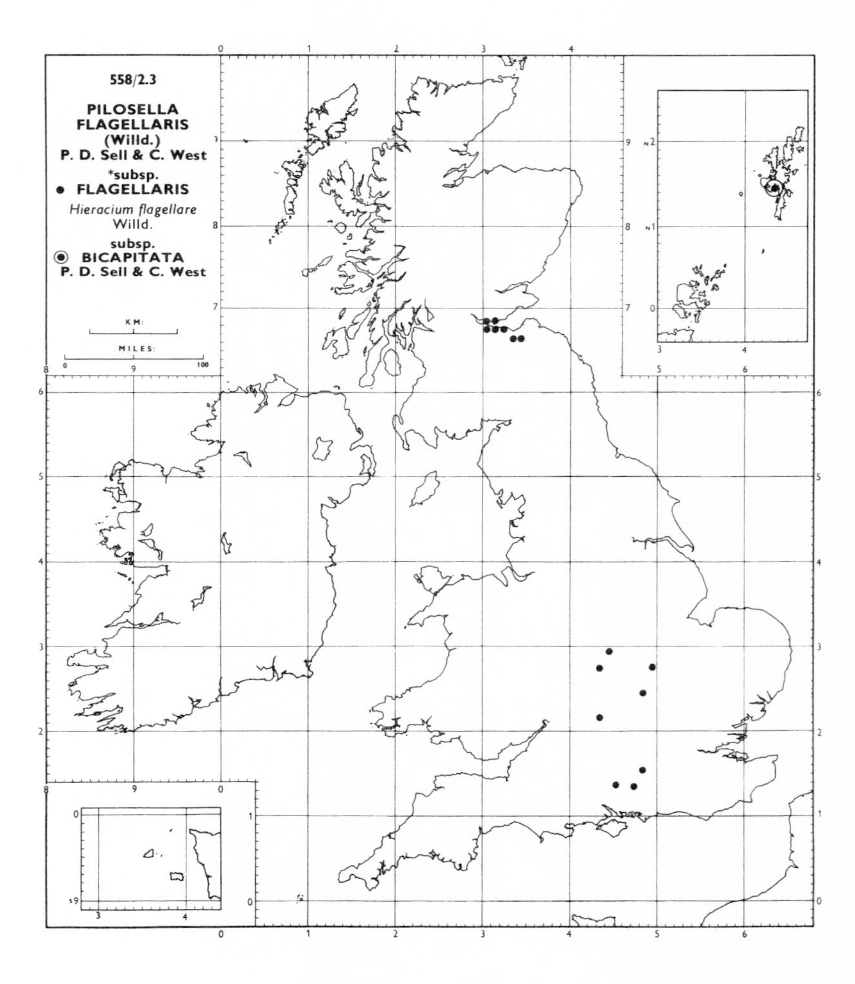

* By P. D. Sell and C. West.

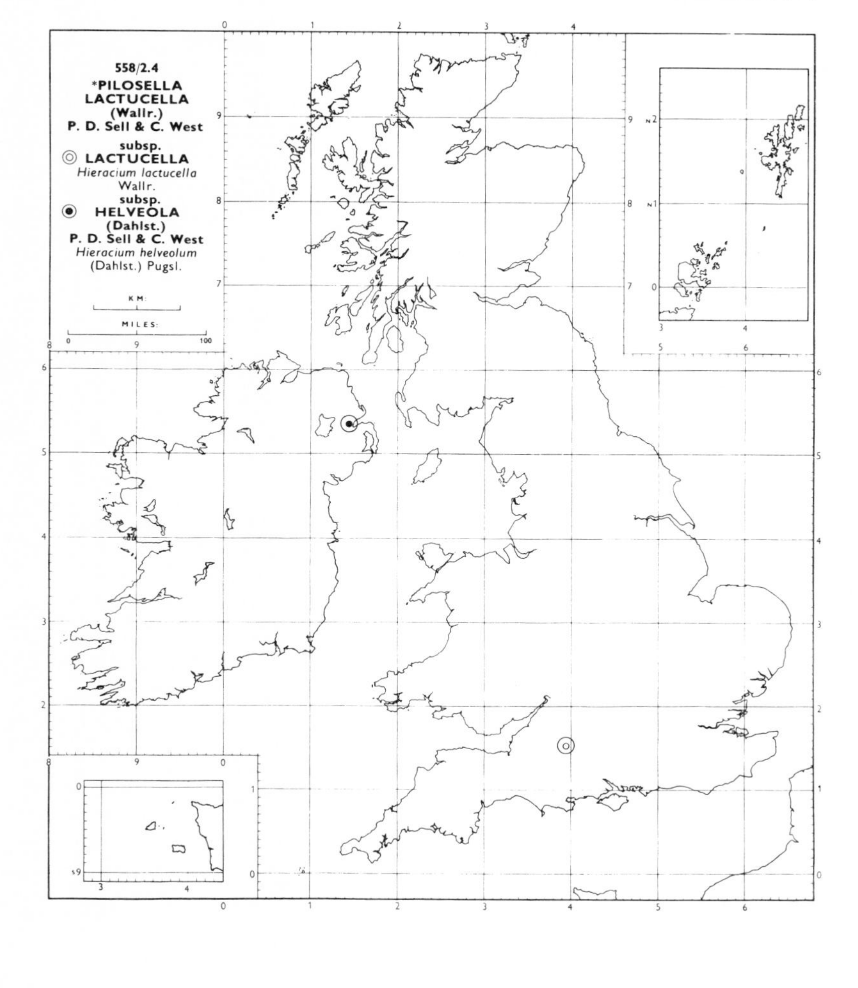

P. lactucella (Wallr.) P. D. Sell & C. West subsp. **lactucella** (*Hieracium lactucella* Wallr.) was discovered in 1904 in a pasture north-east of Keevil, Wiltshire (8), where it was presumably introduced. It has not been seen there since. It is a native of much of continental Europe. **P. lactucella** subsp. **helveola** (Dahlst.) P. D. Sell & C. West (*Hieracium helveolum* (Dahlst.) Pugsl.) was formerly naturalized at Cave Hill, Belfast (H. 39), where its origin was unknown. It is native in Scandinavia.

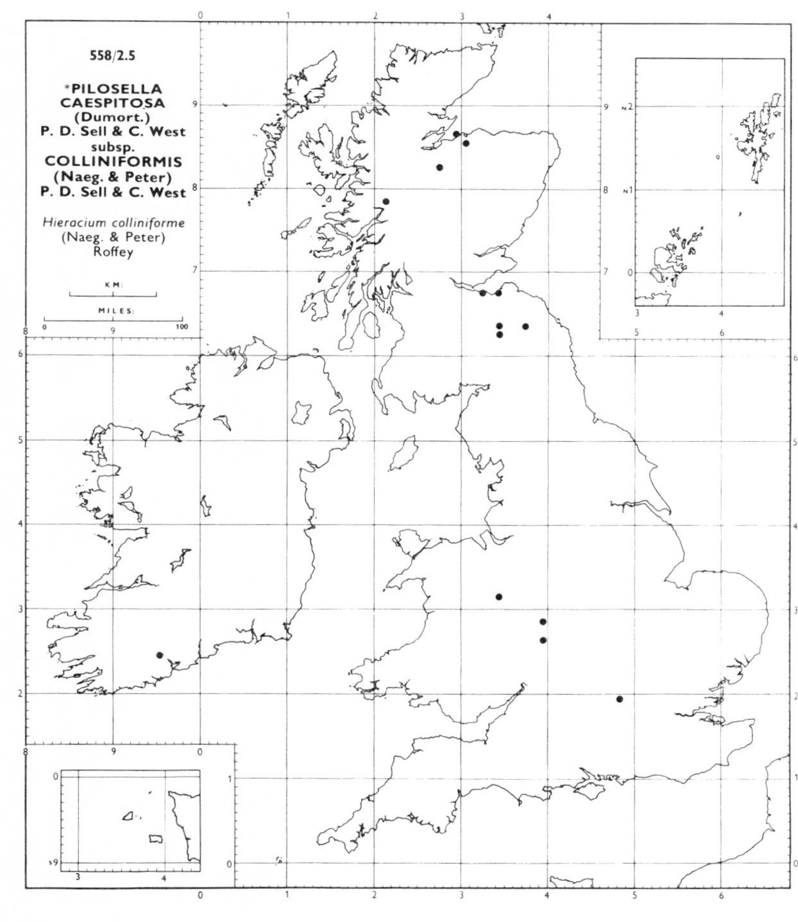

P. caespitosa (Dumort.) P. D. Sell & C. West subsp. **colliniformis** (Naeg. & Peter) P. D. Sell & C. West (*Hieracium colliniforme* (Naeg. & Peter) Roffey) occurs in a few widely scattered localities, mainly on railway banks and old walls, in England and Scotland, with a single record from Ireland, the banks of the River Lee at Cork (H. 5). It is naturalized on the dunes at Culbin, Moray (95). In all its localities it is presumably introduced. It also occurs in Germany and Sweden, and is naturalized in North America.

P. aurantiaca (L.) C. H. & F. W. Schultz (*Hieracium aurantiacum* L.) is naturalized on railway and roadside banks, walls and other grassy and waste places throughout Great Britain. In Ireland there are records only from Galway (H. 17) in 1891 and 1894, Meath (H 22) in 1965, Tyrone (H. 36) in 1948, and Down (H. 38) in 1963. The most widespread type is subsp. **brunneocrocea** (Pugsl.) P. D. Sell & C. West (*Hieracium brunneocroceum* Pugsl.), subsp. **aurantiaca** being of very local occurrence. This species is probably under-collected. *P. aurantiaca* is a native of central Europe, but is widely naturalized elsewhere in Europe, and in North America. At Cromdale, Moray (95), *P. aurantiaca* has hybridized freely with *P. officinarum* and there are two records from Angus (90). Such hybrids are also found occasionally in and around gardens.

P. praealta (Vill. ex Gochnat) C. H. & F. W. Schultz is naturalized in a few places, mainly on railway banks in southern and central England, with an isolated locality in West Lothian (84). Subsp. **praealta** (*Hieracium praealtum* Vill. ex Gochnat) occurs in a number of localities in south Devon (3), and at Elstree, Middlesex (21), Hanslope, Buckingham (24) and Plumley, Cheshire (58). It is widespread in continental Europe from Belgium and France to Russia. Subsp. **arvorum** (Naeg. & Peter) P. D. Sell & C. West (*Hieracium arvorum* (Naeg. & Peter) Pugsl.) occurs at Great Bedwyn, Wiltshire (7), Hungerford, Berkshire (22), and Carriden, Bo'ness, W. Lothian (84). It is native in Austria, Germany and Russia. Subsp. **spraguei** (Pugsl.) P. D. Sell & C. West (*Hieracium spraguei* Pugsl.) is restricted to a small area of Hertfordshire and Buckinghamshire (20, 24). It has not been identified with any continental form, but does not appear to be native.

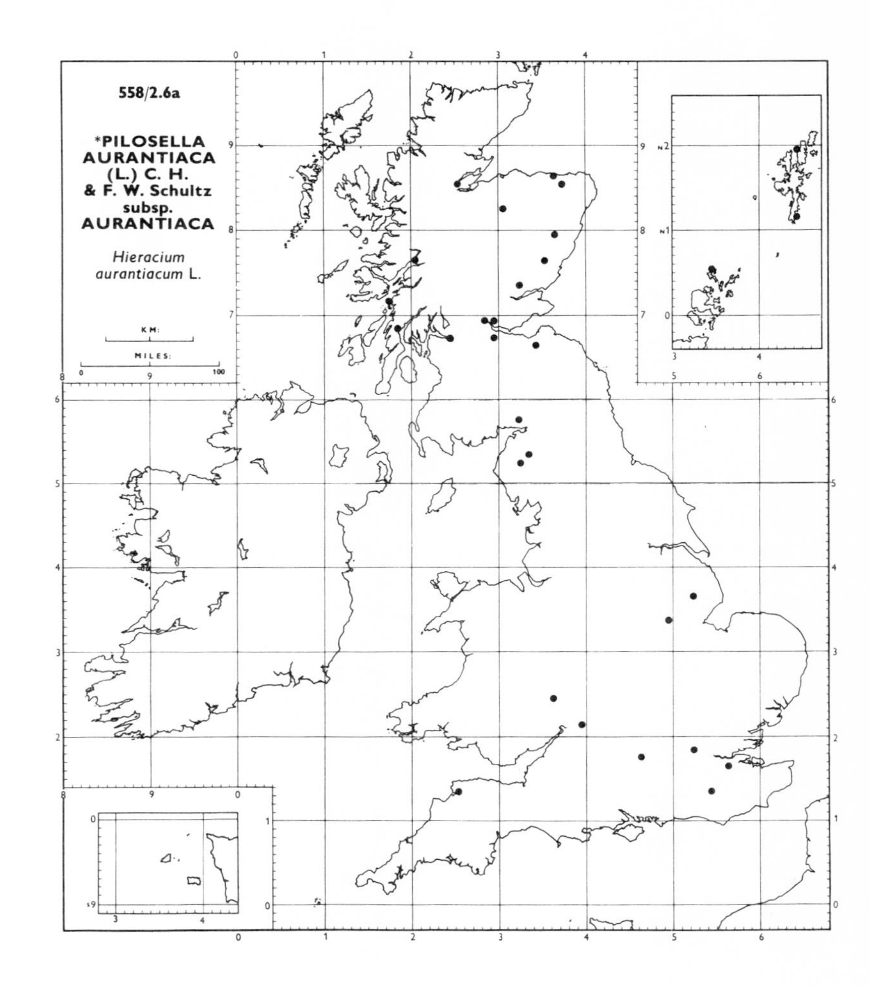

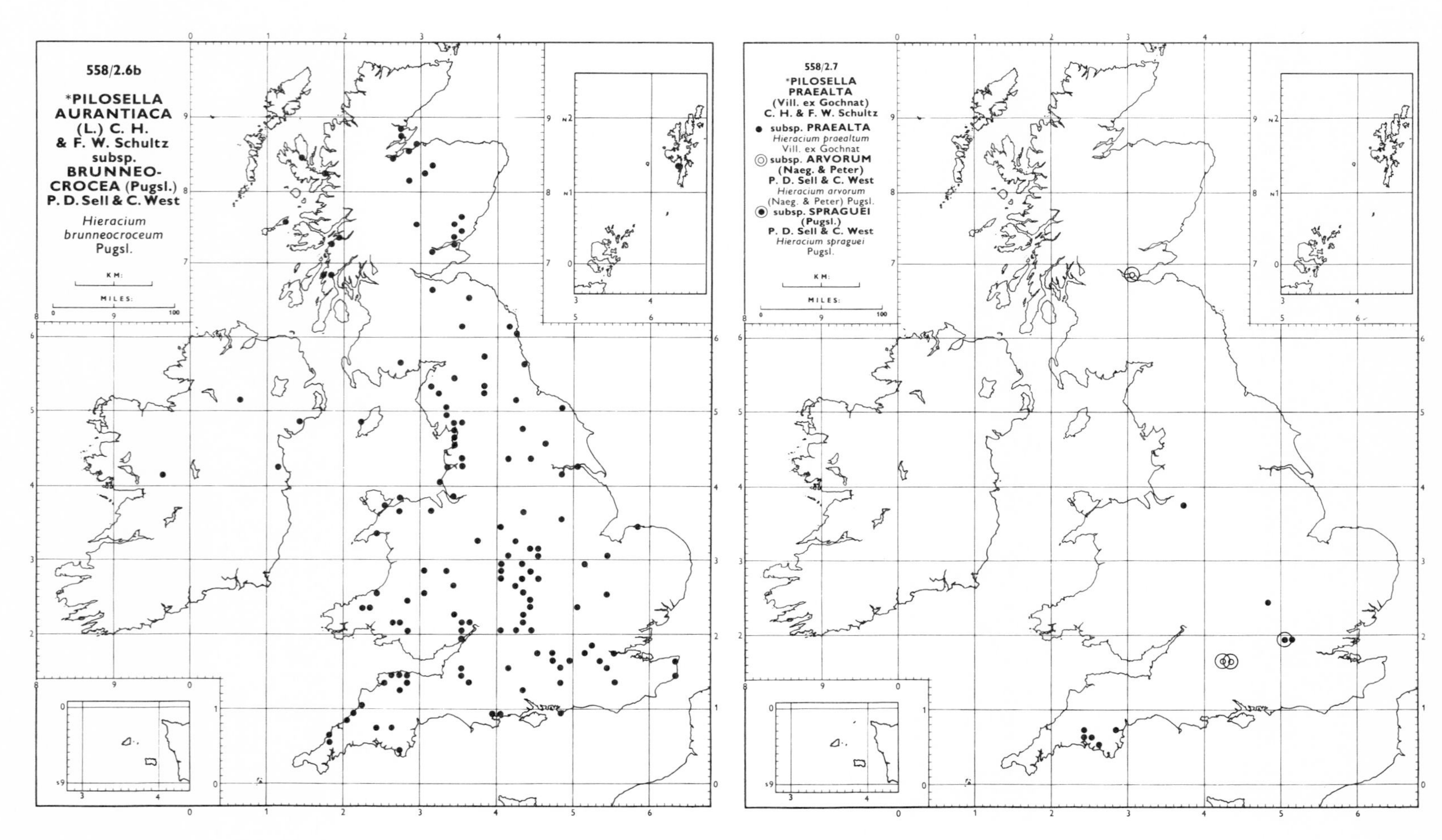

560/2 Taraxacum palustre (Lyons) DC.
560/3 Taraxacum spectabile Dahlst.
560/4 Taraxacum laevigatum (Willd.) DC.

The three maps are based on field records received. Of the four aggregate species recognized by Clapham, Tutin and Warburg (1962), *T. laevigatum* is the most easily distinguished, with its very small flowers and swollen appendage at the tip of the inner involucral bracts. *T. laevigatum* aggregate has consequently been well recorded and the map shows clearly that the species is most frequent on dry soils, largely confined to sand-dunes in the west. Records from v.-cs 79, 80, 92, 96, 100, H. 1, 2 and 20 have not been localized.

T. officinale Weber, *T. palustre* and *T. spectabile* are not so readily separated, as there are many intermediate apomictic species. The maps reflect this difficulty and suggest that there are differences of opinion between recorders about where to draw the line between them. *T. spectabile* particularly is very much under-recorded: it is intermediate between the other two and probably occurs throughout the British Isles in wetter and less disturbed habitats than *T. officinale*. Personal observation suggests that it even replaces *T. officinale* as a plant of roadsides in the far west. Otherwise, as far as is known, *T. officinale* occurs throughout the range of the aggregate mapped in the *Atlas*, p. 300.

Records of *T. palustre* from v.-cs 21, 82, 86, 99, 109 and H. 36 have not been traced.

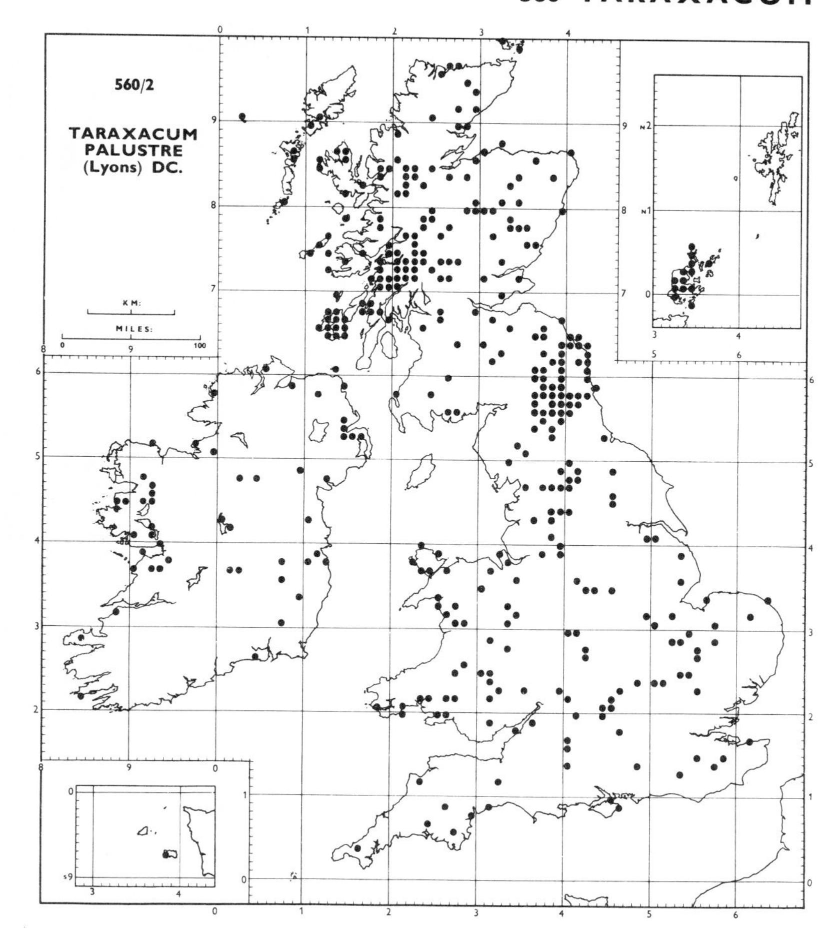

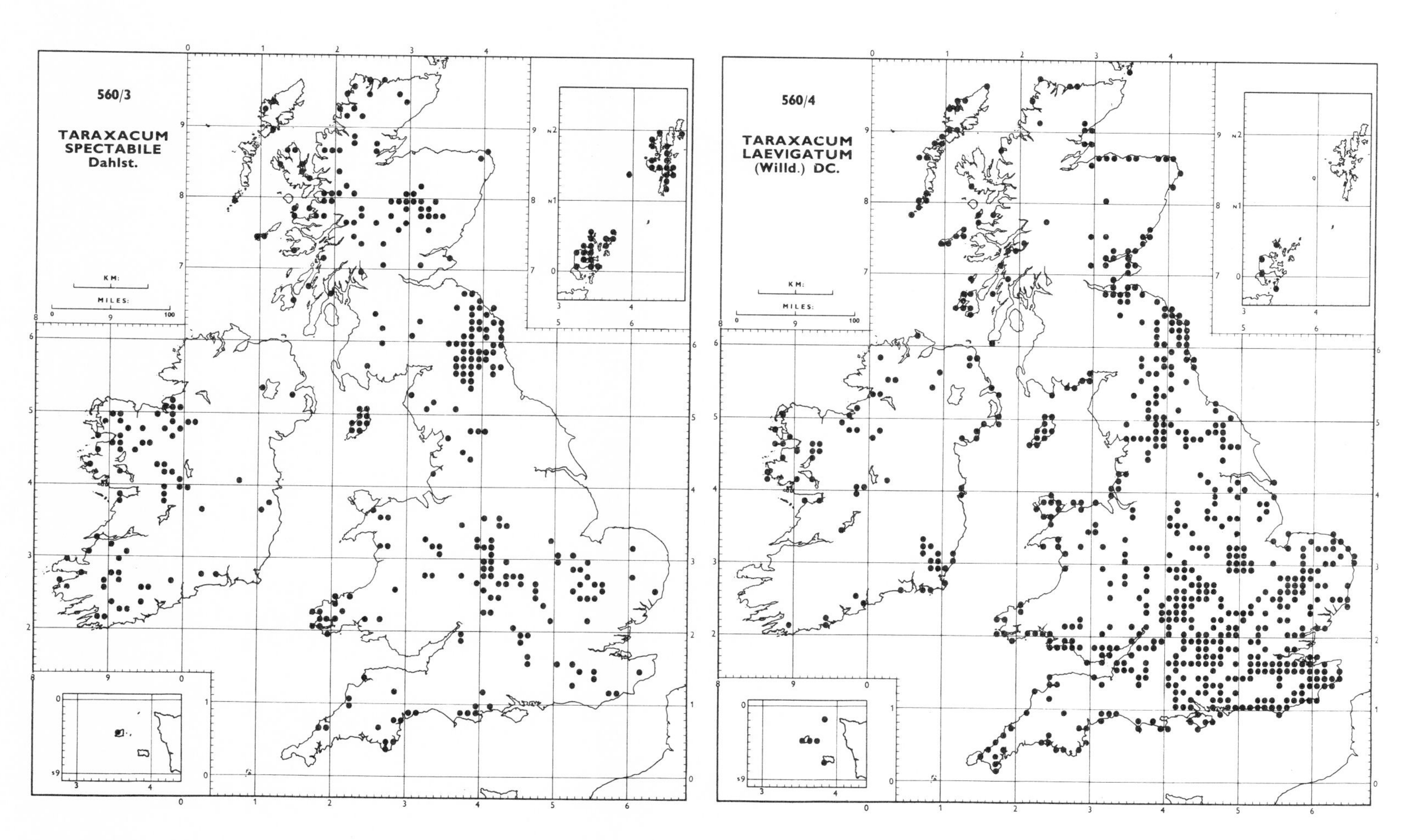

577/5 × 1

POTAMOGETON × FLUITANS Roth

P. lucens × natans

577 POTAMOGETON L.

These maps have been prepared solely from data supplied by J. E. Dandy, who has generously made available the data he has collected with Sir George Taylor during the last thirty years. Though the hybrids are always sterile (except perhaps *P.* × *zizii*), they readily spread by vegetative means, and some are known from areas where one and sometimes both the putative parents are absent.

ABD, BEL, BIRM, BM, CGE, CMM, DBN, DBY, DSY, E, GL, GLR, K, LCN, LDS, LSR, MANCH, NMW, OXF, PTH, RAMM, TCD, WAR

577/5 × 1 **Potamogeton × fluitans** Roth = **P. lucens × natans**

This hybrid is surprisingly rare considering the range of overlap of the two parents (see *Atlas*, pp. 305–6).

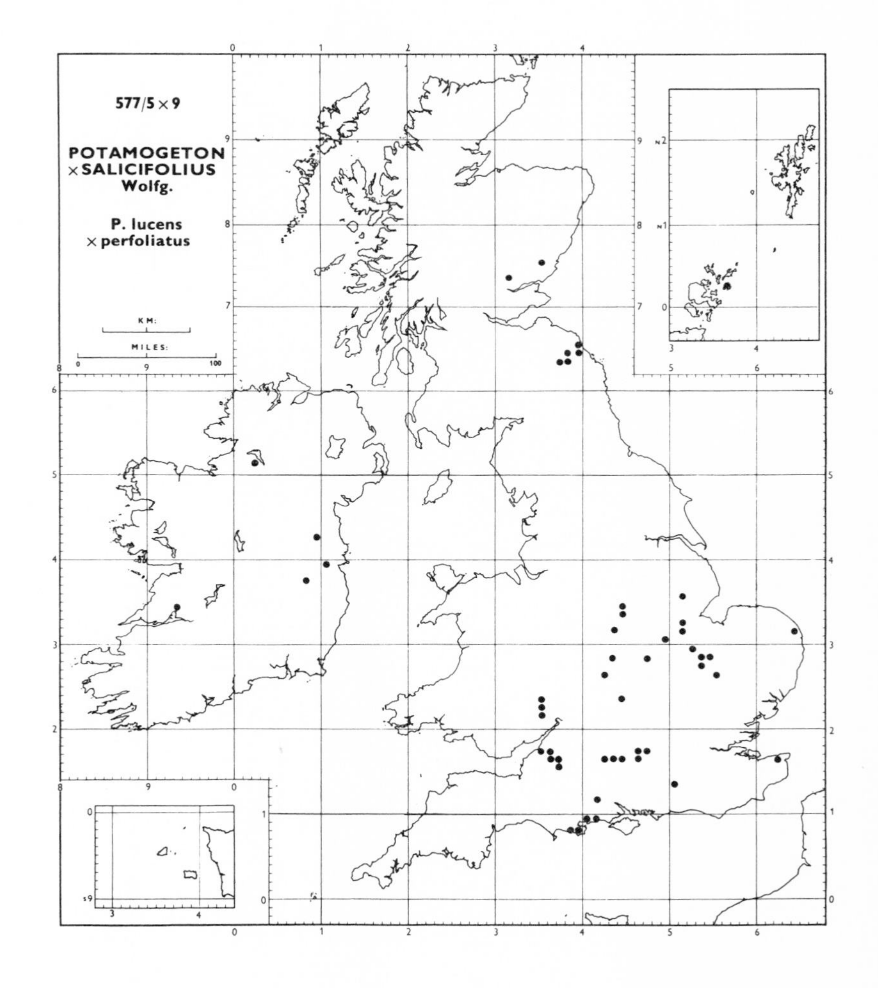

577/5 × 9 **Potamogeton × salicifolius** Wolfg. = **P. lucens × perfoliatus**

Scattered throughout the British Isles within the range of the parents, rarely occurring locally in the absence of one or both.

577/6×1 Potamogeton ×sparganifolius Laest. ex Fries
=P. gramineus×natans

Limited generally to the area in which the more restricted of the two parents, *P. gramineus*, occurs (*Atlas*, p. 306). The record from v.-c. 67 has not been confirmed.

577/6×5 Potamogeton ×zizii Koch ex Roth
=P. gramineus×lucens

One of the most widespread hybrids of *Potamogeton* in Britain. It appears to be partially fertile and is considered by some botanists to justify specific rank. It occurs throughout the range of *P. gramineus* and frequently outside the known range of *P. lucens*. Recorded in error from Kent and the Hebrides.

577/6×9 Potamogeton ×nitens Weber
=P. gramineus×perfoliatus

Similar in distribution to *P.×zizii*, but occurring occasionally well outside the present range of *P. gramineus*, notably in north Devon (4).

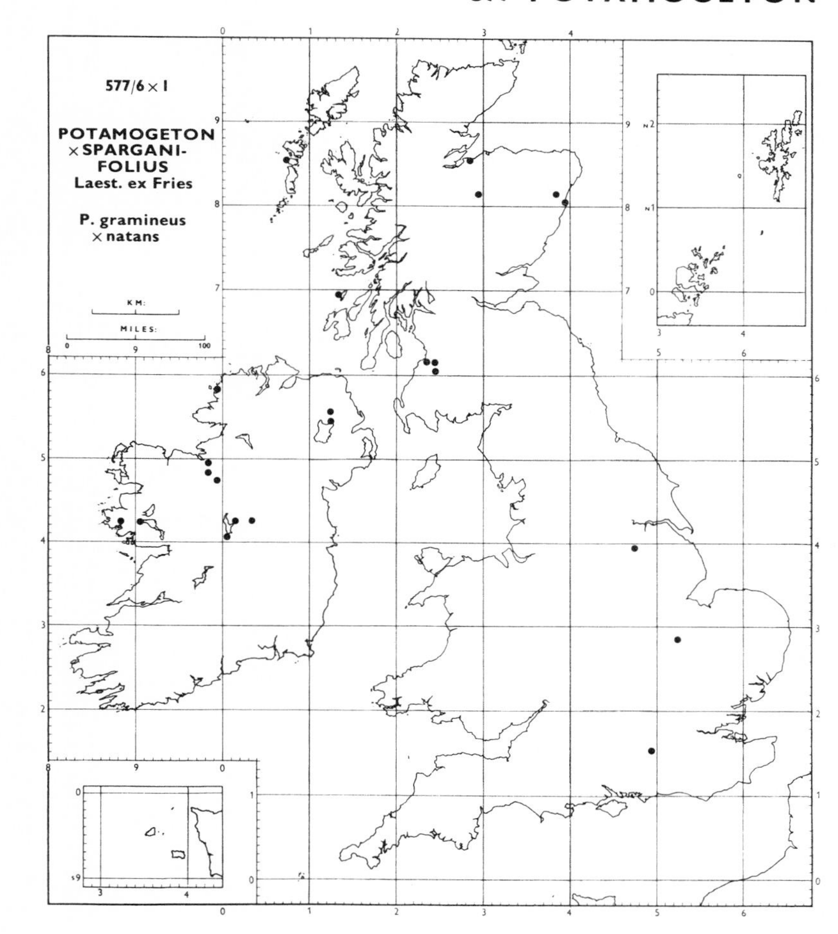

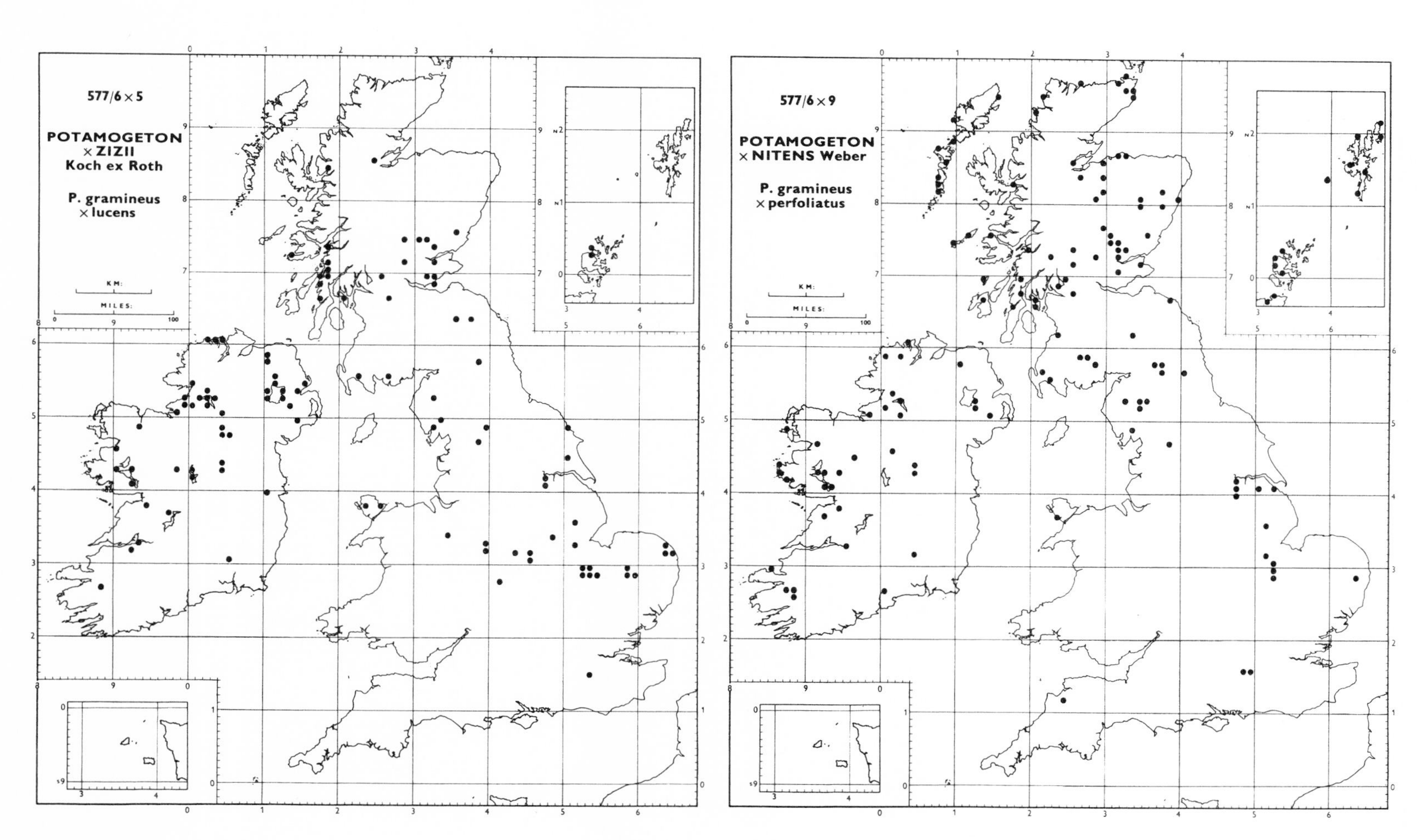

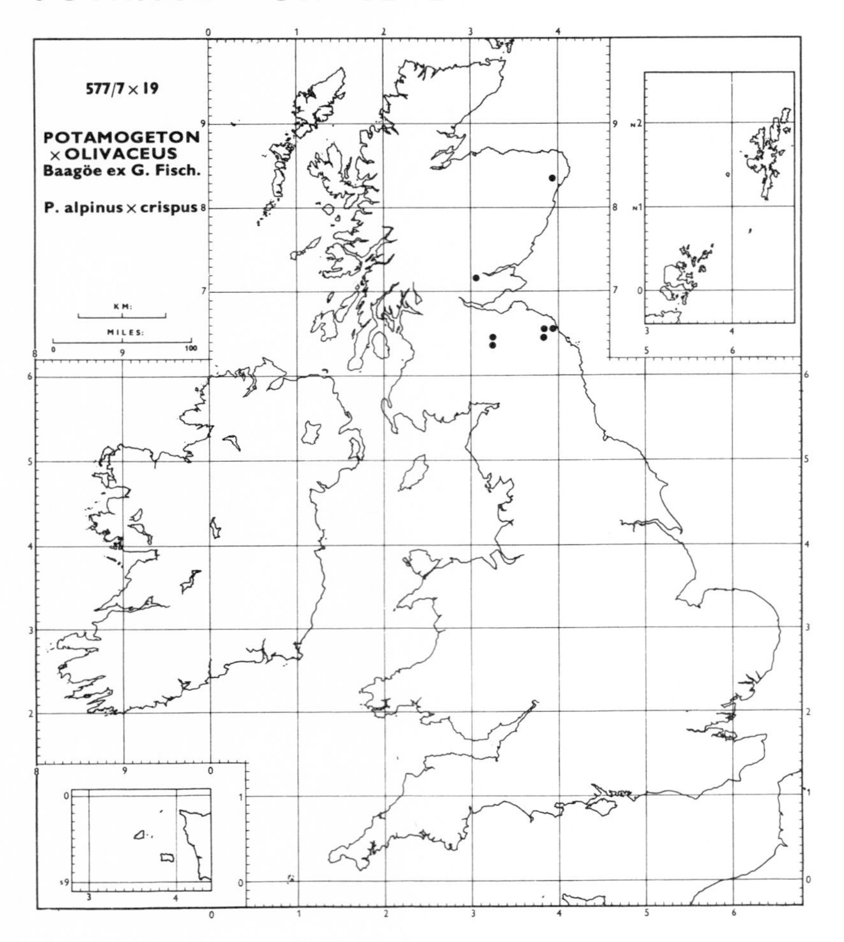

577/7×19 Potamogeton ×olivaceus Baagöe ex G. Fisch.
=P. alpinus×crispus

This hybrid is restricted to a few eastern rivers in Scotland, though the overlap of the two parents is much wider (see *Atlas*, pp. 306, 309).

577/19×9 Potamogeton ×cooperi (Fryer) Fryer
=P. crispus×perfoliatus

Both parents are widespread throughout the waterways of lowland Britain. The hybrid is scattered in this area, though curiously unrecorded from southern Ireland and scarce in England south of the River Thames. →

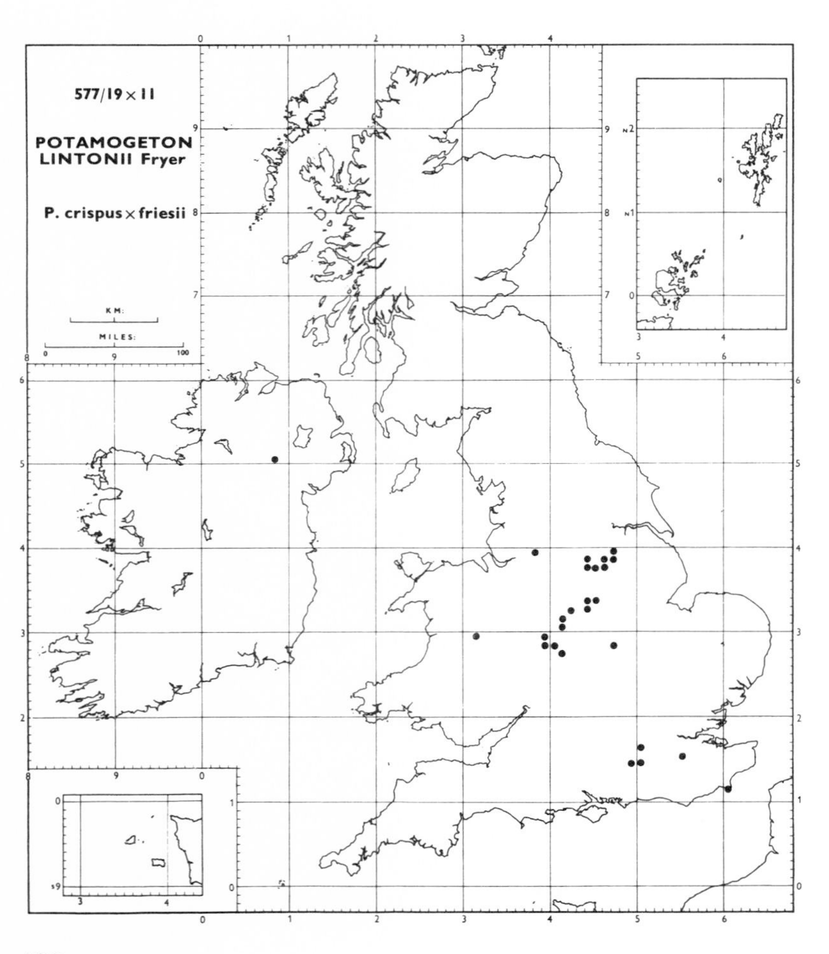

577/19×11 Potamogeton ×lintonii Fryer
= P. crispus×friesii

This hybrid is scattered throughout the British Isles, and occurs frequently in the absence of *P. friesii*, the more local of the two parents (see *Atlas*, p. 307). It is probable that the hybrid has spread along canals vegetatively.

577/20×21 Potamogeton ×suecicus K. Richt.
=P. filiformis×pectinatus

The occurrence of this hybrid in the Rivers Wharfe and Ure in Yorkshire (64, 65) is surprising as the nearest locality for one of the parents, *P. filiformis*, is over 100 miles away in Berwickshire (81) (see *Atlas*, p. 309). →

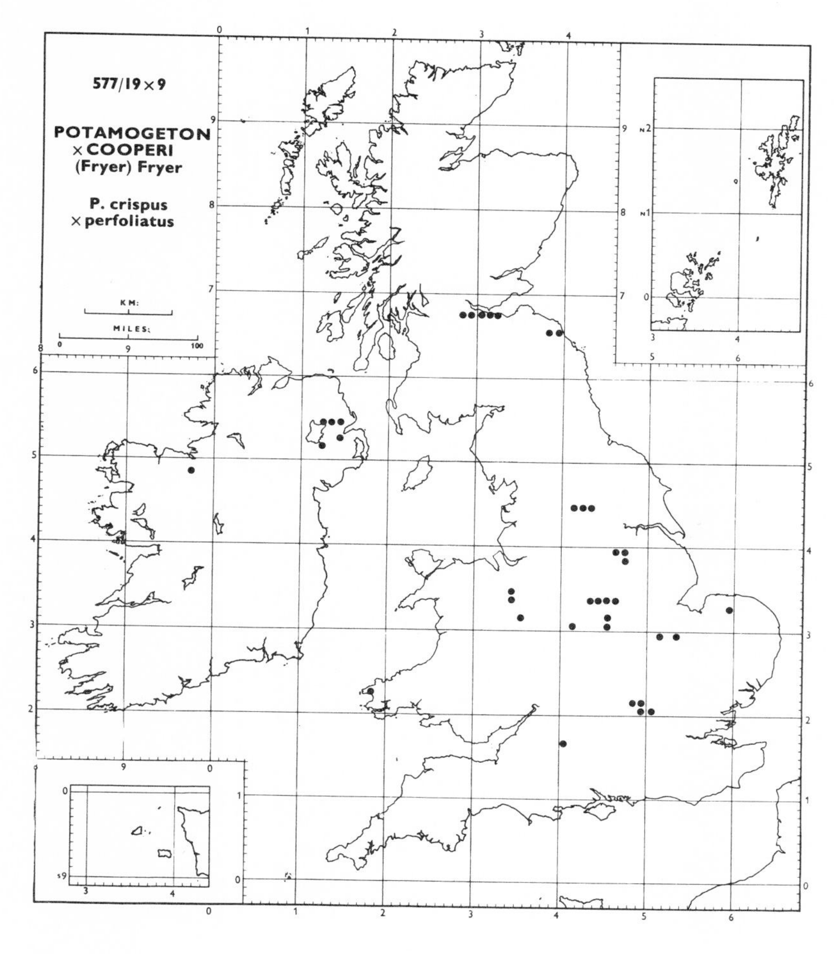

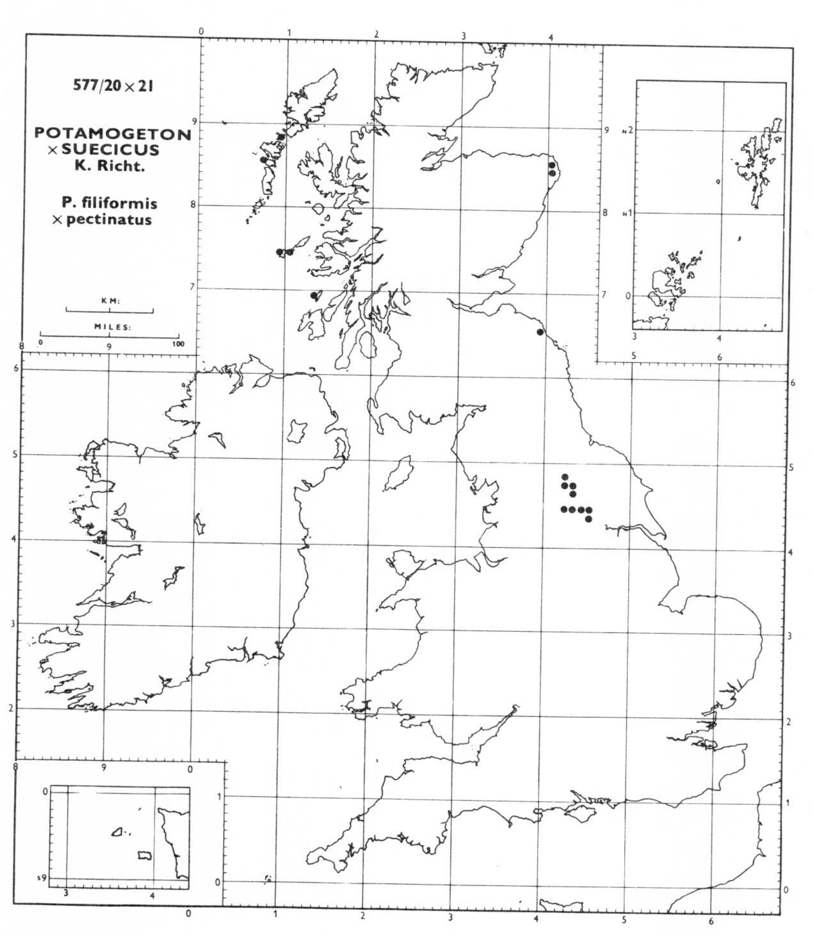

The following *Potamogeton* hybrids are very rare and have not been mapped.

577/3×6 Potamogeton ×billupsii Fryer
=P. coloratus×gramineus

Recorded with certainty only from Benwick, Cambridgeshire (29), and from Benbecula, Outer Hebrides (110). Reported also from Sweden.

577/7×5 Potamogeton ×nerviger Wolfg.
=P. alpinus×lucens

Recorded from the River Fergus, Clare (H. 9). Known also from Russia (Lithuania). (cf. Sell, 1967).

577/7×6 Potamogeton ×nericius Hagstr.
=P. alpinus×gramineus

Recorded from the River Don, south Aberdeenshire (92). Reported also from Norway and Sweden.

577/7×8 Potamogeton ×griffithii A. Benn.
=P. alpinus×praelongus

Known only from Llyn Anafon, Caernarvonshire (49), and from lakes in Ardnamurchan and Moidart, Inverness (97).

557/7×9 Potamogeton ×prussicus Hagstr.
=P. alpinus×perfoliatus

Recorded from Colonsay, South Ebudes (102), and from Benbecula, Outer Hebrides (110). Reported also from Germany and Norway.

577/9×8 Potamogeton ×cognatus Aschers. & Graebn.
=P. perfoliatus×praelongus

Recorded from the North Idle Drain and Double Rivers in north Lincolnshire (54) and from Loch Borralie, west Sutherland (108). Reported also from Germany, Denmark, Norway and Russia.

577/13×16 Potamogeton ×grovesii Dandy & Taylor
=P. pusillus×trichoides

Known only from a dyke near Palling, east Norfolk (27). *P. trinervius* G. Fisch., from Germany, has wrongly been supposed to be this hybrid. (cf. Sell, 1967.)

577/15×1 Potamogeton ×variifolius Thore
=P. berchtoldii×natans

Recorded from the Glenamoy River, west Mayo (H. 27). Known also from France. (cf. Sell, 1967.)

577/15×3 Potamogeton ×lanceolatus Sm.
=P. berchtoldii×coloratus

Recorded from Burwell Fen, Cambridgeshire (29), the Afon Lligwy, Anglesey (52), the Caher River, Clare (H. 9), and the Clonbrock and Grange Rivers in north-east Galway (H. 17). Not known outside the British Isles.

577/18×11 Potamogeton ×pseudofriesii Dandy & Taylor
=P. acutifolius×friesii

Known only from a single locality near Strumpshaw, east Norfolk (27).

577/18×15 Potamogeton ×sudermanicus Hagstr.
=P. acutifolius×berchtoldii

Recorded from ditches near Arne, Dorset (9). Known also from Sweden.

577/19×5 Potamogeton ×cadburyae Dandy & Taylor
=P. crispus×lucens

Known only from a pool near Nuneaton, Warwickshire (38).

577/19×8 Potamogeton ×undulatus Wolfg.
=P. crispus×praelongus

Recorded from Llyn Hilyn, Radnorshire (43) and from the Six Mile Water and River Lagan in Antrim (H. 39). Known also from Russia (Lithuania), Germany and Denmark. (cf. Sell, 1967).

577/19×16 Potamogeton ×bennettii Fryer
=P. crispus×trichoides

Known only from the Forth and Clyde Canal in Lanarkshire (77) and Stirlingshire (86).

605/9 × 8 Juncus × diffusus Hoppe
=J. effusus × inflexus

BFT, BM, CGE, LCN: Floras
This hybrid is readily recognized and rarely occurs in the absence of its parents. It is found more or less throughout the range of overlap of *J. effusus* and *J. inflexus* (see *Atlas*, p. 319). Records for v.-cs 79, 87, H. 7 and 30 have not been traced. It is widespread elsewhere in Europe.

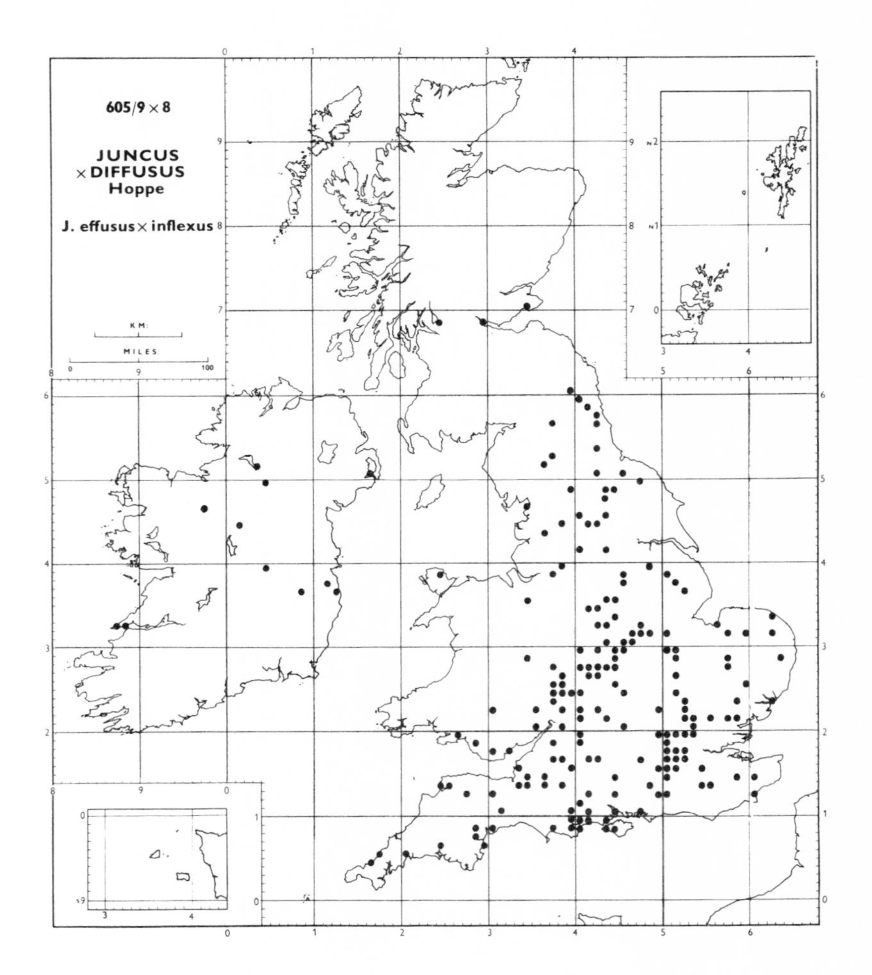

606/2 × 1 Luzula × borreri Bromf. ex Bab.
=L. forsteri × pilosa

CGE: Floras
This sterile hybrid occurs throughout the area in which the distributions of its two parents overlap (see *Atlas*, p. 323). There are very few records from the Continent of Europe. A recent account of this taxon was given by Ebinger (1962).

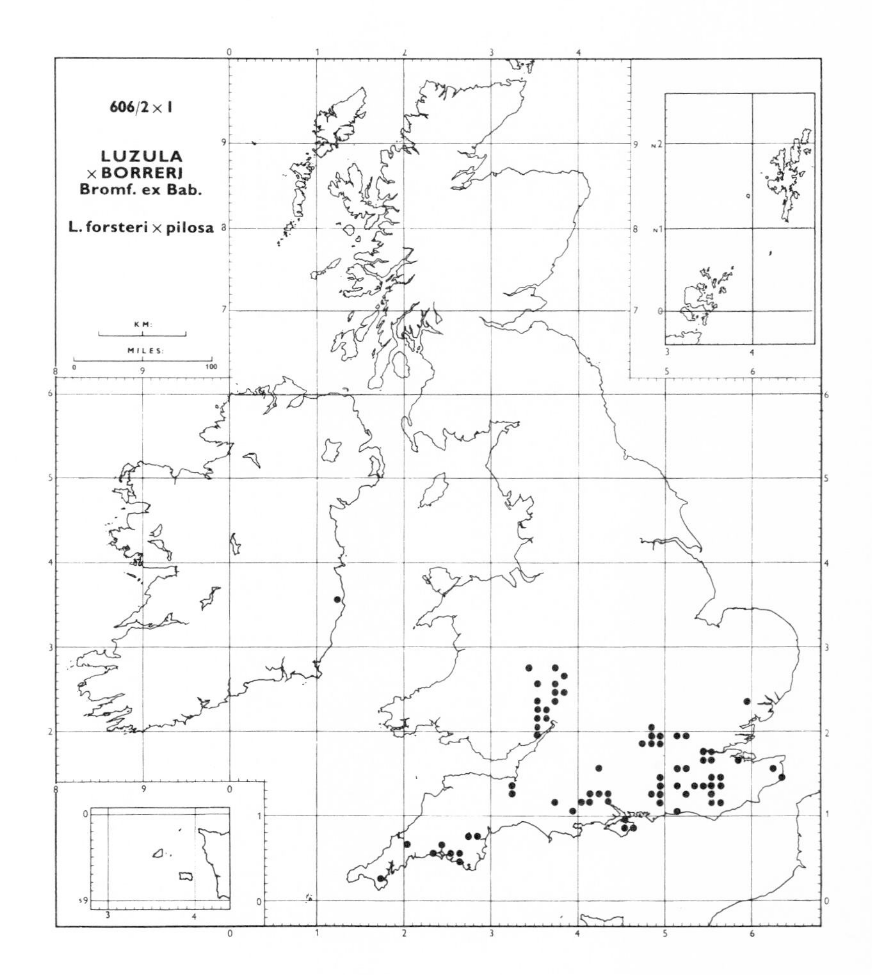

636/1b Gymnadenia conopsea (L.) R. Br.
subsp. **densiflora** (Wahlenb.) G. Camus, Bergon & A. Camus

BFT, BM, CGE, K

This map is based on material seen by the editor, or by P. F. Hunt and V. S. Summerhayes. Only specimens in which the lower leaves exceeded 15 mm. in width and in which the flowering spikes were over 10 cm. long were included.

Subsp. *densiflora* generally occurs in wetter habitats than subsp. **conopsea**, and is the characteristic plant of East Anglian fens. As far as is known, subsp. *conopsea* occurs throughout the range of the species as shown in the *Atlas*, p. 336. Both subsp. *conopsea* and subsp. *densiflora* occur throughout the lowlands of Europe.

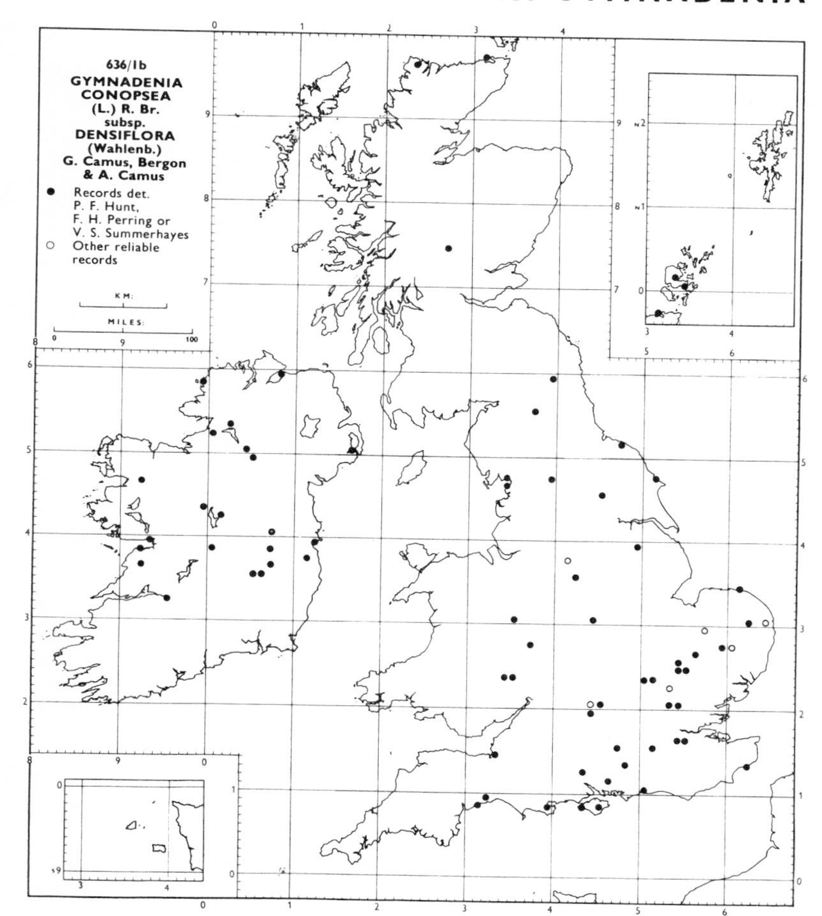

643/1 Dactylorhiza fuchsii (Druce) Soó
Dactylorchis fuchsii (Druce) Vermeul.
subsp. **okellyi** (Druce) Soó
subsp. **hebridensis** (Wilmott) Soó

These maps are based in the main upon herbarium material at Kew determined by P. F. Hunt and V. S. Summerhayes. Other records from reliable sources have been included. Both these subspecies are confined to the British Isles as far as is known. Subsp. **fuchsii** occurs throughout the range of the species (see map in *Atlas*, p. 340) except in the Hebrides (v.-cs 103, 104 and 110) where it is replaced by subsp. *hebridensis*, and in the north and west of Ireland where subsp. *okellyi* is sometimes more frequent than subsp. *fuchsii*, especially in the cracks and crevices of bare limestone pavement. A record of subsp. *okellyi* from Down (H. 38) has not been traced.

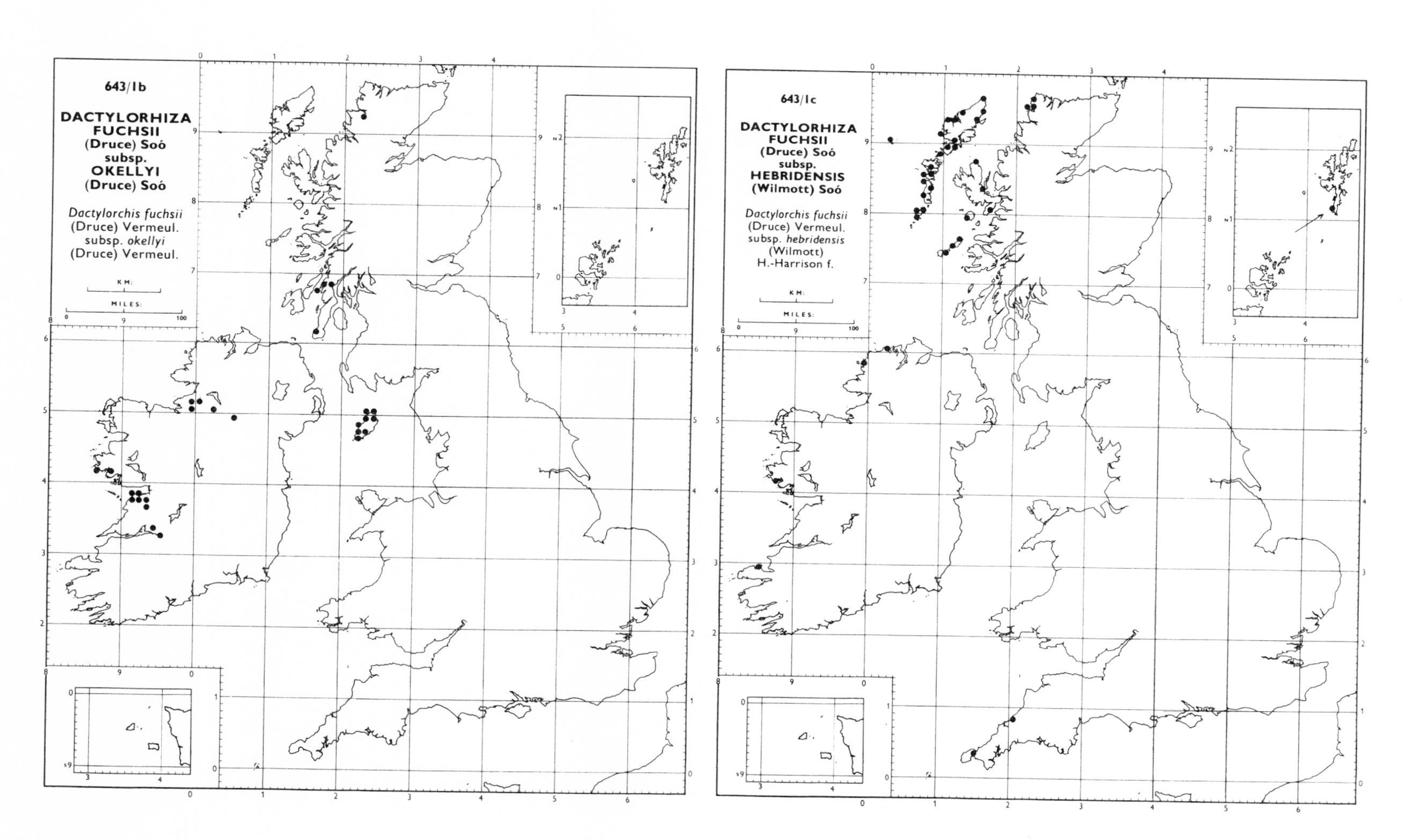

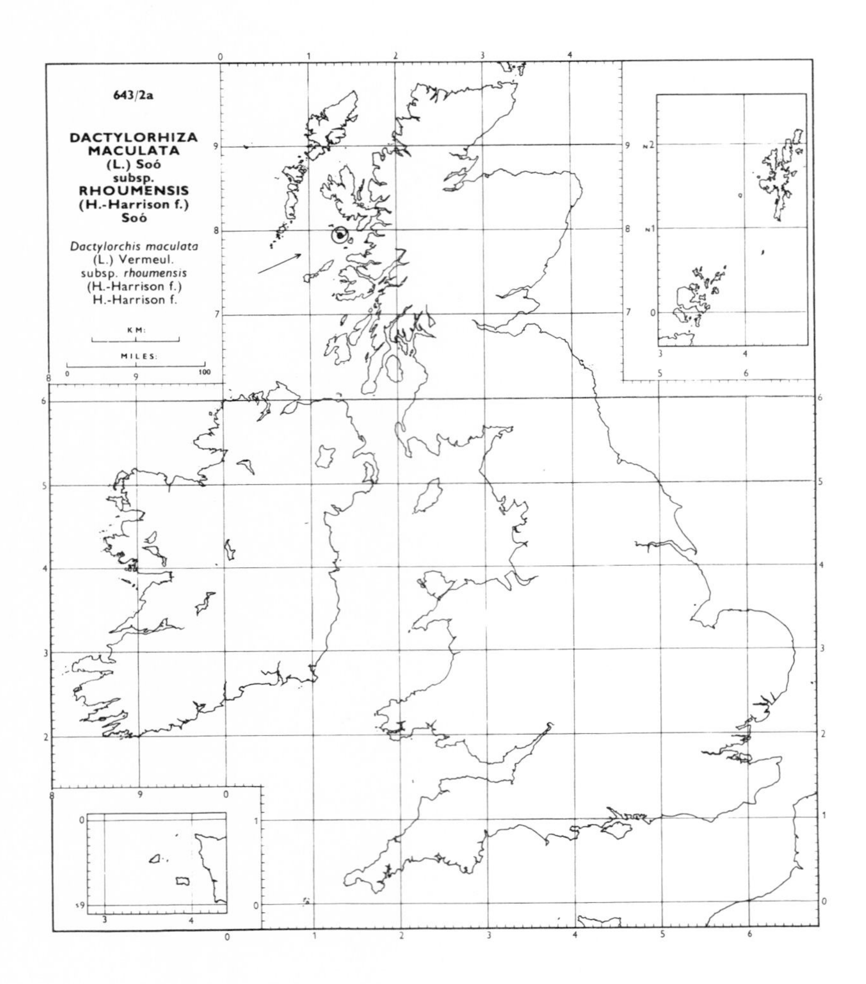

643/2a Dactylorhiza maculata (L.) Soó
subsp. **rhoumensis** (H.-Harrison f.) Soó
Dactylorchis maculata (L.) Vermeul.
subsp. *rhoumensis* (H.-Harrison f.) H.-Harrison f.

This interesting taxon, endemic to Rhum, has the leaves, labellum and spur resembling *D. maculata* (2n=80), but with the chromosome number of *D. fuchsii* (2n=40).

643/3 Dactylorhiza incarnata (L.) Soó
Dactylorchis incarnata (L.) Vermeul.
subsp. **incarnata**
subsp. **pulchella** (Druce) Soó
subsp. **cruenta** (O. F. Muell.) P. D. Sell
subsp. **coccinea** (Pugsl.) Soó
subsp. **gemmana** (Pugsl.) P. D. Sell
subsp. **ochroleuca** (Boll) P. F. Hunt & Summerh.

BM, CGE, K

These six maps are based on material at Kew determined by P. F. Hunt and V. S. Summerhayes, and on records published or determined by J. Heslop Harrison. They are possibly complete for the rarer taxa, but undoubtedly the commoner subsp. *incarnata*, subsp. *pulchella* and subsp. *coccinea* are very much under-recorded (cf. distribution of the species, *Atlas*, p. 341). However, the maps do indicate the differences in geographical distribution and ecological preference of the three taxa.

Subsp. *incarnata*, with flesh-pink flowers, is the most frequent variant in fens and other wet base-rich habitats in south England and East Anglia. In the north and west it is widespread, but less plentiful. It is widespread in continental Europe. Subsp. *pulchella* consists of a variable series of populations characterized by having magenta or purplish flowers. It replaces subsp. *incarnata* in acid *Sphagnum* bogs in the north and west. However, populations referable to subsp. *pulchella*, or intermediate between subsp. *pulchella* and subsp. *incarnata*, are scattered throughout the British Isles, often in association with subsp. *incarnata*. It is probable that similar populations occur in *Sphagnum* bogs on the Continent. Subsp. *cruenta* differs from all other subspecies of *D. incarnata* in having the leaves spotted on both sides, because of which it is sometimes regarded as a distinct species. In the British Isles the typical form is restricted to the neighbourhood of lakes in the western part of the Irish limestone plain (H. 9, 17, 26). Some individuals in populations of subsp. *pulchella* have fine spotting on the leaves and seem to link that subspecies with subsp. *cruenta*. In continental Europe its main range is in Fennoscandia between 55° and 65°N. extending into Siberia to east of Lake Baikal; there are also six or seven isolated localities in the Alps (Heslop Harrison, 1950). Subsp. *coccinea*, distinguished by its small stature and ruby-red or crimson-red flowers, is the common form of calcareous dune-slacks. It is apparently endemic. Subsp. *gemmana* is in the British Isles recorded from a number of widely scattered marsh localities. It is larger in all its parts than any other subspecies of *D. incarnata*. It is also widespread in the marshes of continental Europe. Subsp. *ochroleuca*, distinguished by its creamy-white flowers and lack of anthocyanin in the whole of the plant, is known in Britain only from calcareous fens in East Anglia (27, 29) and south Wales (44). Homogeneous colonies of it are rare. It is widespread in calcareous fens in continental Europe.

The taxonomic treatment followed is that of Heslop Harrison (1954). Although most populations of *D. incarnata* in the British Isles can conveniently be placed into six groups, it must be realized that variable populations exist within which any of the characters of the ecological races may appear together, intergrading or in atypical combinations. (Nomenclature, cf. Sell, 1967.)

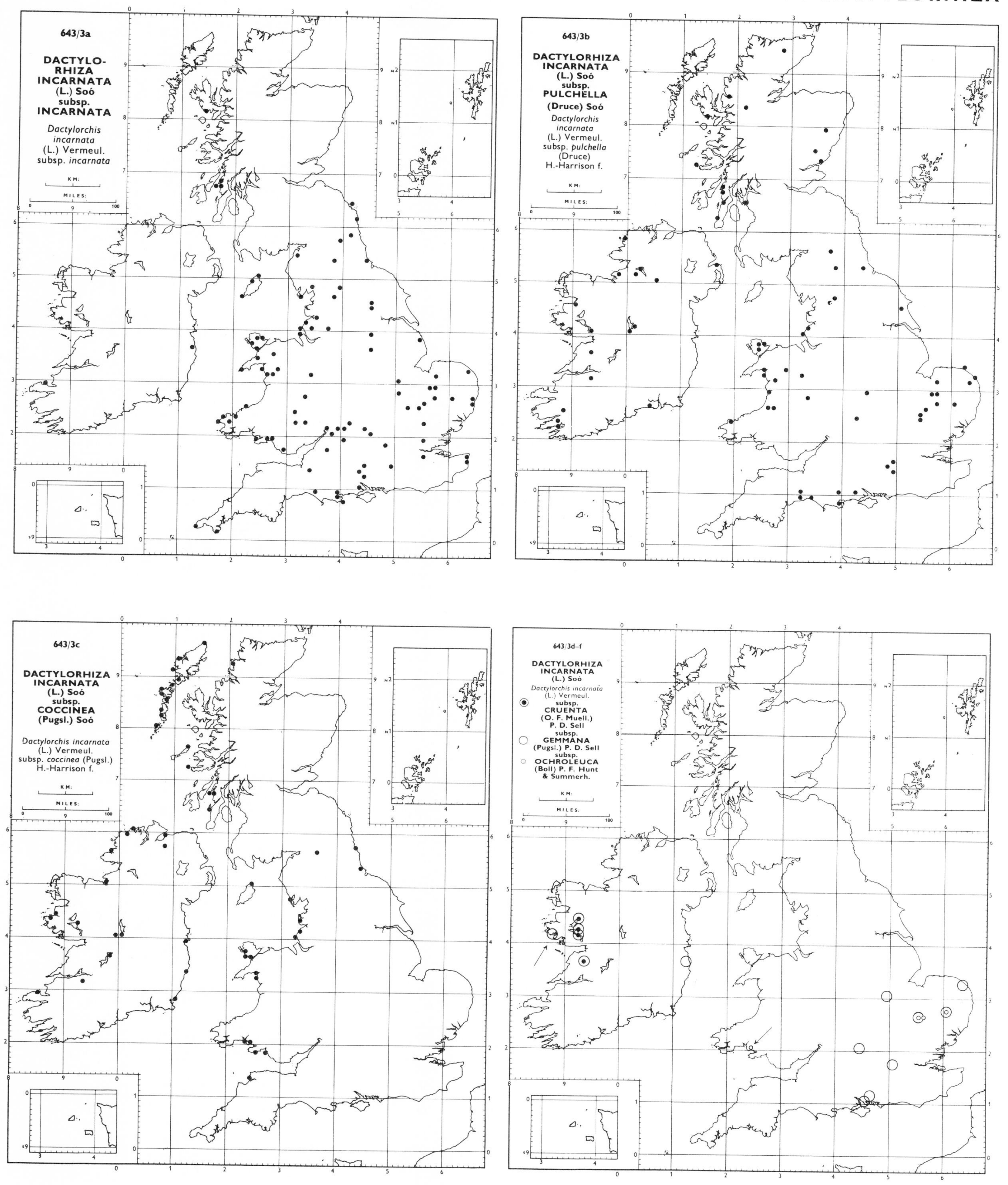
643/3a
DACTYLORHIZA INCARNATA (L.) Soó subsp. INCARNATA
Dactylorchis incarnata (L.) Vermeul. subsp. incarnata
KM:
MILES:
643/3b
DACTYLORHIZA INCARNATA (L.) Soó subsp. PULCHELLA (Druce) Soó
Dactylorchis incarnata (L.) Vermeul. subsp. pulchella (Druce) H.-Harrison f.
643/3c
DACTYLORHIZA INCARNATA (L.) Soó subsp. COCCINEA (Pugsl.) Soó
Dactylorchis incarnata (L.) Vermeul. subsp. coccinea (Pugsl.) H.-Harrison f.
643/3d–f
DACTYLORHIZA INCARNATA (L.) Soó
Dactylorchis incarnata (L.) Vermeul.
subsp. CRUENTA (O. F. Muell.) P. D. Sell
subsp. GEMMANA (Pugsl.) P. D. Sell
subsp. OCHROLEUCA (Boll) P. F. Hunt & Summerh.

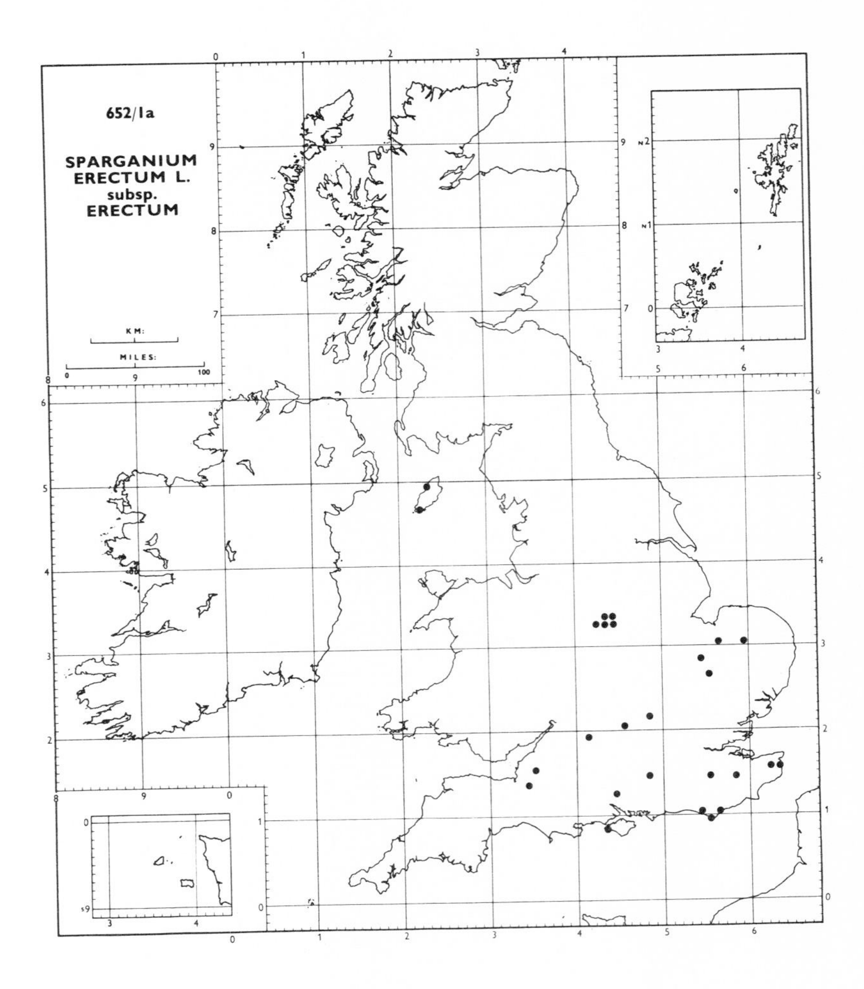

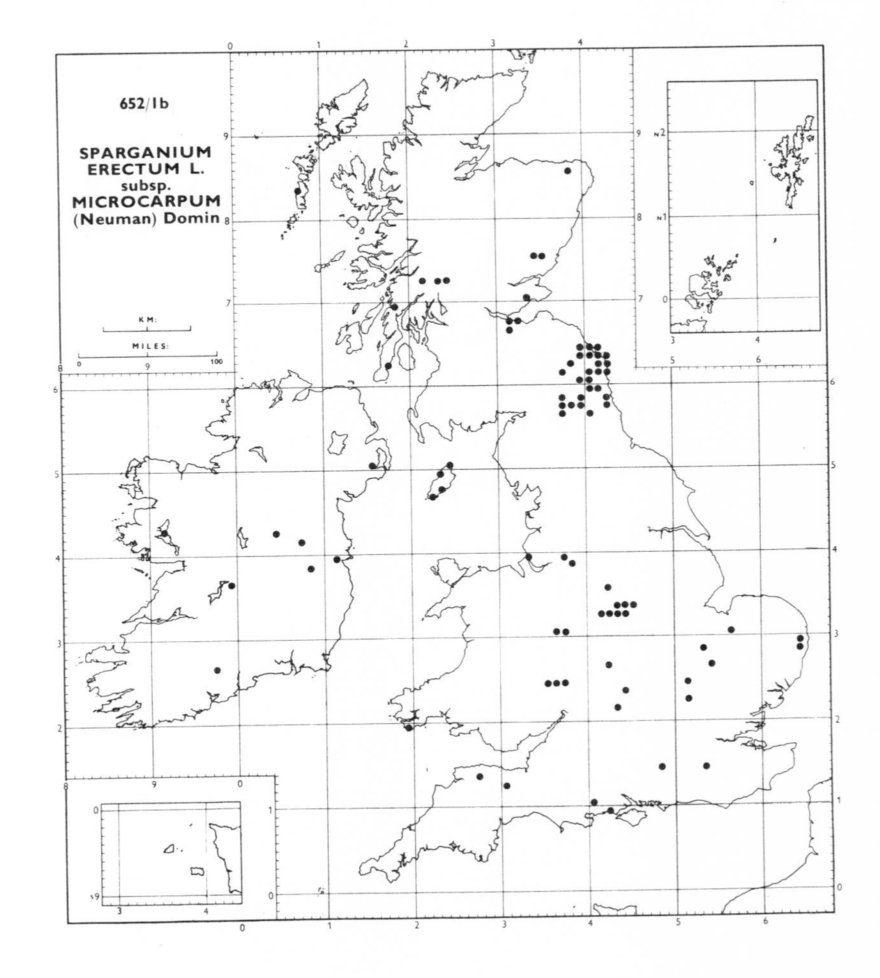

652/1 Sparganium erectum L.
subsp. **erectum**
subsp. **microcarpum** (Neuman) Domin
subsp. **neglectum** (Beeby) Schinz & Thell.
subsp. **oocarpum** (Čelak.) Domin

BM, CGE, DBN, K
These maps are based on material identified by the editor and C. D. K. Cook. Identifications have to be made on ripe fruit, which does not mature until late in the season. As much herbarium material is collected early it is immature and the subspecies cannot be determined. For the same reason field records have been scarce. There is conspicuous observer bias owing to the activities of Cook in Leicestershire (55) and G. A. Swan in Northumberland (67, 68). However, the maps, though very incomplete, do reflect the different distribution of at least the two more widespread subspecies, *microcarpum* and *neglectum*.

Subsp. *erectum* occurs from south Sweden and Finland to the south Mediterranean and eastwards to central Siberia. Subsp. *microcarpum* extends from the Arctic circle southwards to north Africa and eastwards to Siberia. Subsp. *neglectum* is found between south Sweden and north Africa and eastwards to the Caucasus. Subsp. *oocarpum* has been under-recorded in Europe, but is known from Turkey and north Africa. For a full account of these subspecies see Cook (1961).

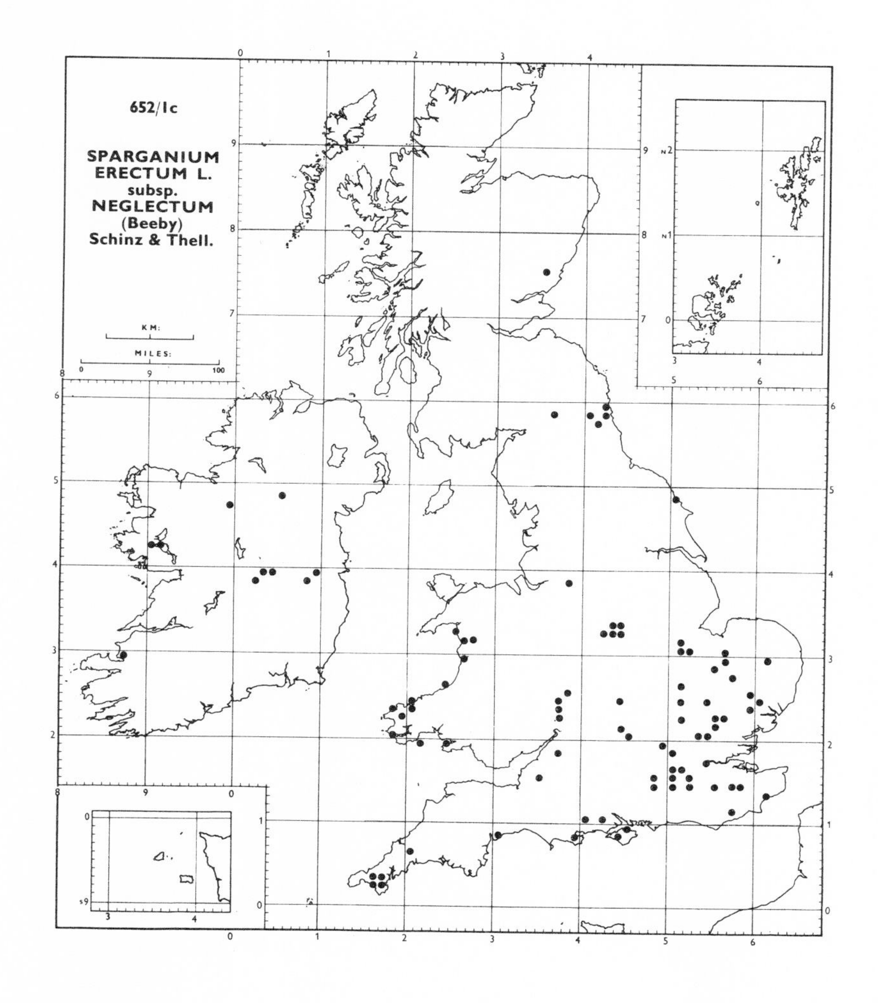

655/8×6 Scirpus ×carinatus Sm.
=S. lacustris × triquetrus

655/9×6 Scirpus ×kuekenthalianus Junge
=S. tabernaemontani × triquetrus

BM, CGE, K: Floras

It is not surprising that these hybrids of *S. triquetrus* with the closely related *S. lacustris* and *S. tabernaemontani* were for a long time all included under the same name, *S.×carinatus*. However, *S.× kuekenthalianus* may be distinguished by the asperous glumes and the more glaucous stems, which are more slender than in *S.× carinatus*. For further details see Lousley (1931).

The distribution of these hybrids in England is limited to the area from which the rarer parent, *S. triquetrus*, has been recorded (see, *Atlas*, p. 348). On the banks of the River Thames (17, 21) *S.* × *carinatus* is more frequent than *S. triquetrus*, and *S. lacustris* is absent from the immediate vicinity. At Aylesford, Kent (15), and in west Sussex (13) *S.×kuekenthalianus* persists, but *S. triquetrus* can no longer be found. The apparent absence of *S.×kuekenthalianus* from Ireland is strange as both parents are reported growing together on the banks of the River Shannon below Limerick (H. 8).

These hybrids are widely reported from central Europe, but the details are uncertain because of past confusion as to their separate identity.

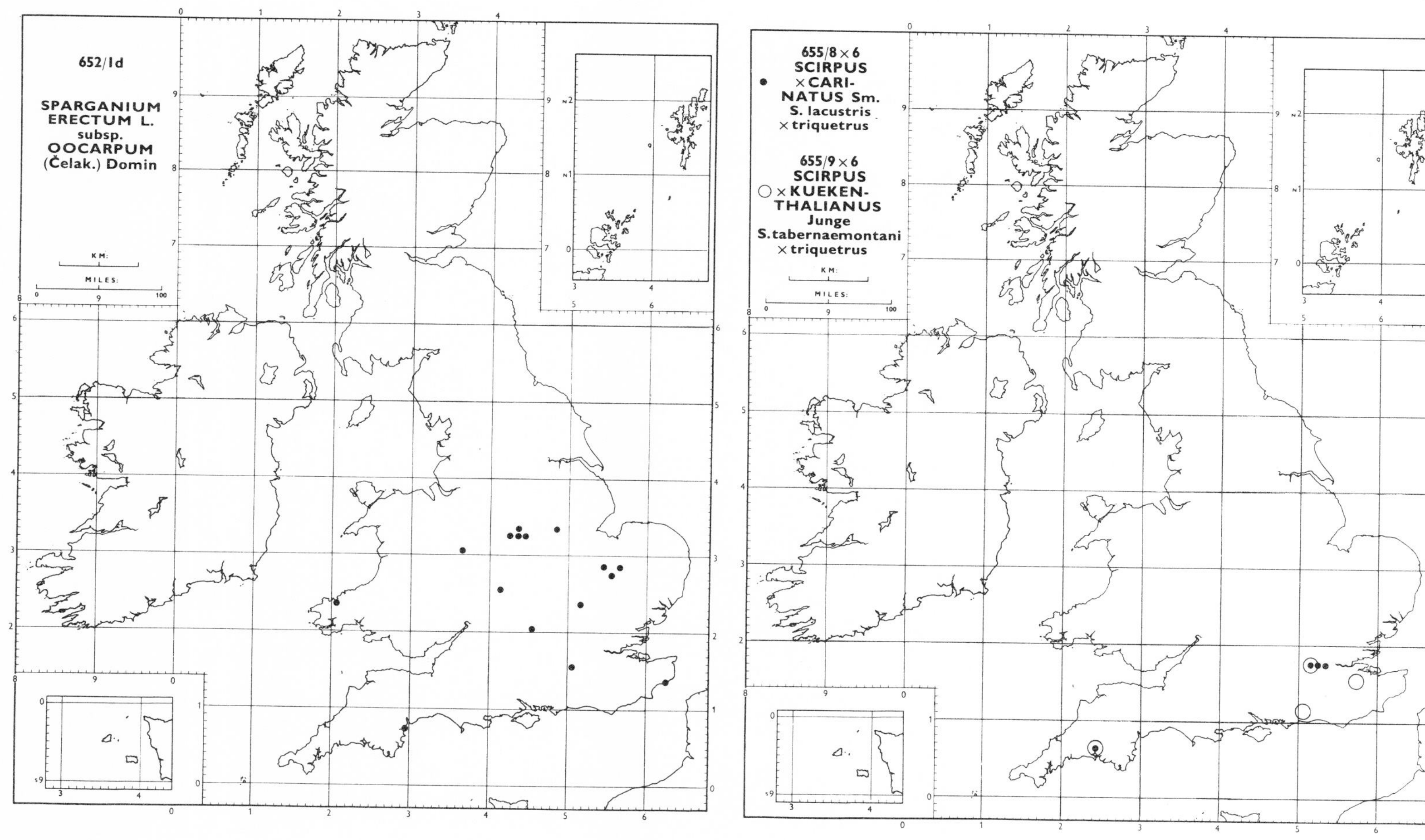

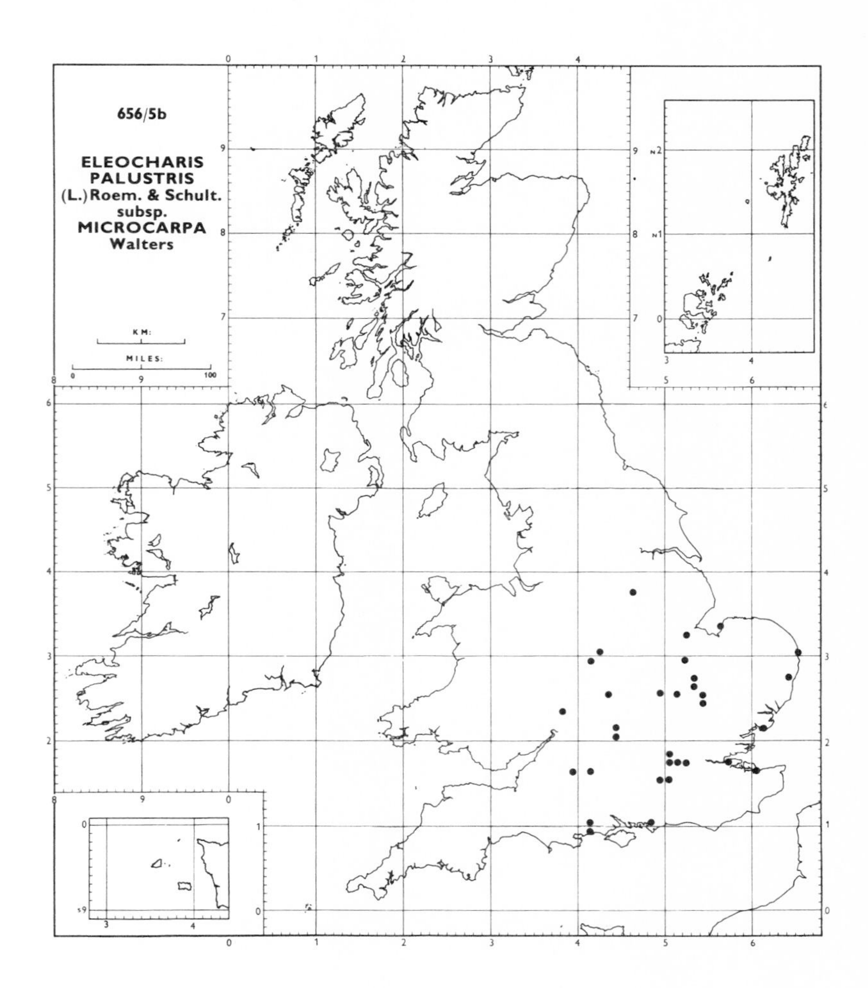

656/5b Eleocharis palustris (L.) Roem. & Schult. subsp. **microcarpa** Walters

BM, CGE, K

This map is based on material determined by S. M. Walters. It may not be complete, but the subspecies is apparently confined to south and east England and the west midlands. It is widespread in Europe, but apparently absent from north Scandinavia, and extends further south and east than subsp. **vulgaris** Walters. As far as is known, the distribution of subsp. *vulgaris* is shown by the map in the *Atlas*, p. 351. For a full account of *E. palustris*, see Walters (1949c).

(To avoid ambiguity, the names of the two subspecies here used are as in Walters (1949c). The question of the use of the nominate subspecies in this case is complex and disputed.)

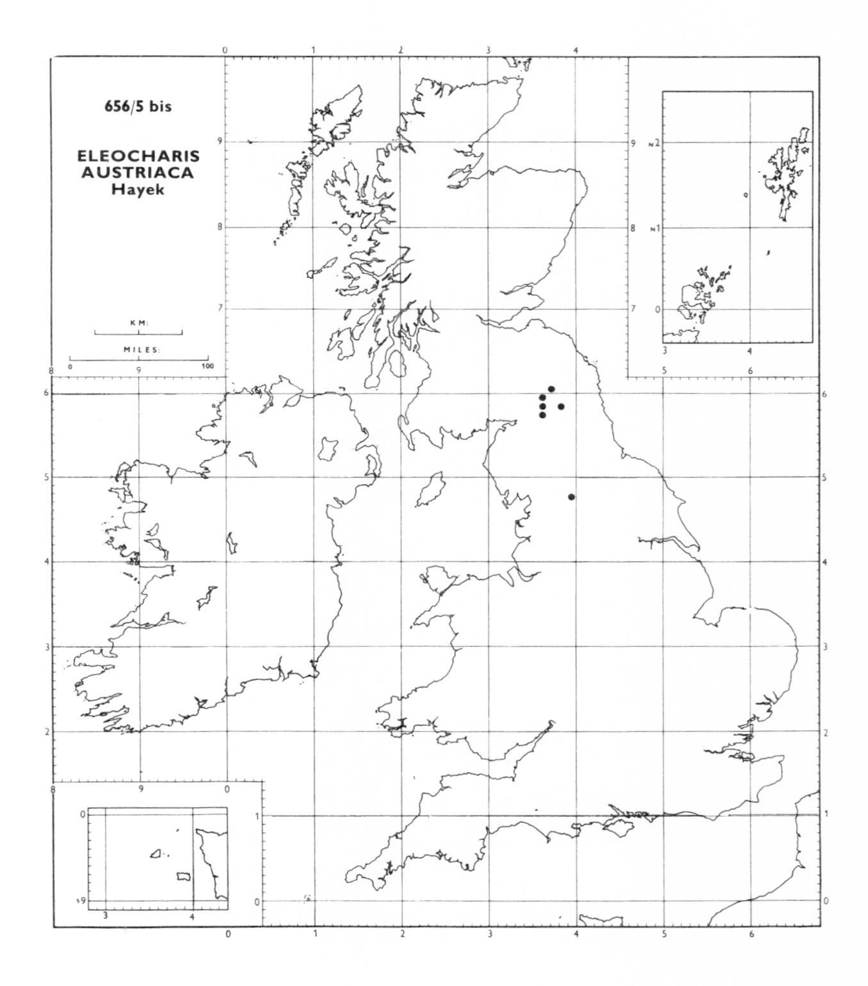

656/5 bis Eleocharis austriaca Hayek

CGE, K

This easily recognized species of the *E. palustris* group was first discovered in Yorkshire (64) by N. Y. Sandwith in 1947, but it was not identified until 1960. Six British localities are now known in Yorkshire, Northumberland and Cumberland (64, 67 and 70), all of which are by upland rivers. It is a widespread species of the mountains of Europe. For a full account of the species see Walters (1963).

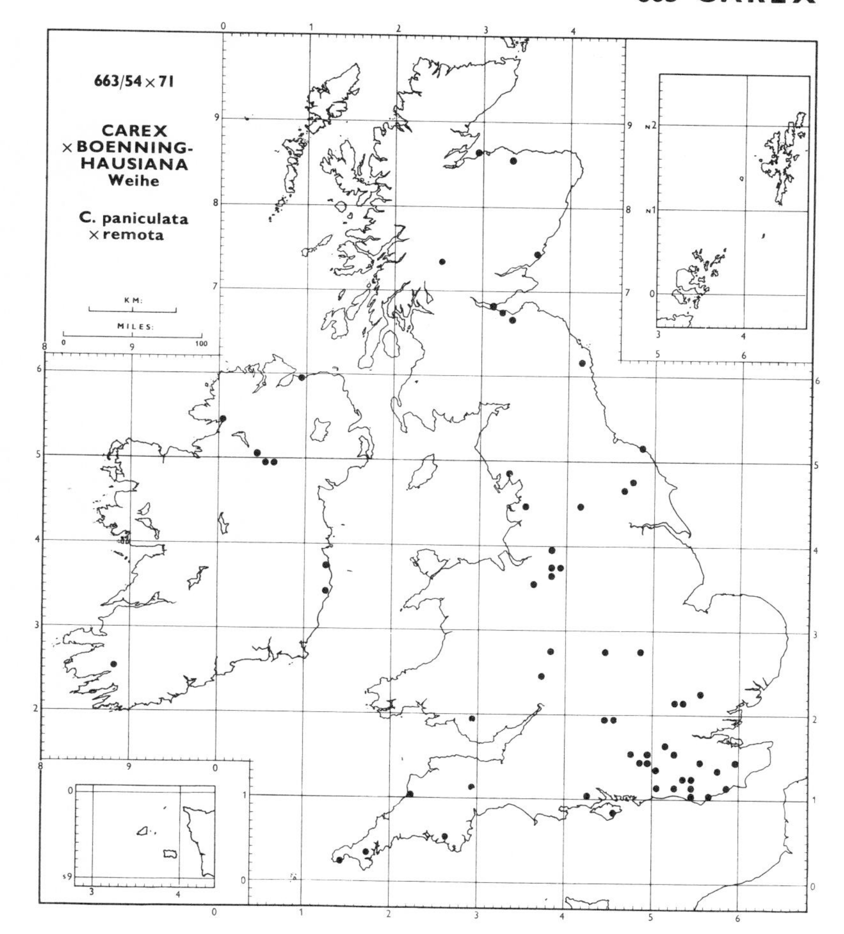

663/54×71 **Carex** × **boenninghausiana** Weihe
=**C. paniculata** × **remota**

BM, CGE, K, NMW, Lousley, Wallace: Floras
This widespread hybrid occurs throughout the area of overlap of the two parents (see *Atlas*, pp. 365, 369), and occasionally apparently in the absence of *C. paniculata*. It is known from much of continental Europe.

Records from v.-cs 23, 25, 40, 67, 75 and 97 have not been traced.

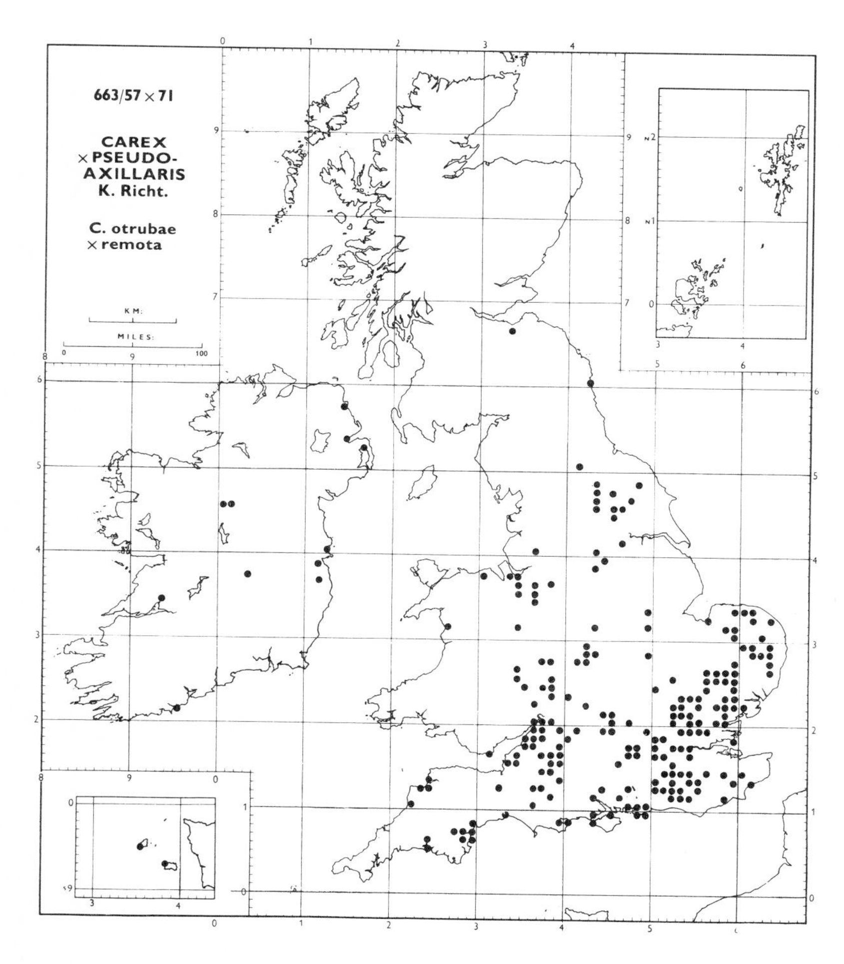

663/57×71 **Carex** ×**pseudoaxillaris** K. Richt.
=**C. otrubae** × **remota**

BM, CGE, K: Floras
This hybrid occurs in scattered localities throughout the range of overlap of the two parents (see *Atlas*, pp. 366, 369) and is particularly frequent on heavy clays where the habitats of the parents are contiguous. It is widely recorded from the Continent of Europe.

Records from v.-cs 1 and 60 have not been traced: that from v.-c. 69 was probably an error.

GRAMINEAE

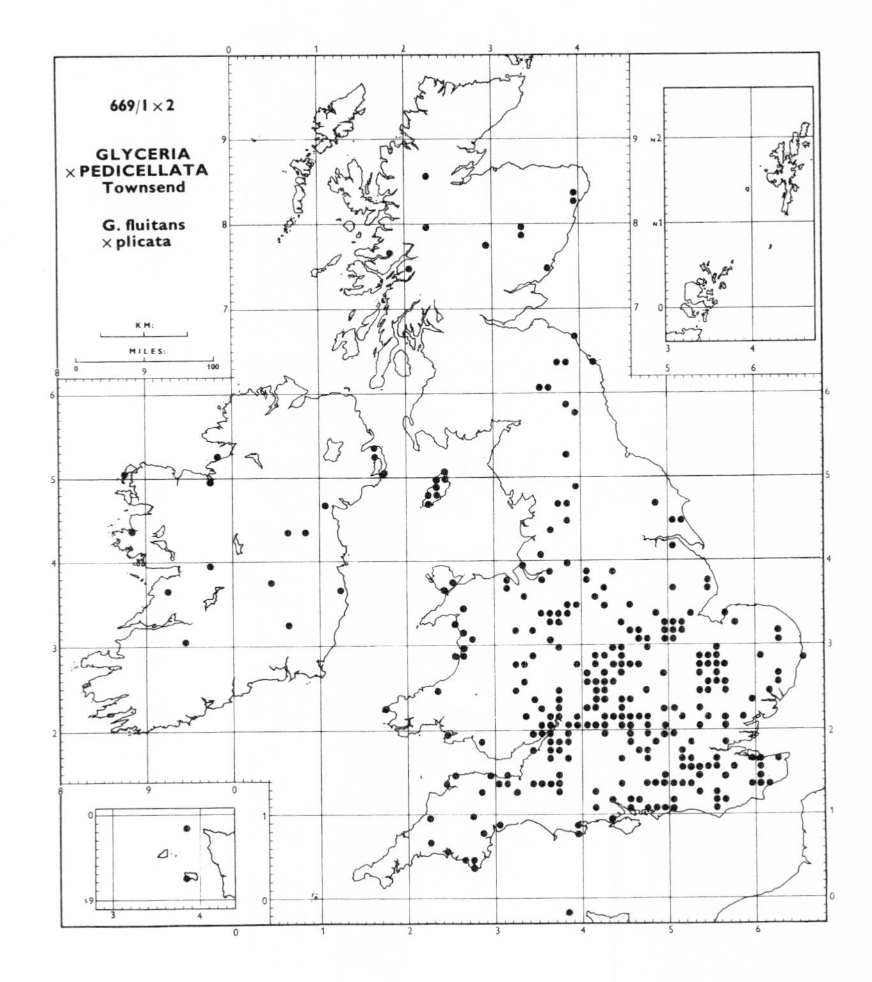

669/1 × 2 **Glyceria ×pedicellata** Townsend **=G. fluitans × plicata**

BFT, CGE, K, NMW

This hybrid occurs throughout the area of overlap of the two parents (see *Atlas*, p. 372) and occasionally in the absence of either. It is sterile but spreads strongly, vegetatively. The map is based on material seen by C. E. Hubbard and S. M. Walters, and on field records from reliable recorders. It is apparently widespread on the Continent of Europe.

For an account of this taxon, see Borrill (1956).

670/9 **Festuca tenuifolia** Sibth.

BFT, BM, CGE, DBN, NMW, TCD

The characters which distinguish this species have not always been clearly understood in the past, so that this map is based only on material seen by the editor, C. E. Hubbard, A. Melderis and a few other recorders known to be reliable. Although widespread on suitable soils in Great Britain, its apparent rarity in the west, and particularly in Ireland, seems to be a reality: no material has been found in the Irish herbaria examined.

This grass is scattered throughout Europe and has been introduced into north-east North America.

670/1 × 671/1 **×Festulolium loliaceum** (Huds.) P. Fourn. **=Festuca pratensis × Lolium perenne**

CGE, DBN, K: Floras

This widespread hybrid occurs almost throughout the area of overlap of the two parents (see *Atlas*, pp. 373, 375), being particularly frequent in permanent pastures on the heavy clays. It is known from much of continental Europe.

Records from v.-cs 35 and 54 have not been traced, whilst that from H. 34 was an error.

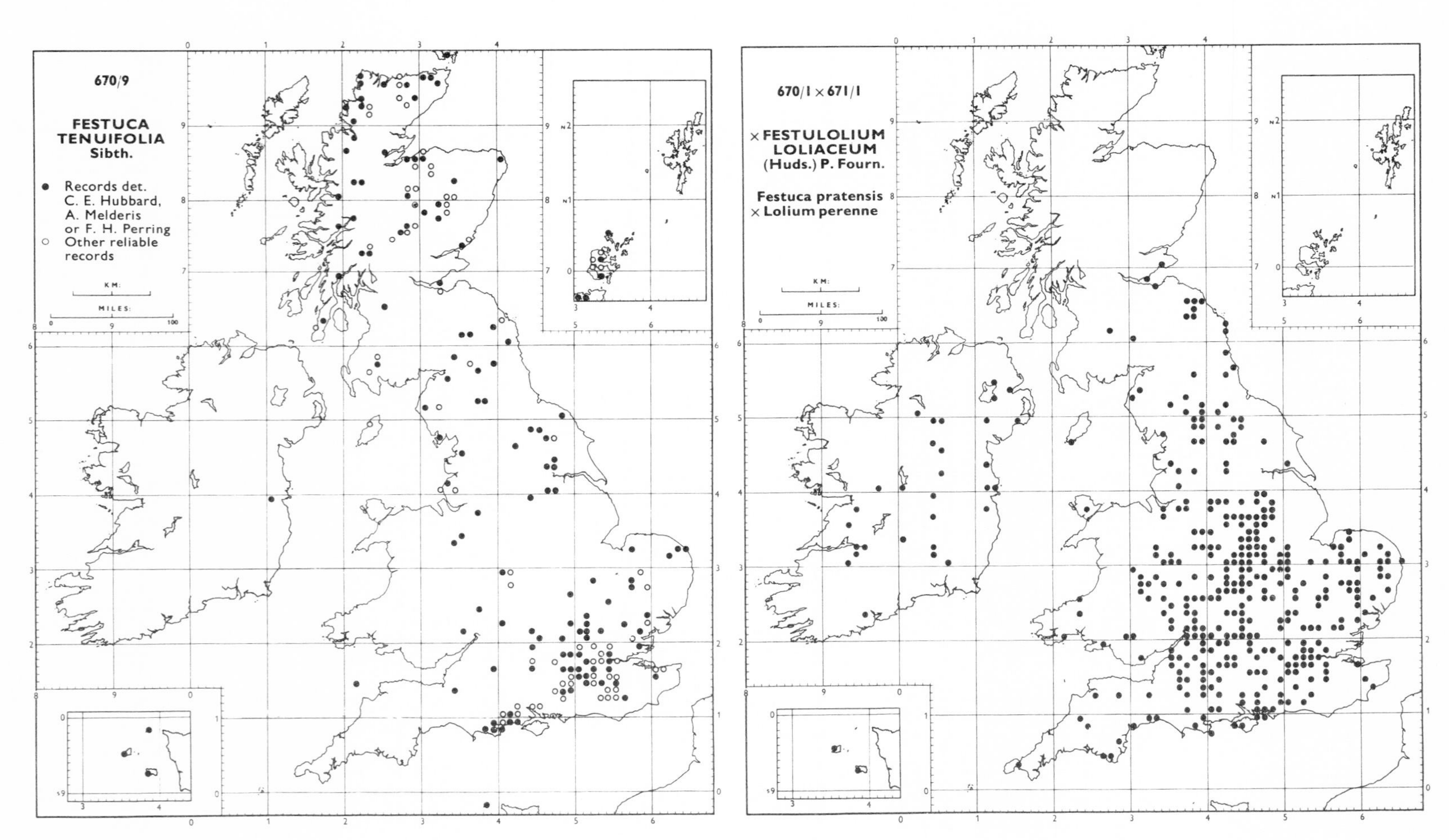

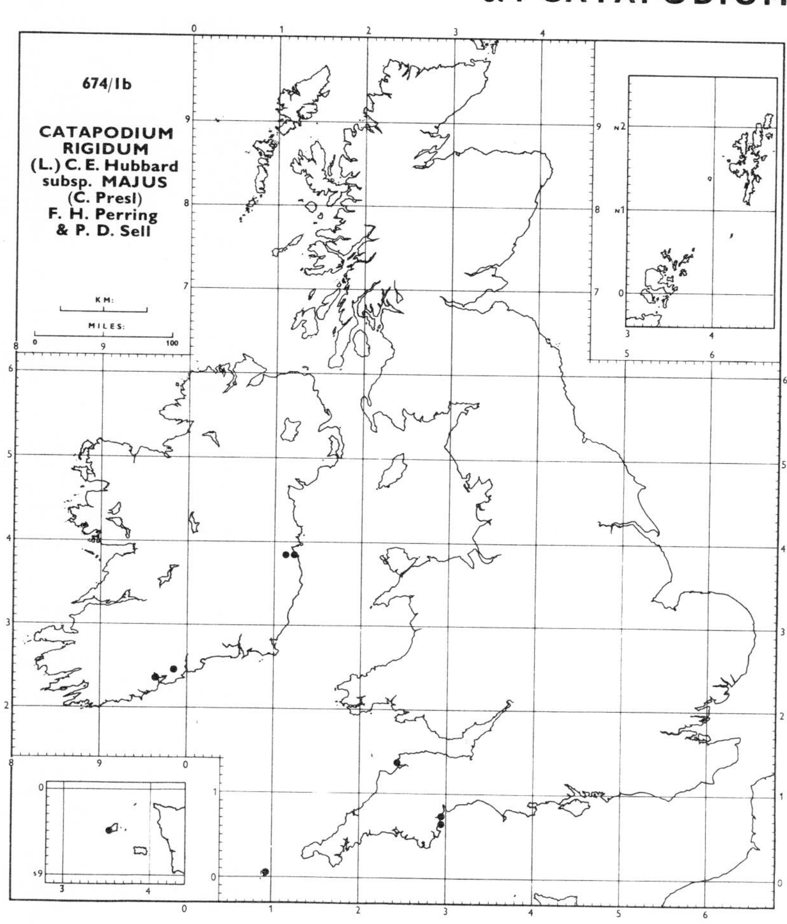

674/1b Catapodium rigidum (L.) C. E. Hubbard
subsp. **majus** (C. Presl) F. H. Perring & P. D. Sell

BM, CGE, DBN

Subsp. *majus* can be distinguished from subsp. *rigidum* by its taller habit, wider leaves and open pyramidal inflorescence. It occurs in the south and west of the British Isles, often near the sea. There is some doubt about its status in this country, but it is certainly a native of south-west Europe and the Mediterranean region. (cf. Sell, 1967.)

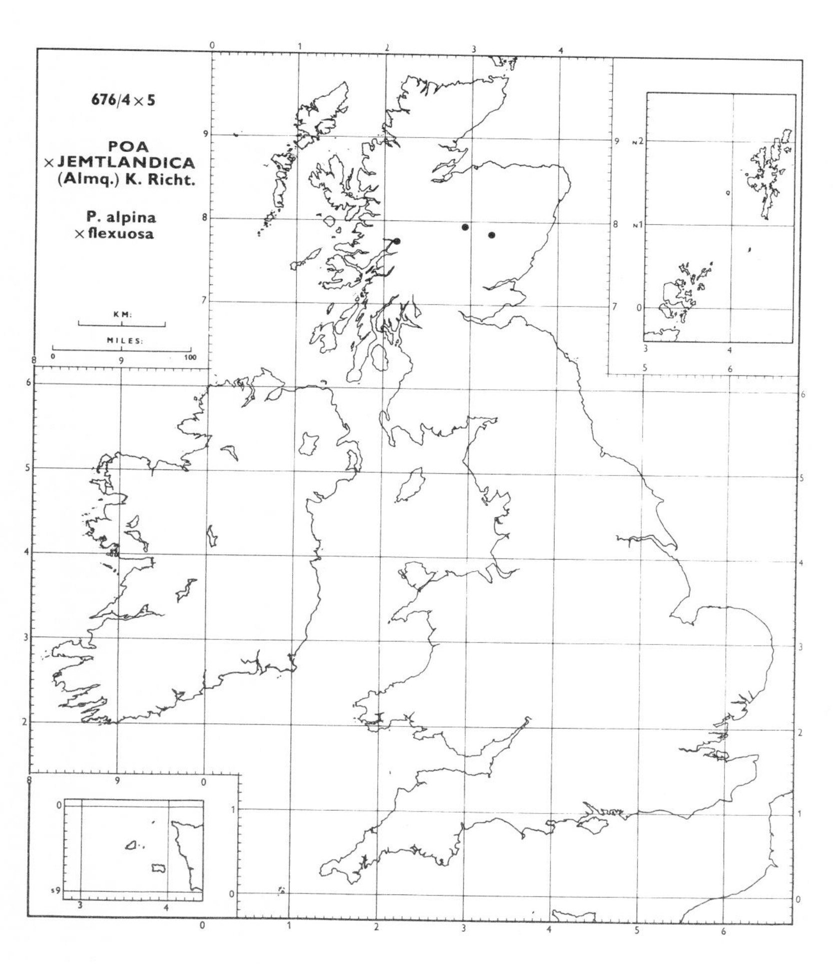

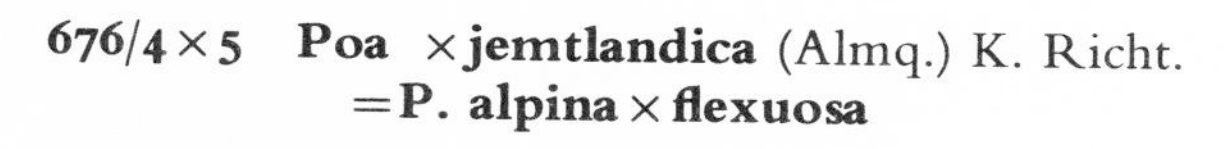
676/4×5 Poa ×jemtlandica (Almq.) K. Richt.
=P. alpina×flexuosa

BM, CGE, E, NMW, OXF, Lousley, Wallace

This viviparous hybrid is very rare on damp stony slopes and rock ledges on Ben Nevis (97), Lochnagar (92) and the Cairntoul (92). It also occurs in Scandinavia and Iceland.

676/10 **Poa pratensis** L.

It has not been possible to include a map of this species, and the map of the aggregate in the *Atlas*, p. 381, is unsatisfactory as there is evidence to show that *P. pratensis* is replaced by *P. angustifolia* in south-east England, and by *P. subcaerulea* in north Scotland.

676/11 **Poa angustifolia** L.

BFT, BM, CGE, K, NMW: Floras

The map is based upon material seen by the editor, C. E. Hubbard and D. M. Barling, who has made a special study of this species (see Barling, 1959). Field and literature records from reliable sources have also been included, and the map is reasonably complete.

It is a species of dry grassland and wall-tops which flowers about a month earlier than *P. pratensis*. In Europe it is widely recorded and extends into south-west Asia.

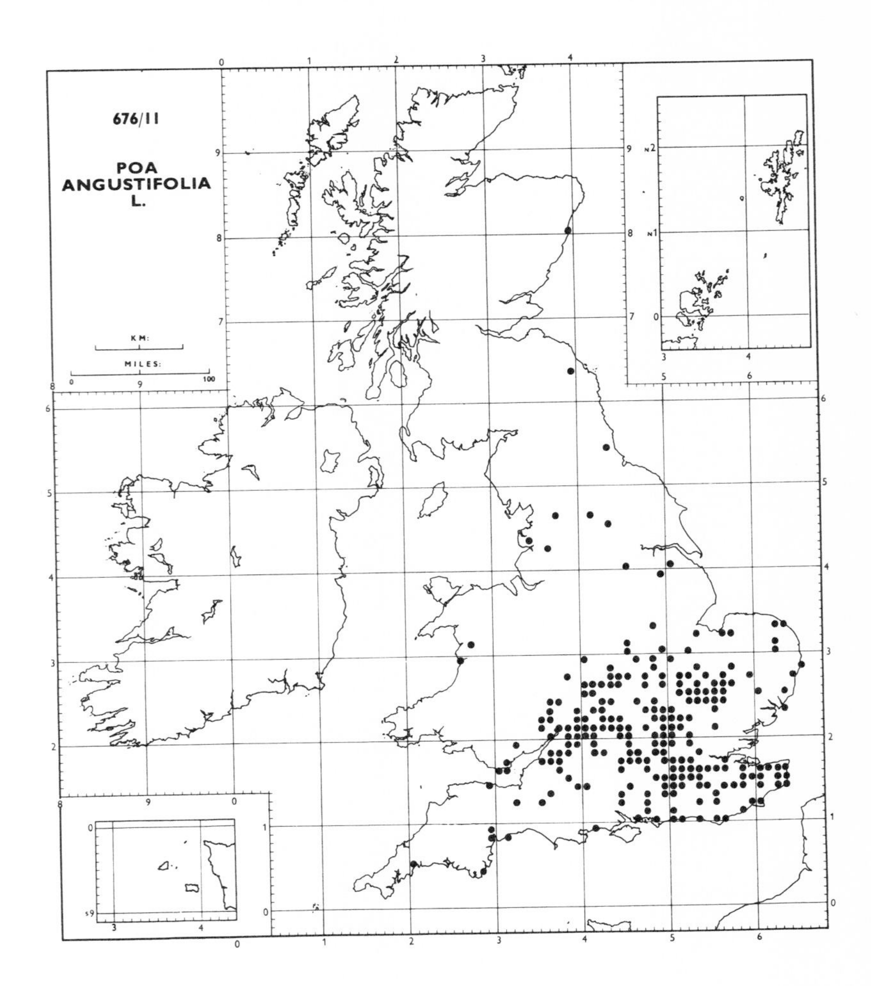

676/12 **Poa subcaerulea** Sm.

BM, CGE, DBN, E, K, NMW, OXF, SHY

The map is based on material seen by the editor, C. E. Hubbard, A. Melderis and D. M. Barling, who has made a study of this species (see Barling, 1962). It is a variable species, but it can be distinguished by the low number of its panicle branches, never exceeding four at the lowest node, and by having three veins on both upper and lower glumes. It is essentially a plant of moist habitats, being common in the hill-country of Scotland and Wales, where it is undoubtedly under-recorded. A very distinct form occurs in sand-dunes all round the British coast.

Outside the British Isles this species has been recorded from Fennoscandia, Denmark, northern European U.S.S.R. and Iceland. It is probably at its southern limit in the British Isles.

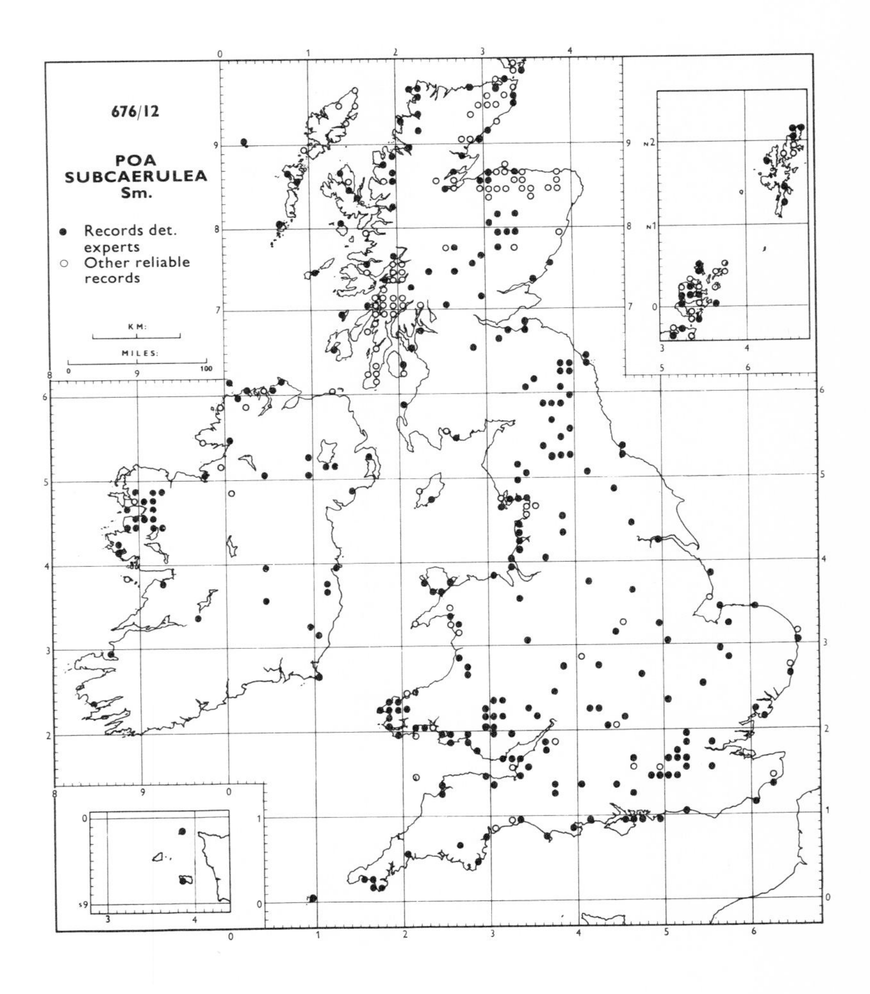

677/1b Catabrosa aquatica (L.) Beauv.
subsp. **minor** (Bab.) F. H. Perring & P. D. Sell
C. aquatica var. *littoralis* Parn.

BM, CGE, DBN, E, K

The map is based on herbarium material seen by the editor. Subsp. *minor* differs from subsp. *aquatica* in having shorter culms, leaves and panicles, and in the spikelets being only 1-flowered. It occurs on wet, open sand by the sea at various places on the north and west coasts of Britain. It is almost certainly underrecorded, particularly in the Western Isles (102, 103, 110), where the species is abundant (see *Atlas*, p. 382). No specimens have been seen from outside Britain. (cf. Sell, 1967.)

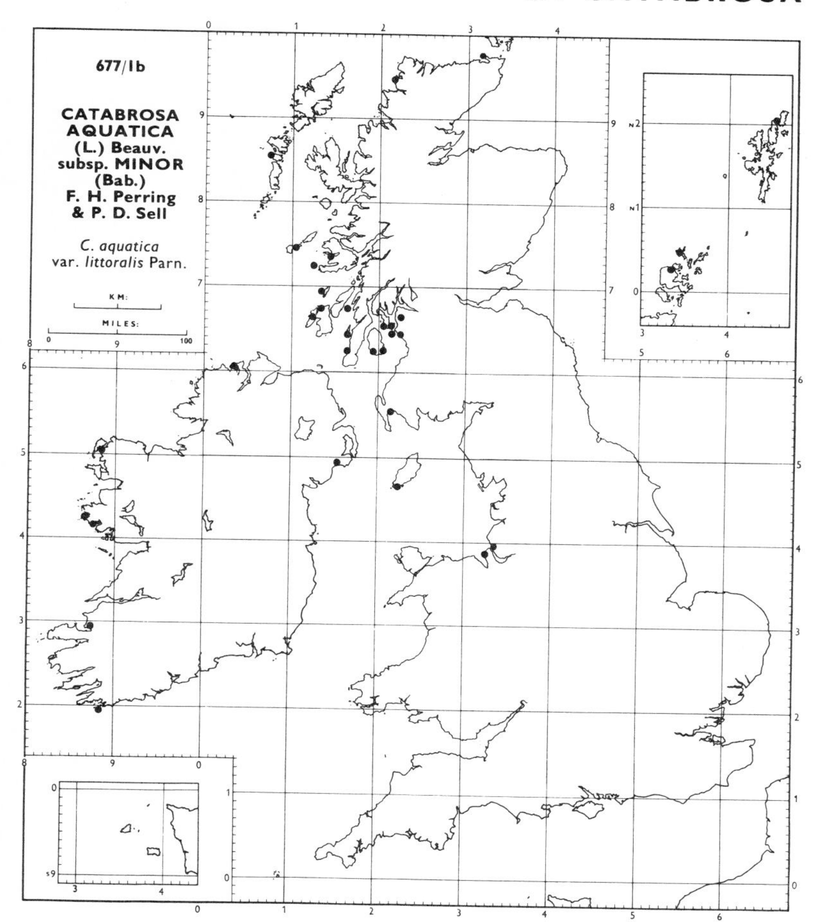

683/11 Bromus ferronii Mabille

BM, CGE, K, NMW: Floras

This coastal segregate of *Bromus mollis* L. is specially frequent in the south and west. It is recorded from western France and Spain and may occur elsewhere. It should be looked for in southern Ireland. The record from east Sussex (14) is probably an error. The map is based upon herbarium and literature records, to which field records from reliable sources have been added.

683/12 Bromus thominii Hardouin

BFT, BM, CGE, DBN, K, NMW, TCD: Floras

This segregate of *Bromus mollis* L. is widespread in the lowlands of the British Isles, and occurs elsewhere in western Europe from France to Scandinavia. The species is probably commoner than the map suggests, particularly in Ireland. All records from reliable sources have been accepted.

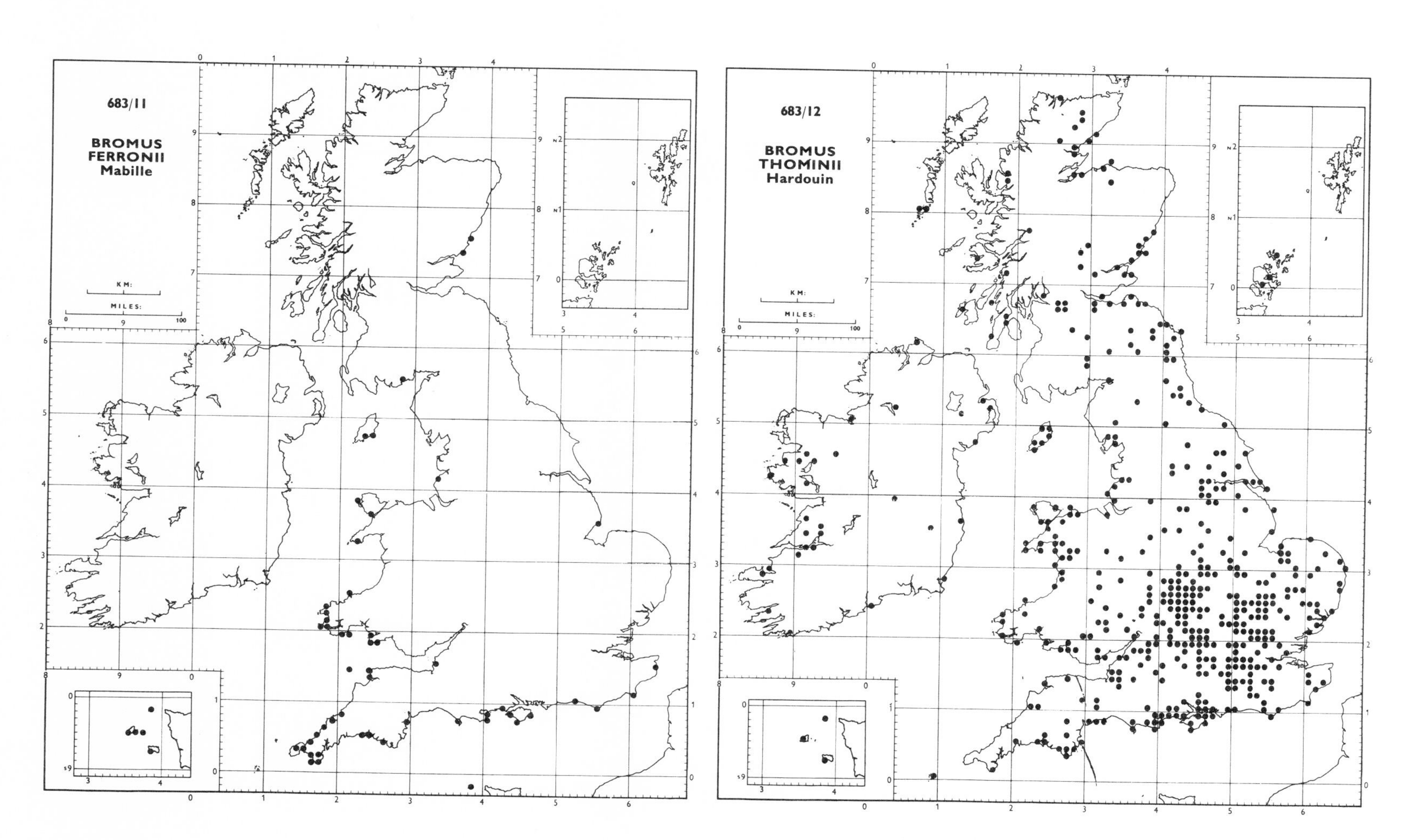

699/1 × 700/1 × Ammocalamagrostis baltica (Schrad.) P. Fourn.
= Ammophila arenaria × Calamagrostis epigejos

BM, CGE

This rare hybrid occurs naturally in a few places where *Calamagrostis epigejos* is found on or near the coast. It is planted as a dune stabilizer in several places on the East Anglian coast because of its aggressive vegetative habit of growth. It occurs on the sandy coasts of western Europe from north-east France to Scandinavia.

701/5 × 703/1 × Agropogon littoralis (Sm.) C. E. Hubbard
= Agrostis stolonifera × Polypogon monspeliensis

BM, CGE, K

This rare intergeneric hybrid is found on a few muddy salt-marshes in southern England from Dorset to Norfolk, where *Polypogon monspeliensis* is regarded as native (see *Atlas*, p. 399). It has also been recorded occasionally as a casual. Elsewhere it is known only from France.

716/2 × 1 Spartina × townsendii H. & J. Groves
= S. alterniflora × maritima

K

This hybrid is the sterile F_1 between *S. alterniflora* and *S. maritima*. It has $2n=62$. The map is based on material determined by C. E. Hubbard. The map in the *Atlas*, p. 405, probably shows the distribution of the amphidiploid plants in which $2n=120$, 122 or 124. This also was previously known as *S.* × *townsendii*, but is at present without a name. For details, see Marchant, 1963.

716/2 bis Spartina glabra Muhl.
S. alterniflora var. *glabra* (Muhl.) Fernald

K

This plant has been recently recognized as occurring in Britain (Marchant, 1963). It was probably first introduced into this country by F. W. Oliver in 1924.

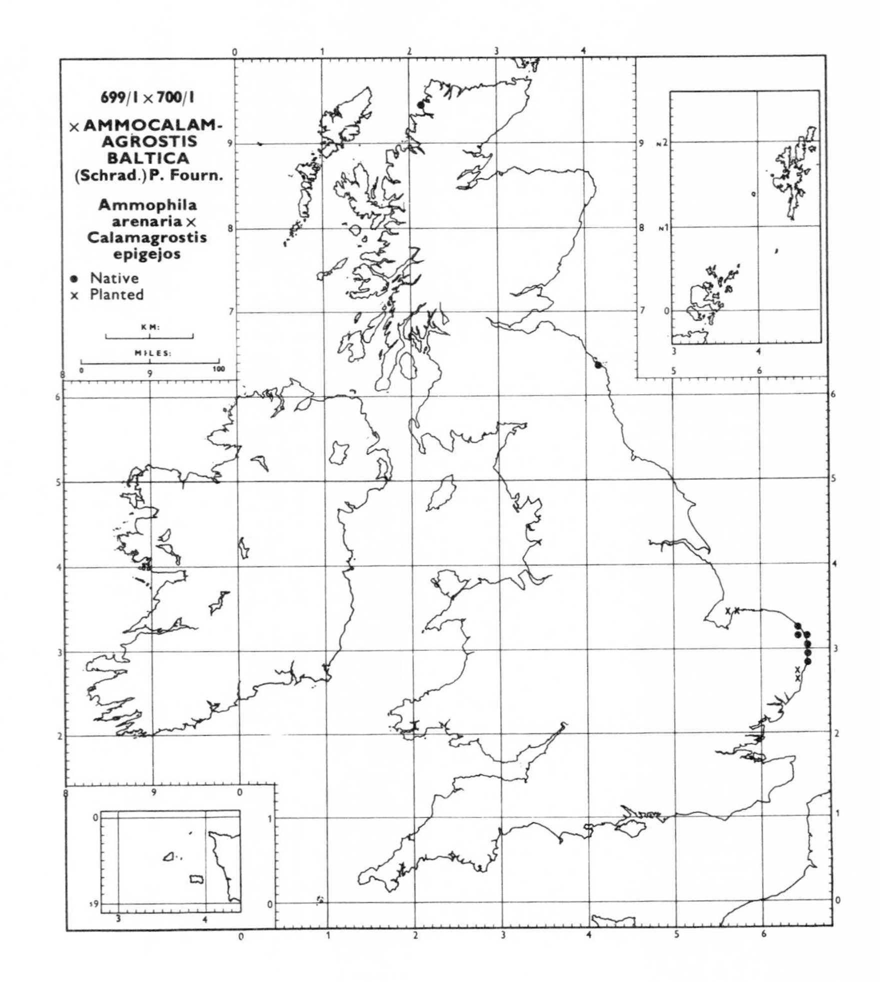

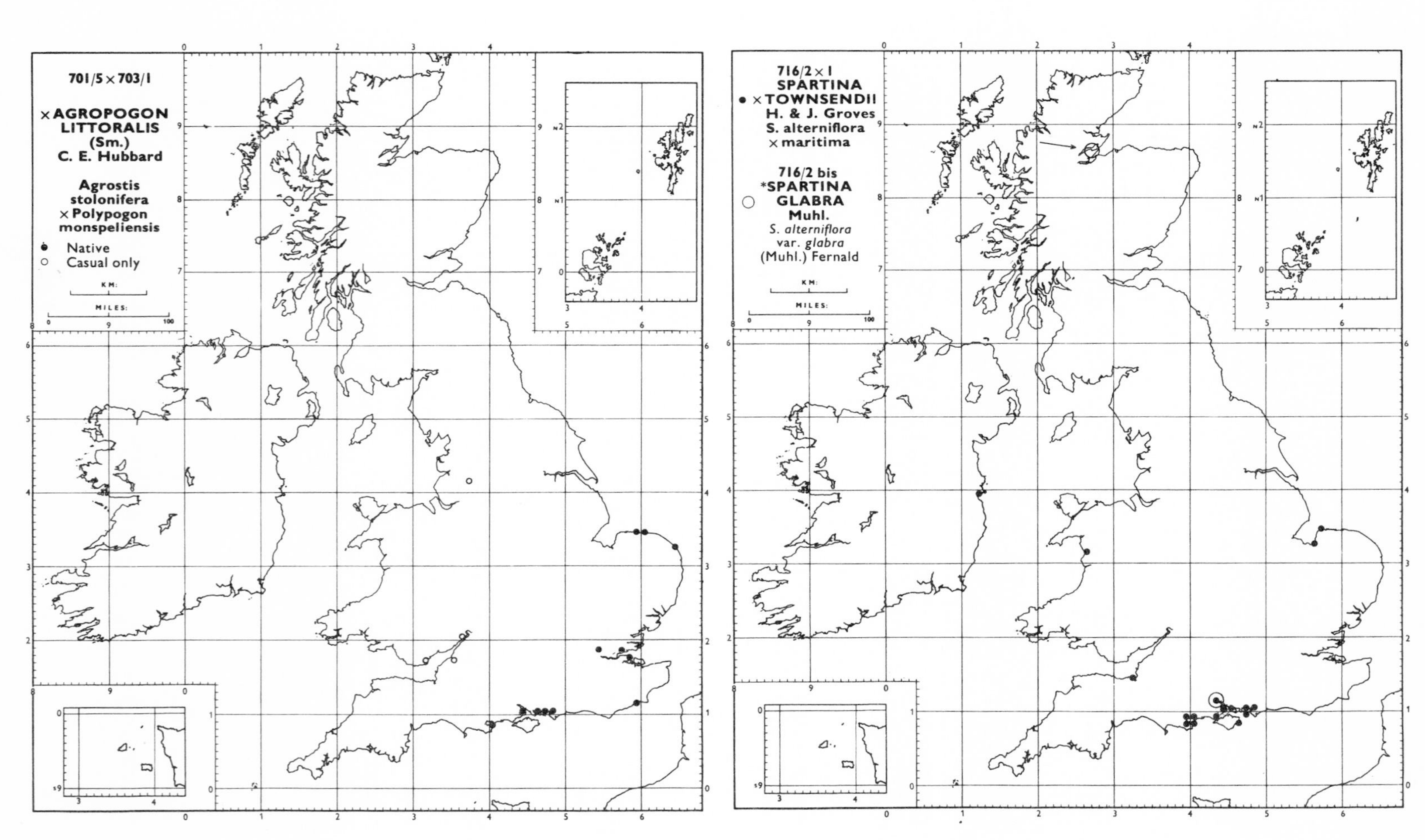

Appendix

KEY TO HERBARIUM ABBREVIATIONS

PUBLIC

ABD ABERDEEN: Department of Botany of the University
ABS ABERYSTWYTH: Department of Botany of the University College of Wales
BEL BELFAST: Ulster Museum and Art Gallery
BFT BELFAST: The Queen's University Natural History and Geological Museum
BIRA BIRMINGHAM: City Museum and Art Gallery
BIRM BIRMINGHAM: Department of Botany of the University
BM LONDON: Department of Botany of the British Museum (Natural History)
BRIST BRISTOL: Department of Botany of the University
CGE CAMBRIDGE: Botany School of the University
CLE CARLISLE: Corporation Museum and Art Gallery
CMM BRADFORD: Cartwright Memorial Museum
DBN DUBLIN: National Museum of Ireland
DBY DERBY: County Borough Museum and Art Gallery
DHM DURHAM: Department of Botany of the University
DSY DEWSBURY: County Borough Museum
E EDINBURGH: Royal Botanic Garden
EXR EXETER: Department of Botany of the University
GL GLASGOW: Department of Botany of the University
GLR GLOUCESTER: City Museum
HAMU NEWCASTLE UPON TYNE: Hancock Museum
HLU HULL: Department of Botany of the University
HWB HARROW: Butler Museum
K KEW: Royal Botanic Gardens
KLE KEELE: Department of Biology of the University of North Staffordshire
LCN LINCOLN: City and County Museum
LDS LEEDS: Department of Botany of the University
LIV LIVERPOOL: City Museum
LIVU LIVERPOOL: Department of Botany of the University
LSR LEICESTER: City Museum and Art Gallery
LTN LUTON: Public Museum and Art Gallery
LTR LEICESTER: Department of Botany of the University
MACO MARLBOROUGH: Marlborough College
MANCH MANCHESTER: Department of Botany of the University
MNE MAIDSTONE: Museum and Art Gallery
NMW CARDIFF: Department of Botany of the National Museum of Wales
NOT NOTTINGHAM: Natural History Museum, Wollaton Hall
NWH NORWICH: City Museum
OXF OXFORD: Department of Botany of the University
PTH PERTH: City Art Gallery and Museum
QMC LONDON: Queen Mary College
RAMM EXETER: Royal Albert Memorial Museum
SHD SHEFFIELD: Department of Botany of the University
SHY SHREWSBURY: Borough Public Library, Museum and Art Gallery
SLBI SOUTH NORWOOD: South London Botanical Institute
SPN SOUTHAMPTON: Department of Botany of the University
STA ST ANDREWS: Department of Botany of the University
TCD DUBLIN: School of Botany of the University
UCNW BANGOR: Department of Botany of University College of North Wales
WAR WARWICK: The Museum

PRIVATE

Graham = R. A. Graham, now in possession of R. M. Harley, Department of Botany, The University, Bristol 8
Lousley = J. E. Lousley, 7 Penistone Road, London S.W.16
Roberts = R. H. Roberts, Quinton, Belmont Road, Bangor, Caernarvon
Sandwith = N. Y. Sandwith, now in Rijksherbarium, Leiden (**L**)
Simpson = N. D. Simpson, Maesbury, 3 Cavendish Road, Bournemouth, Hants
Wallace = E. C. Wallace, 2 Strathearn Road, Sutton, Surrey

Bibliography

ALLEN, D. E. (1954). Variation in *Peplis portula* L. *Watsonia*, **3**, 85–91.

ALLEN, D. E. (1961). A new variety of *Valerianella locusta* (L.) Betcke. *Watsonia*, **5**, 45–6.

ALLEN, D. E. (1967). The taxonomy and nomenclature of the radiate variants of *Senecio vulgaris* L. *Watsonia*, **6**, 280–2.

BACKHOUSE, J. (1856). *A Monograph of the British Hieracia*. York.

BAKER, H. G. (1950). The inheritance of certain characters in crosses between *Melandrium dioicum* and *M. album*. *Genetica*, **25**, 126–56.

BAKER, H. G. (1955). *Geranium purpureum* Vill. and *G. robertianum* L. in the British Flora.—I. *Geranium purpureum*. *Watsonia*, **3**, 160–7.

BAKER, H. G. (1956). *Geranium purpureum* Vill. and *G. robertianum* L. in the British Flora.—II. *Geranium robertianum*. *Watsonia*, **3**, 270–9.

BALL, P. W., and HEYWOOD, V. H. (1962). The taxonomic separation of the cytological races of *Kohlrauschia prolifera* (L.) Kunth *sensu lato*. *Watsonia*, **5**, 113–16.

BARLING, D. M. (1959). Biological studies in *Poa angustifolia*. *Watsonia*, **4**, 147–68.

BARLING, D. M. (1962). Studies in the Biology of *Poa subcaerulea* Sm. *Watsonia*, **5**, 163–73.

BORRILL, M. (1956). A biosystematic study of some *Glyceria* species in Britain. 1. Taxonomy. *Watsonia*, **3**, 291–8.

BRADSHAW, M. E. (1962). The distribution and status of five species of the *Alchemilla vulgaris* L. aggregate in Upper Teesdale. *J. Ecol.*, **50**, 681–706.

BRADSHAW, M. E. (1963a). Studies on *Alchemilla filicaulis* Bus., *sensu lato*, and *A. minima* Walters.—Introduction, and I. Morphological variation in *A. filicaulis, sensu lato*. *Watsonia*, **5**, 304–20.

BRADSHAW, M. E. (1963b). Studies on *Alchemilla filicaulis* Buser, *sensu lato* and *A. minima* Walters. II. Cytology of *A. filicaulis, sensu lato*. *Watsonia*, **5**, 321–6.

BRADSHAW, M. E. (1964). Studies on *Alchemilla filicaulis* Bus. *sensu lato* and *A. minima* Walters. III. *Alchemilla minima*. *Watsonia*, **6**, 76–81.

BRADSHAW, M. E., and WALTERS, S. M. (1961). A Russian *Alchemilla* in South Scotland. *Watsonia*, **4**, 281–2.

BRENAN, J. P. M., and SIMPSON, N. D. (1949). The results of two botanical journeys in Ireland in 1938–9. *Proc. Roy. Irish Acad.*, **52**, B, 57–84.

CLAPHAM, A. R., TUTIN, T. G., and WARBURG, E. F. (1962). *Flora of the British Isles*. Ed. 2. Cambridge.

COOK, C. D. K. (1961). *Sparganium* in Britain. *Watsonia*, **5**, 1–10.

COOMBE, D. E. (1961). *Trifolium occidentale*, a new species related to *T. repens*. *Watsonia*, **5**, 68–87.

CULLEN, J. (1961). *A taxonomic revision of* Anthyllis vulneria *L. in Europe*. Unpublished. Thesis Univ. Liverpool.

DAKER, M. G. (1963). Cytotaxonomic studies on *Fumaria officinalis*. *Proc. B.S.B.I.*, **5**, 168–9.

DANDY, J. E. (1958). *List of British Vascular Plants*. London.

DRUCE, G. C. (1897). *The Flora of Berkshire*. Oxford.

DRUCE, G. C. (1926). *The Flora of Buckinghamshire*. Arbroath.

DRUCE, G. C. (1928). *British Plant List*. Ed. 2. Arbroath.

EBINGER, J. E. (1962). *Luzula × borreri* in England. *Watsonia*, **5**, 251–4.

FEINBRUN, N. and STEARN, W. T. (1964). Typification of *Lycium barbarum* L., *L. afrum* L. and *L. europaeum* L. *Israel J. Bot.*, **12**, 114–23.

FRASER, J. (1927). Menthae Britannicae. *Rep. Bot. Exch. Club Brit. Isles*, **8**, 213–47.

GRAHAM, R. A. (1958). Mint Notes. VIII. A new Mint from Scotland. *Watsonia*, **4**, 119–21.

GREEN, P. S. (1954). *Stellaria nemorum* L. subspecies *glochidisperma* Murbeck in Britain. *Watsonia*, **3**, 122–6.

GREGORY, E. S. (1912). *British Violets*. Cambridge.

GREGORY, E. S. (1918). Some notes on British Violets, with additional localities. *Rep. Bot. Exch. Club Brit. Isles*, **5**, 148a–g.

HANBURY, F. J. (1904). *Hieracium* in Babington, C.C. *Manual of British Botany*. Ed. 9. London.

HANBURY, F. J. (Ed.) (1925). *The London Catalogue of British Plants*. Ed. 11. London.

HANBURY, F. J., and MARSHALL, E. S. (1899). *Flora of Kent*. London.

HASKELL, G. (1961). Genetics and distribution of British *Rubi*. *Genetica*, **32**, 118–33.

HESLOP HARRISON, J. (1950). *Orchis cruenta* Müll. in the British Islands. *Watsonia*, **1**, 366–75.

HESLOP HARRISON, J. (1954). A synopsis of the Dactylorchids of the British Isles. *Ber. Geobot. Forsch. Zurich*, **1953**, 53–82.

HESLOP HARRISON, Y. (1953). *Nuphar intermedia* Ledeb., a presumed relict hybrid, in Britain. *Watsonia*, **3**, 7–25.

HOWARD, H. W., and LYON, A. G. (1950). The identification and distribution of the British Watercress species. *Watsonia*, **1**, 228–33.

HOWARD, H. W., and LYON, A. G. (1951). Distribution of the British Watercress species. *Watsonia*, **2**, 91–2.

HYDE, H. A., and WADE, A. E. (1962). *Welsh Ferns*. Ed. 4. Cardiff.

HYLANDER, N. (1943). Die Grassameneinkömmlinge Schwedisher Parke. *Symb. Bot. Upsal.*, **7** (**1**), 106–274.

LAMB, J. G. D. (1964). On the possible occurrence of *Erica mackaiana* Bab. in Co. Mayo. *Irish Nat. J.*, **14**, 213–14.

LINTON, W. R. (1905). *An Account of the British Hieracia*. London.

LOUSLEY, J. E. (1931). The *Schoenoplectus* group of the genus *Scirpus* in Britain. *J. Bot.*, **69**, 151–63.

MARCHANT, C. J. (1963). Corrected chromosome numbers for *Spartina × townsendii* and its parent species. *Nature*, **199**, 929.

MARSDEN-JONES, E. M., and TURRILL, W. B. (1954). *British Knapweeds*. London.

PERRING, F. H. (1960). Report on the survey of *Arctium* L. agg. in Britain, 1959. *Proc. B.S.B.I.*, **4**, 33–7.

PERRING, F. H. (1962). Hints on the determination of some critical species, microspecies, subspecies, varieties and hybrids in the British Flora. *Proc. B.S.B.I.*, **4**, 359–83.

PERRING, F. H., and WALTERS, S. M. (Ed.) (1962). *Atlas of the British Flora*. London and Edinburgh.

PRAEGER, R. Ll. (1921). *Equisetum litorale* Kühlew. *Irish Nat.*, **30**, 145.

PRAEGER, R. Ll. (1929). Report on recent additions to the Irish Fauna and Flora (Terrestrial and Freshwater). *Proc. Roy. Irish Acad.*, **39**, B, 1–94.

PRAEGER, R. Ll. (1930). *Equisetum litorale* Kühlew. (*E. arvense × limosum*). *J. Bot., Lond.*, **68**, 250.

PRAEGER, R. Ll. (1932). Some noteworthy plants found in or reported from Ireland. *Proc. Roy. Irish Acad.*, **41**, B, 95–124.

PRAEGER, R. Ll. (1934a). A contribution to the Flora of Ireland. *Proc. Roy. Irish Acad.*, **42**, B, 55–86.

PRAEGER, R. Ll. (1934b). Irish Junipers. *Irish Nat. J.*, **5**, 58.

PRAEGER, R. Ll. (1939). A further contribution to the Flora of Ireland. *Proc. Roy. Irish Acad.*, **45**, B, 231–54.

PRAEGER, R. Ll. (1946). Additions to the knowledge of the Irish Flora 1939–1945. *Proc. Roy. Irish Acad.*, **51**, B, 27–51.

PRITCHARD, N. M. (1959). *Gentianella* in Britain. I. *G. amarella*, *G. anglica* and *G. uliginosa*. *Watsonia*, **4**, 169–92.

PRITCHARD, N. M. (1960). *Gentianella* in Britain. II. *Gentianella septentrionalis* (Druce) E. F. Warburg. *Watsonia*, **4**, 218–37.

PROCTOR, M. C. F. (1957). Variation in *Helianthemum canum* (L.) Baumg. in Britain. *Watsonia*, **4**, 28–40.

PUGSLEY, H. W. (1912). The Genus *Fumaria* in Britain. *J. Bot., Lond.*, **50** Suppl., 1–76.

PUGSLEY, H. W. (1930). A revision of the British Euphrasiae. *J. Linn. Soc. London (Bot.)*, **48**, 467–544.

PUGSLEY, H. W. (1948). A Prodromus of the British Hieracia. *J. Linn. Soc. London* (Bot.), **54**.

RAVEN, P. H. (1963). *Circaea* in the British Isles. *Watsonia*, **5**, 262–72.

RIDDELSDELL, H. J. (1939). *Rubus* in MARTIN, W. K., and FRASER, G. T. (Ed.) *Flora of Devon*. Arbroath. 228–88.

RILSTONE, F. (1952). Rubi from Dartmoor to the Land's End. *Watsonia*, **2**, 151–62.

ROBSON, N. K. B. (1957). *Hypericum maculatum* Crantz. *Proc. B.S.B.I.*, **2**, 237–8.

ROFFEY, J. (1925). *Hieracium* in *The London Catalogue of British Plants*. Ed. 11. London. 26–30.

ROGERS, W. M. (1900). *Handbook of British Rubi*. London.

SELL, P. D. (1963a). Taxonomic and nomenclatural notes on European *Fumaria* species. *Feddes Rep.*, **68**, 174–8.

SELL, P. D. (1963b). Notes on the European species of *Scleranthus*. *Feddes Rep.*, **68**, 167–9.

SELL, P. D. (Ed.) (1967). Taxonomic and nomenclatural notes on the British Flora. *Watsonia*, **6**, 292–318.

SELL, P. D., and WEST, C. (1955). Notes on British Hieracia-I. *Watsonia*, **3**, 233–6.

SELL, P. D., and WEST, C. (1956). A Shetland 'Endemic' *Hieracium*. *Proc. Bot. Soc. Brit. Isles*, **2**, 79.

SELL, P. D., and WEST, C. (1962). Notes on British Hieracia. II. The species of the Orkney Islands. *Watsonia*, **5**, 215–23.

SELL, P. D., and WEST, C. (1965). A revision of the British species of *Hieracium* Section *Alpestria* [Fries] F. N. Williams. *Watsonia*, **6**, 85–105.

SMITH, A. J. E. (1963). Variation in *Melampyrum pratense* L. *Watsonia*, **5**, 336–67.

STYLES, B. T. (1962). The taxonomy of *Polygonum aviculare* and its allies in Britain. *Watsonia*, **5**, 177–214.

TUTIN, T. G., *et al.* (1964). *Flora Europaea*. **1**, Cambridge.

VALENTINE, D. H. (1946). The Butterbur in Yorkshire. *Naturalist*, **1946**, 45–6.

VALENTINE, D. H. (1947). The distribution of the sexes in Butterbur. *N. W. Naturalist*, **1947**, 111–14.

WADE, A. E. (1958). The history of *Symphytum asperum* Lepech. and *S.* × *uplandicum* Nyman in Britain. *Watsonia*, **4**, 117–18.

WALTERS, S. M. (1949a). *Alchemilla vulgaris* L. agg. in Britain. *Watsonia*, **1**, 6–18.

WALTERS, S. M. (1949b). *Aphanes microcarpa* (Boiss. et Reut.) Rothm. in Britain. *Watsonia*, **1**, 163–9.

WALTERS, S. M. (1949c). Biological Flora of the British Isles: *Eleocharis* R. Br. *J. Ecol.*, **37**, 192–206.

WALTERS, S. M. (1952). *Alchemilla subcrenata* Buser in Britain. *Watsonia*, **2**, 277–8.

WALTERS, S. M. (1953). *Montia fontana* L. *Watsonia*, **3**, 1–6.

WALTERS, S. M. (1963). *Eleocharis austriaca* Hayek, a species new to the British Isles. *Watsonia*, **5**, 329–35.

WATSON, W. C. R. (1958). *Handbook of the Rubi of Great Britain and Ireland*. Cambridge.

WEBB, D. A. (1954). Notes on four Irish Heaths. *Irish Nat. J.*, **11**, 187–90.

WEBB, D. A. (1955). Biological Flora of the British Isles. *Erica mackaiana* Bab. (*E. mackayi* Hook. ined.). *J. Ecol.*, **43**, 319–30.

WEBB, D. A. (1956). A new subspecies of *Pedicularis sylvatica*. *Watsonia*, **3**, 239–41.

WHITE, F. B. (1898). *The Flora of Perthshire*. Edinburgh.

WILLIAMS, F. N. (1902). *Prodromus Florae Britannicae*. Brentford. **1**, 76–184.

YEO, P. F. (1956). Hybridization between diploid and tetraploid species of *Euphrasia*. *Watsonia*, **3**, 253–69.

Index

Arabic numerals refer to the text. Bold numerals refer to maps.